AF606558

ADVANCES IN LIPID RESEARCH

Volume 26

Sphingolipids
Part B: Regulation and Function of Metabolism

Advances in Lipid Research

Volume 26

Sphingolipids Part B: Regulation and Function of Metabolism

Edited by

ROBERT M. BELL
Department of Biochemistry
Duke University Medical Center
Durham, North Carolina

YUSUF A. HANNUN
Departments of Medicine and Cell Biology
Duke University Medical Center
Durham, North Carolina

ALFRED H. MERRILL, JR.
Department of Biochemistry
Emory University School of Medicine
Atlanta, Georgia

SERIES EDITORS

RICHARD J. HAVEL
Cardiovascular Research Institute
University of California, San Francisco
San Francisco, California

DONALD M. SMALL
Department of Biophysics
Boston University
Boston, Massachusetts

ACADEMIC PRESS, INC.
A Division of Harcourt Brace & Company

San Diego New York Boston London Sydney Tokyo Toronto

This book is printed on acid-free paper. ∞

Academic Press, Inc.
1250 Sixth Avenue, San Diego, California 92101-4311

United Kingdom Edition published by
Academic Press Limited
24–28 Oval Road, London NW1 7DX

International Standard Serial Number: 0065-2849

International Standard Book Number: 0-12-024926-X

PRINTED IN THE UNITED STATES OF AMERICA
93 94 95 96 97 98 BB 9 8 7 6 5 4 3 2 1

CONTENTS

PART I
Enzymes of Sphingolipid Metabolism

Sphingomyelinases

Matthew W. Spence

Neutral Sphingomyelinase

Subroto Chatterjee

Ceramidases: Enzymology and Metabolic Roles

Daniel F. Hassler and Robert M. Bell

Sphingosine Kinase: Properties and Cellular Functions

Benjamin M. Buehrer and Robert M. Bell

Sphingosine-Phosphate Lyase

Paul P. Van Veldhoven and Guy P. Mannaerts

PART II
Trafficking and Topology of Sphingolipid Metabolism

Intracellular Transport of Ceramide and Its Metabolites at the Golgi Complex: Insights from Short-Chain Analogs

Anne G. Rosenwald and Richard E. Pagano

Ganglioside Metabolism—Topology and Regulation

Konrad Sandhoff and Gerhild van Echten

Truncated Ceramide Analogs as Probes for Sphingolipid Biosynthesis and Transport

Dieter Jeckel and Felix Wieland

Molecular Approaches to Studying the Intracellular Trafficking of Glycosphingolipids

William W. Young, Jr.

PART III
Functional and Dysfunctional Effects of Sphingolipid Metabolism

Metabolic Effects of Inhibiting Glucosylceramide Synthesis with PDMP and Other Substances

Norman S. Radin, James A. Shayman, and Jin-Ichi Inokuchi

Fumonisins and Other Inhibitors of *de Novo* Sphingolipid Biosynthesis

Alfred H. Merrill, Jr., Elaine Wang, David G. Gilchrist, and Ronald T. Riley

Extracellular Sialidases

Charles C. Sweeley

Sphingolipids with Inositolphosphate-Containing Head Groups

Robert L. Lester and Robert C. Dickson

Sphingomyelinase D: A Pathogenic Agent Produced by Bacteria and Arthropods

A. P. Truett III and L. E. King, Jr.

Sphingolipid-like Molecule Linked to CD45, a Protein Tyrosine Phosphatase

Akiko Takeda

PART IV
Comparative and Developmental Biology

Glycosphingolipids of the Invertebrata as Exemplified by a Cestode Platyhelminth, *Taenia crassiceps*, and a Dipteran Insect, *Calliphora vicina*

Roger D. Dennis and Herbert Wiegandt

Developmental Changes of Glycosphingolipid Composition of Epithelia of Rat Digestive Tract

Jean-François Bouhours, Danièle Bouhours, and Gunnar C. Hansson

CONTRIBUTORS

Numbers in parentheses indicate the pages on which the authors' contributions begin.

Robert M. Bell (49, 59), Department of Biochemistry and Section of Cell Growth, Regulation, and Oncogenesis, Duke University Medical Center, Durham, North Carolina 27710

Danièle Bouhours (353), Institut National de la Santé et de la Recherche Médicale, Unité 76, F-75739 Paris, France

Jean-François Bouhours (353), Institut National de la Santé et de la Recherche Médicale, Unité 76, F-75739 Paris, France

Benjamin M. Buehrer (59), Department of Biochemistry and Section of Cell Growth, Regulation, and Oncogenesis, Duke University Medical Center, Durham, North Carolina 27710

Subroto Chatterjee (25), Department of Pediatrics, Lipid Research Arteriosclerosis Division, Johns Hopkins University School of Medicine, Baltimore, Maryland 21205

Roger D. Dennis (321), Physiologisch–Chemisches Institut, Philipps Universität—Marburg, W-3550 Marburg, Germany

Robert C. Dickson (253), Department of Biochemistry, University of Kentucky College of Medicine, Lexington, Kentucky 40536

David G. Gilchrist (215), Department of Plant Pathology, University of California, Davis, Davis, California 95616

Gunnar C. Hansson (353), Department of Medical Biochemistry, University of Göteborg, S-413 90 Göteborg, Sweden

Daniel F. Hassler (49), Department of Biochemistry and Section of Cell Growth, Regulation, and Oncogenesis, Duke University Medical Center, Durham, North Carolina 27710

Jin-Ichi Inokuchi (183), Seikagaku Corporation, Tokyo Research Institute, Higashiyamato, Tokyo 207, Japan

Dieter Jeckel (143), Institut für Biochemie I, Universität Heidelberg, D-6900 Heidelberg, Germany

L. E. King, Jr. (275), Division of Dermatology, Vanderbilt University Medical Center, Nashville, Tennessee 37232

Robert L. Lester (253), Department of Biochemistry, University of Kentucky College of Medicine, Lexington, Kentucky 40536

Guy P. Mannaerts (69), Katholieke Universiteit Leuven, Fakulteit Geneeskunde, Campus Gasthuisberg, Afdeling Farmakologie, B-3000 Leuven, Belgium

Alfred H. Merrill, Jr. (215), Department of Biochemistry, Emory University School of Medicine, Atlanta, Georgia 30322

Richard E. Pagano (101), Department of Embryology, Carnegie Institution of Washington, Baltimore, Maryland 21210

Norman S. Radin (183), Department of Internal Medicine and Mental Health Research Institute, University of Michigan Medical Center, Ann Arbor, Michigan 48109

Ronald T. Riley (215), Toxicology and Mycotoxins Research Unit, USDA-ARS, Athens, Georgia 30613

Anne G. Rosenwald (101), Department of Embryology, Carnegie Institution of Washington, Baltimore, Maryland 21210

Konrad Sandhoff (119), Institut für Organische Chemie und Biochemie, Universität Bonn, D-5300 Bonn 1, Germany

James A. Shayman (183), Department of Internal Medicine, University of Michigan Medical Center, Ann Arbor, Michigan 48109

Matthew W. Spence (3), Atlantic Research Centre, and Departments of Paediatrics and Biochemistry, Dalhousie University, Halifax, Nova Scotia, Canada B3H 4H7

Charles C. Sweeley (235), Department of Biochemistry, Michigan State University, East Lansing, Michigan 48824

Akiko Takeda (293), Department of Pathology, Roger Williams Medical Center—Brown University, Providence, Rhode Island 02908

A. P. Truett III (275), Division of Dermatology, Vanderbilt University Medical Center, Nashville, Tennessee 37232

Gerhild van Echten (119), Institut für Organische Chemie und Biochemie, Universität Bonn, D-5300 Bonn 1, Germany

Paul P. Van Veldhoven (69), Katholieke Universiteit Leuven, Fakulteit Geneeskunde, Campus Gasthuisberg, Afdeling Farmakologie, B-3000 Leuven, Belgium

Elaine Wang (215), Department of Biochemistry, Emory University School of Medicine, Atlanta, Georgia 30322

Herbert Wiegandt (321), Physiologisch–Chemisches Institut, Philipps Universität—Marburg, W-3550 Marburg, Germany

Felix Wieland (143), Institut für Biochemie I, Universität Heidelberg, D-6900 Heidelberg, Germany

William W. Young, Jr. (161), Departments of Biological and Biophysical Sciences and Biochemistry Health Sciences Center, University of Louisville, Louisville, Kentucky 40292

PREFACE

The sphingolipids were aptly named in 1884 by their discoverer, J. L. W. Thudichum, after the riddle of the ancient Greek Sphinx which guarded the city of Thebes. The structural complexity of the sphingolipids now exceeds 400 distinct molecules; this portends important functions. The first of these functions is the role of glycosphingolipids on the cell surface as receptors and as modulators of growth factor receptors. In 1986, we reported on the possible role of sphingosine, a breakdown product of complex sphingolipids, on cellular regulation and its inhibitory effects on protein kinase C. These findings marked the beginning of a pathway leading to the discovery of a variety of intracellular sphingolipid metabolites that function as lipid second messengers/lipid mediators. The first bona fide sphingolipid cycle, "the sphingomyelin cycle," was discovered in 1989; ceramide appears to be a second messenger. Since then, a variety of sphingosine derivatives, including *N,N*-dimethylsphingosine, sphingosine phosphate, and sphingosine phosphorylcholine, have been reported to be bioactive.

Thus, the answer to the sphingolipid riddle posed long ago by Thudichum appears, in part, to reflect the fact that complex sphingolipids serve as reservoirs for the production of bioactive molecules that regulate cellular responses, which encompasses the regulation of cell growth and differentiation, mitogenesis, calcium mobilization, regulation of protein kinases and phosphatases, and apoptosis. These studies have led to reinterpretations of the significance of findings 20 years old or older on the metabolism of the sphingolipids and to renewed interest and understanding of the biosynthesis, catabolism, and role of breakdown products in cellular regulation.

These findings offer promise for new molecular insights into tumor growth and metastasis (glycosphingolipids become dominant tumor antigens), neural functions, diabetes, obesity, hypoxic injury, atherosclerosis, and inflammatory disorders including psoriasis. Defects in these metabolic pathways, as well as their disruption by toxic agents in the environment, should emerge as causes of diseases. As recent data define novel pathways and pinpoint targets for potential therapeutic intervention, these findings place the sphingolipids and their roles in cells in parallel with the more advanced studies on the glycerolipids. Still, much remains to be done to define the extent of the set of intracellular bioactive lipids, second messengers, and lipid mediators derived from the sphingolipids. Our intuition suggests that an extensive number of sphingolipid metabolites may emerge. The field may be poised in a manner similar to the early days of the eicosanoids, as novel metabolic pathways that generate and attenuate these messengers await discovery and clarification. Most functions of the cellular sphingolipids remain to be elucidated. The role of sphingolipids as dietary components represents another frontier for future research.

We are grateful to Dr. Donald Small for suggesting the subject matter for Volumes 25 and 26 of *Advances in Lipid Research* and to the authors of the 32 chapters that constitute these two volumes for the many new ideas contained in their contributions. The authors appreciate the encouragement of their colleagues, students, and postdocs in planning and editing these volumes. We anticipate further exciting discoveries that will advance and clarify this new field of research.

ROBERT M. BELL
YUSUF A. HANNUN
ALFRED H. MERRILL, JR.

Part I

ENZYMES OF SPHINGOLIPID METABOLISM

ADVANCES IN LIPID RESEARCH, VOL. 26

Sphingomyelinases

MATTHEW W. SPENCE*

Atlantic Research Centre and
Departments of Paediatrics and Biochemistry
Dalhousie University
Halifax, Nova Scotia,
Canada B3H 4H7

I. Introduction

A. Sphingomyelin Structure, Tissue Distribution, and Functions

Sphingomyelin is a naturally occurring membrane lipid consisting of two aliphatic lipid components, sphingosine and a long-chain fatty acid, linked with a polar head group, phosphocholine (Thudicum, 1962; Rouser *et al.*, 1953; Shapiro and Flowers, 1962; Barenholz and Thompson, 1980; Barenholz and Gatt, 1982). Sphingosine is a class of 18- and 20-carbon amine diols of which the most common is 4-*trans*-sphingenine (1,3-dihydroxy-2-amino-4-octadecane). Lesser amounts of the sphinganine (1,3-dihydroxy-2-amino-octadecane) and 4-hydroxysphinganine (phytosphingosine; 1,3,4-trihydroxy-2-amino octadecane) have been observed in many mammalian tissues, together with trace amounts of other sphingosine bases. The fatty acyl group is in amide linkage to the sphingosine and varies in length from 12 to over 30 carbons in length, but is chiefly

* Present address: Alberta Heritage Foundation for Medical Research, Edmonton, Alberta, Canada T5J 3S4.

16:0, 18:0, 22:0, 24:0, and 24:1 (Rouser *et al.*, 1972; Barenholz and Thompson, 1980; White, 1974; O'Brien and Sampson, 1965). Species with polyenoic very long-chain fatty acids, and with 2-hydroxy fatty acids have been described (Robinson *et al.*, 1992).

The sphingosine and the amide-linked fatty acid form the lipid portion of the sphingomyelin molecule that is called ceramide. The various permutations in the sphingosine and fatty acyl chains lead to heterogeneity even between different leaflets of the same membrane (Boegheim *et al.*, 1983). As over one-half of the fatty acyl chains of ceramide are greater than 20 carbons in length, and the sphingosine hydrocarbon tail is generally only 15 carbons, the ceramide molecule exhibits marked hydrocarbon chain-length asymmetry (Levin *et al.*, 1985). This leads to mixed, interdigitated chain packing in bilayers that is more ordered than the noninterdigitated layers formed by other lipids with more symmetrical acyl chains. The phosphocholine, attached to the hydroxyl at position 1 of ceramide, is identical to the phosphocholine of phosphatidylcholine, and the latter may be the metabolic source of the phosphocholine in sphingomyelin (Diringer *et al.*, 1972; Ullman and Radin, 1974; Marggraf *et al.*, 1981; Voelker and Kennedy, 1982; Eppler *et al.*, 1987). Despite the similar polar phosphocholine moiety, the two lipids differ in physical properties, with sphingomyelin tending to confer a more rigid structure on membranes due to its higher proportion of saturated acyl chains and intermolecular hydrogen bonding between the amide bond and the 3-hydroxyl (Barenholz and Thompson, 1980).

Sphingomyelin is ubiquitously distributed in all membranes of mammalian cells and in serum lipoproteins (White, 1974; Barenholz and Gatt, 1982). In general, its distribution parallels that of cholesterol, and, in most cells, there is an increasing gradient of sphingomyelin and cholesterol relative to other lipids from the nuclear membrane through the various organelles to the cell surface (Barenholz and Thompson, 1980; White, 1974; Rouser *et al.*, 1968). Approximately 70 to 90% of the cell sphingomyelin is in the plasma membrane in hepatocytes (Koval and Pagano, 1991), skin fibroblasts (Lange *et al.*, 1989), and C_6 glioma cells (Cook *et al.*, 1988). Where examined, most of the plasma membrane sphingomyelin is in the extracytoplasmic leaflet of the plasma membrane (Slotte and Bierman, 1988; Allan and Quinn, 1988; Kolesnick, 1991). Sphingomyelin levels tend to increase with age in different organs, and may also vary with environmental influences such as diet (Barenholz and Gatt, 1982; Myher *et al.*, 1981).

Sphingosine is synthesized from palmitoyl-CoA and serine by serine palmitoyltransferase (Brady and Koval, 1958; Williams *et al.*, 1984). The product, 3-ketosphinganine is reduced to sphinganine by a microsomal

NADPH-dependent reductase (Stoffel *et al.*, 1968; Stoffel, 1970). The 4-*trans* double bond is probably introduced following N-acylation to form ceramide (Ong and Brady, 1973). At least three different pathways of ceramide biosynthesis have been described: acyl-CoA-dependent acylation of the amino group (Sribney, 1966; Morell and Radin, 1970; Ullman and Radin, 1972; Narimatsu *et al.*, 1986; Kishimoto and Kawamura, 1979), acyl-CoA-independent acylation via a reduced pyridine-nucleotide-dependent pathway that also forms α-hydroxy fatty acid-containing ceramides (Kishimoto and Kawamura, 1979), and the condensation of free fatty acid and sphingosine in the reversal of the ceramidase reaction (Yavin and Gatt, 1969; Narimatsu *et al.*, 1986). The latter reaction is generally thought to be of limited physiological significance (Stoffel *et al.*, 1980).

The final reaction is the addition of phosphocholine to ceramide. The principal route of synthesis is the direct transfer of phosphocholine from phosphatidylcholine to ceramide to form sphingomyelin and diacylglycerol. First described in whole cells and homogenates and tissue subcellular fractions (Diringer *et al.*, 1972; Ullman and Radin, 1974), more recent work has localized this enzyme activity (phosphatidylcholine:ceramide phosphocholine transferase) to the Golgi (Futerman *et al.*, 1990; Koval and Pagano, 1991; Kobayashi *et al.*, 1992) and to the plasma membrane (Marggraf *et al.*, 1981; van den Hill *et al.*, 1985). Other pathways that have been described include the formation of sphingomyelin from CDP-choline and ceramide in a reaction analogous to phosphatidylcholine biosynthesis (Sribney and Kennedy, 1958; Stoffel and Melzner, 1980), and the acylation of sphingosylphosphocholine by acyl-CoA (Brady *et al.*, 1965; Fujino *et al.*, 1968). While these latter two pathways have been demonstrated under special conditions, they are probably not major pathways of sphingomyelin biosynthesis (Stoffel and Melzner, 1980; Voelker and Kennedy, 1982).

Like all membrane lipids, sphingomyelin appears to serve a wide variety of physiological functions, in addition to a structural role. Its preferential association with cholesterol has been demonstrated in tissue and organelle lipid profiles (Barenholz and Thompson, 1980; White, 1974; Rouser *et al.*, 1968); in physicochemical studies of lipid, and lipid–protein mixtures in artificial systems (Demel *et al.*, 1977; Yeagle and Young, 1986; Gronberg and Slotte, 1990; Kanfer *et al.*, 1991; McIntosh *et al.*, 1992); and in various pathological states such as atherosclerotic lesions (Barenholz and Gatt, 1982) and the inherited lipidosis, Niemann-Pick disease (Spence and Callahan, 1989). In the sphingomyelinase-deficient forms of this disease, sphingomyelin and cholesterol accumulate; in the cholesterol-processing-deficient forms of the disease, cholesterol and sphingomyelin also accumulate (Spence and Callahan, 1989). Not only is there a physical association between these lipids, but coincidentally, and perhaps related causally, there is a metabolic relationship. Depletion of plasma membrane sphingomyelin in cultured cells by exogenous sphingomyelinase leads to an increase in acyl-CoA:cholesterol acyltransferase activity (Slotte and Bierman, 1988; Byers *et al.*, 1992) and a reduced activity of hydroxy-methylglutaryl-CoA reductase activity (Gupta and Rudney, 1991). Sphingomyelinase- treated low-density lipid (LDL) is taken up avidly by macrophages (Xu and Tabas, 1991). Conversely, excess cholesterol can alter sphingomyelin catabolism. Cholesterol loading of cultured fibroblasts, or inherited disorders leading to the accumulation of cholesterol in fibroblasts cause inhibition of sphingomyelin degradation (Mazière *et al.*, 1983, 1984). Further, cholesterol can inhibit sphingomyelin degradation *in vitro* (Mazière *et al.*, 1981).

The close metabolic relationship between sphingomyelin and phosphatidylcholine is evident in the synthesis of sphingomyelin (Voelker and Kennedy, 1982). Interestingly, the relative molar proportions of the two vary markedly in membranes, with marked effects on the physical properties as demonstrated in artificial lipid mixtures (Barenholz and Thompson, 1980; Sankaram and Thompson, 1990). However, the sum of the two is relatively constant in the plasma membrane at more than 50% of the total phospholipids (Barenholz and Thompson, 1980).

Sphingomyelin has been implicated in the binding of tetradecanoylphorbol-13-acetate (Esumi and Fujiki, 1983), and has been reported to associate preferentially with several proteins, including glucagon (Epand *et al.*, 1982), 5′-nucleotidase (Widnell and Unkeless, 1968), ATPase (Sood *et al.*, 1972), sea anemone toxin (Linder *et al.*, 1977), apolipoprotein A-III (Stoffel *et al.*, 1974), and acetylcholinesterase (Cohen and Barenholz, 1984). Sphingomyelin has also been implicated as an inhibitor, *in vitro*, of the diacylglycerol-stimulated activities of human platelet phosphatidylinositol phosphodiesterase, rat intestinal phospholipase A_2, and rat liver phospholipase A (Dawson *et al.*, 1985).

The role of sphingomyelin as a precursor of other biologically important macromolecules, such as ceramide, ceramide phosphate, and sphingosine, has been reviewed (Hannun and Bell, 1989; Merrill and Jones, 1990; Dennis *et al.*, 1991; Kolesnick, 1991) and is discussed in other articles in this volume.

B. Sphingomyelinases

1. *The First Step in Sphingomyelin Catabolism*

At least three possible routes of sphingomyelin degradation have been described. One is hydrolysis by a phospholipase D to form choline and ceramide phosphate. Bacterial (Berheimer *et al.*, 1980) and cabbage

(Heller, 1978) phospholipase D can hydrolyze sphingomyelin; the latter enzyme is more active with glycerolipids. No major degradation by this route has been described in mammalian tissues, nor has base-exchange with sphingomyelin been demonstrated (Kanfer, 1980). The second possible route of degradation is by deacylation to form sphingenylphosphocholine and fatty acid. Small amounts of sphingenylphosphocholine have been isolated from the spleen of a patient with Niemann-Pick disease (Strasberg and Callahan, 1988). In another sphingolipidosis, Krabbe's disease, a similar deacylated sphingolipid, sphingenylgalactose, has been isolated (Suzuki and Suzuki, 1982). The failure to demonstrate deacylation of the parent glycosphingolipid has led to the suggestion that the deacylated intermediate may be formed by synthesis rather than degradation. Similar considerations may apply to sphingenylphosphocholine formation, and, if present at all, deacylation is probably of minor significance.

The third, and probably the major route of sphingomyelin degradation, is a phospholipase C-type reaction to form ceramide and phosphocholine. The reaction is catalyzed by a sphingomyelinase (sphingomyelin phosphodiesterase EC 3.1.4.12; SMase). Inherited deficiencies of SMase lead to accumulation of sphingomyelin in tissues (Spence and Callahan, 1989). Inhibition of SMase in cultured cells by drugs causes sphingomyelin accumulation (Aubert-Tulkens *et al.*, 1979; Albouz *et al.*, 1986; Gosh and Chatterjee, 1987). Treatment of HL-60 cells with vitamin D_3 (Okazaki *et al.*, 1989), interferon α, or tumor necrosis factor (Kim *et al.*, 1991) causes a decrease in sphingomyelin and a reciprocal elevation in ceramide. Injection of labeled dihydrosphingomyelin intravenously in rats and measurement of the metabolic products suggests a reaction sequence: dihydrosphingomyelin $\rightarrow$ dihydroceramide $\rightarrow$ sphinganine (Schneider and Kennedy, 1968). Taken together, all of this evidence is compatible with sphingomyelinase as the principal first step in sphingomyelin degradation.

2. *Method of Measurement*

Radiolabeled sphingomyelin has been used by most investigators studying SMase activity, and the radioactive products are separated from the substrate by precipitation, solvent partition, or adsorption. The labeled substrate may be prepared by reduction of the double bonds in ceramide with ^3H (Rao and Spence, 1976), acylation of sphingosylphosphocholine with labeled fatty acid (Spence *et al.*, 1983), or the incorporation of [^3H]methyl or [^{14}C]methyl from methyl iodide into the methyl group of the choline moiety (Stoffel *et al.*, 1971). The sphingomyelin is isolated from natural sources. If the product of hydrolysis is labeled ceramide, it may be separated from the labeled substrate by a combination of solvent extraction and batch silicic acid absorption (Schneider and Kennedy,

1967). When the choline moiety is labeled, the labeled water-soluble phosphocholine can be separated from the labeled substrate by partition (Spence *et al.*, 1989) or by trichloroacetic acid precipitation of the reaction mixture and extraction with ether (Kanfer *et al.*, 1966).

A variety of artificial substrates with colored or fluorescent products are available. Colored substrates include *N*-trinitrophenylaminolauryl-sphingosylphosphocholine (Gatt *et al.*, 1981), *N*-[10-(1-pyrene)decanoyl]-sphingomyelin (Levade and Gatt, 1987), and 2-hexadecanoylamino-4-nitrophenylphosphocholine (Gal *et al.*, 1980). Fluorescent substrates include bis(4-methylumbelliferyl) phosphate (Jones *et al.*, 1982) and 11-(9-anthroyloxy)-undecanoyl sphingomyelin (Gatt *et al.*, 1981).

II. Mammalian Sphingomyelinases

A. Acid Sphingomyelinase

1. *Tissue Distribution and Function*

Acid SMase, so named because the pH optimum for maximal activity in most tissues and under most reaction conditions ranges between 4 and 5.5, is ubiquitously distributed in every mammalian tissue examined (Barenholz *et al.*, 1966; Kanfer *et al.*, 1966; Spence *et al.*, 1979). The enzyme was first extensively described by Barenholz *et al.* (1966) in brain and by Kanfer *et al.* (1966) in liver, where activity levels are high; subcellular fractionation studies in liver (Weinreb *et al.*, 1968; Fowler, 1969) and cultured neuroblastoma cells (Spence *et al.*, 1982) support a lysosomal location. Other tissues in which activity has been reported include rat and human kidney (Gosh and Chatterjee, 1987); spleen (Schneider and Kennedy, 1967); rat lung, cardiac and psoas muscle, testis, aorta, adrenal gland, adipose tissue (Spence *et al.*, 1979); human, bovine, and porcine seminal fluid (Vanha-Perttula, 1988); human urine (Quintern *et al.*, 1987); cultured fibroblasts (Poulos *et al.*, 1984); bovine and human eye (Noda *et al.*, 1989); and cultured neuroblastoma and glioma cell lines (Chakravarthy *et al.*, 1985). The membrane localization of sphingomyelin and the central role of the lysosome in membrane turnover through endocytosis and autophagy (Duncan and Pratten, 1977; Dean, 1977; de Duve, 1983) suggest that this enzyme may function in the bulk turnover of membrane or lipoprotein-associated sphingomyelin as part of general membrane or lipoprotein turnover. This is supported by the sphingomyelin accumulation in lysosomes observed in the inherited disorders characterized by acid SMase deficiency known as type I Niemann-Pick disease (Spence and Callahan, 1989). However, the observations that, in some circumstances, other plasma membrane SMases or exogenous SMase may degrade the major portion of cell sphingomyelin, indicates that this is not always the major route of degradation (Spence *et al.*, 1983).

There is little change in activity, based on protein, with age in rat brain (Spence and Burgess, 1978), but activity in human whole epidermis gradually decreases with age (Yamamura and Tezuka, 1990). Activity measured subsequently *in vitro* is increased in human leukocytes preincubated for 2 hours with dexamethosone (Nelson *et al.*, 1982), and in cultured human fibroblasts incubated for 1 to 4 days in 2% dimethyl sulfoxide (DMSO) (Sato *et al.*, 1988). Although DMSO affected several lysosomal hydrolase activities similarly, the greatest effect observed was on SMase activity. Several amphiphilic cationic drugs, such as AY-9944, gentamicin, chloroquine, and imipramine, when either administered to experimental animals over several weeks or added to the medium of cultured cells, cause a decrease in SMase activity (Yoshida *et al.*, 1985; Aubert-Tulkens *et al.*, 1979; Albouz *et al.*, 1986). The calmodulin antagonist W7 inhibits activity in C_6 glioma cells (Masson *et al.*, 1989). Activity is also reduced in cultured fibroblasts following cholesterol loading (Mazière *et al.*, 1983).

2. *Purification and Properties*

The enzyme has been purified to varying degrees from human placenta (Pentchev *et al.*, 1977; Jones *et al.*, 1981; Sakuragawa, 1982), brain (Yamanaka *et al.*, 1981), rat liver (Watanabe *et al.*, 1983), and human urine (Quintern *et al.*, 1987). Urine from subjects with peritonitis is a rich source of enzyme (Quintern *et al.*, 1989a) (Table I). Purification protocols vary, but most have included a combination of lectin affinity on Con-A, ion-exchange, hydrophobic supports, dye binding, and, in some cases, affinity chromatography.

Comparisons of the various studies of the properties of crude and purified enzyme preparations are complicated by the requirement for detergents for maximal activity with lipid substrates. However, the following properties are common to most studies, or, where examined in only a few studies, have been similar.

Highest activity levels are obtained with D-*erythro*-sphingomyelin and dihydrosphingomyelin; the L-*erythro* and D,L-*threo* conformers are hydrolyzed at only 20–30% of the rate of the D-*erythro* form (Barenholz *et al.*, 1966). The K_m for sphingomyelin as substrate has been reported to be between 10 and 500 μM. Although sphingomyelin is the preferred substrate, the purified enzyme can hydrolyze phosphatidylcholine, phosphatidylglycerol, and lysophosphatidylglycerol; activity can also be demonstrated with a variety of artificial substrates, as described above. Monovalent and divalent cations, EDTA, and EGTA have no effect on activity;

Table I

The Purification of Acid Sphingomyelinase from Mammalian Tissues

Investigators	Tissue	Fold purification	Specific activity (μmol/hr/mg protein)	Comments
Pentchev *et al.* (1977)	Human placenta	10,300	195	Two bands on SDS-PAGE
Jones *et al.* (1981)	Human placenta	12,000	52	Some minor bonds on SDS-PAGE
Sakuragawa (1982)	Human placenta	32,800	380	Two major protein bands on SDS-PAGE
Watanabe *et al.*(1983)	Rat liver	5,000	3,200	Trace amounts of acid phosphatase and β-galactosidase activity; several minor bands on SDS-PAGE
Yamanaka *et al.* (1981)	Human brain	1,200–1,500	64–88	Other lysosomal enzyme activities. Several minor bands on SDS-PAGE
Quintern *et al.* (1987)	Human urine	3,000	6,900	Pure denatured enzyme obtained following HPLC

reducing agents such as dithiothreitol inactivate the enzyme. Activity can be strongly inhibited by AMP, and, for the purified urine enzyme, by 9-β-D-arabinofuranosyladenine-5′-monophosphate and phosphatidylinositol 4′,5′-bisphosphate (Quintern *et al.*, 1987). Activity of the urine enzyme can be stimulated by mono-, di-, and triacylglycerols and by free fatty acids.

In all cases, highest activity levels are obtained when the sphingomyelin substrate is part of a lipid micelle, and activity levels are dependent on the nature of the other lipids or detergents that form the micelle (Gatt *et al.*, 1973; Gatt, 1982). These complex effects probably relate to a combination of micellar size, shape, charge, and the spacing of the substrate molecules in the micelle.

The purified, glycosylated urine enzyme is a single polypeptide chain with an apparent molecular mass of 72 kDa; the mass reduces to 61 kDa following deglycosylation (Quintern and Sandhoff, 1991), suggesting that as many as five sites may be glycosylated. Size estimates of partially purified enzyme, or by Western blotting with a polyclonal antiserum to the placental enzyme, have suggested size heterogeneity in human tissues, ranging from 70,000 to 120,000 kDa (Spence and Callahan, 1989). The reasons for this are not known.

3. *Activator Proteins*

Activator proteins may play an important role in the action of acid SMase *in vivo*. Alpert and Beaudet (1981) have shown that ApoC-III, an apolipoprotein from human very low density lipoprotein, stimulates the activity of acid SMase of human fibroblasts measured *in vitro*. In subsequent studies (Ahmad *et al.*, 1986), this group has shown that stimulation by ApoC-III is greater than for ApoC-II, and that A-I, A-II, B, C, and E have no effect. The stimulation is a property of the phospholipid-binding, carboxyl-terminal half of the protein and may be due to a breaking down of the lipid substrate matrix into smaller units that are more easily accommodated at the enzyme active site.

Another group of activator proteins are the saposins (O'Brien and Kishimoto, 1991). Prosaposin, a 524-amino acid glycoprotein, is composed of 4–80 amino acid domains, each with nearly identical positioning of cysteine residues, glycosylation sites, and helical regions and flanked by potential proteolytic cleavage sites. Proteolytic cleavage results in four saposins—A, B, C, and D. A point mutation in saposin B has been described in transformed fibroblasts from a subject with the activator-deficient form of the autosomal-recessive inherited human lysosomal storage disease metachromatic leukodystrophy (Kretz *et al.*, 1990). Thus, the biological importance of the saposins is clear. Saposin D appears to stimulate acid SMase specifically (Morimoto *et al.*, 1988); saposins A and C have also been reported to stimulate acid SMase. Whether this is a primary property of saposins A and C, or is a nonspecific detergent-like property, or represents trace-contamination by saposin D remains to be determined. All of the saposins have sequences capable of forming amphipathic helices which may associate as a hydrophobic pocket to accommodate the lipids (O'Brien *et al.*, 1988).

4. *Molecular Studies*

The human acid SMase gene has been localized to chromosomal region 11p15.1–p15.4 using somatic cell hybrids and *in situ* hybridization techniques (Pereira *et al.*, 1991), and the complete nucleotide sequence and structural organization have been described (Quintern *et al.*, 1989b; Schuchman *et al.*, 1991, 1992). The locus is small (approximately 5 kilobases) and is composed of 6 exons, ranging in size from 77 to 773 base pairs (bp) and 5 introns, with a size range from 153 to 1059 bp. Three cDNAs have been isolated, and detailed studies of their base composition suggest they are formed by alternative splicing of the SMase mRNA transcript. The principal cDNA (>90%) contains a 172-bp in-frame exonic sequence (exon 3); the other two do not contain this sequence, but in the more abundant (approximately 10%) of the minor forms, it is replaced by a 40-bp in-frame intronic sequence. The alternative splicing may be due to weak donor sites adjacent to the 172-bp exon 3 sequence. Only the major species causes the transient production of functionally active protein in COS-1 cells (Schuchman *et al.*, 1991). Three other open reading frames have been identified within the genomic sequence that could encode functional proteins of 101, 104, and 158 amino acid residues (Schuchman *et al.*, 1992). Two are in the opposite transcriptional orientation to SMase, and the predicted proteins share no homology. The third is in the same transcriptional orientation and coding phase with the SMase gene. The significance of this is not known.

5. *Niemann-Pick Disease*

a. Pathophysiology. The Niemann-Pick group of diseases can be divided into two broad subgroups (Spence and Callahan, 1989). Both are characterized by varying degrees of hepatospengomegaly, psychomotor retardation, and the formation of storage or foam cells in bone marrow and tissues. There are variably increased amounts of sphingomyelin, cholesterol, glycosphingolipids, and bis(monoacylglycero)phosphate in visceral organs. Clinically, both broad subgroups can be separated into acute, subacute, and chronic forms which vary in age of onset, severity of symptomatology, and life expectancy. The type I subgroup has a deficiency of acid SMase, generally less than 10% of normal levels; sphingomyelin and cholesterol are the principal lipids stored. In the type II group, acid SMase may be normal, or reduced by as much as 80%, and the reduction in activity may not affect all tissues. A reduction in cholesterol esterification has been demonstrated in some patients (Pentchev *et al.*, 1985; Vanier *et al.*, 1991; Liscum and Faust, 1987; Byers *et al.*, 1989; Sidhu *et al.*, 1992), and the reduced activity levels of acid SMase in cultured fibroblasts from some patients can be increased to normal levels by removing the lipoprotein fraction from the culture media (Thomas *et al.*, 1989). Taken together, these observations suggest that, in the type II patient, the variably decreased levels of SMase activity are secondary to some other primary process.

Cross-reactive immunological material to acid SMase has been present in crude homogenates of cultured fibroblasts from type I subjects that have been examined (Jobb and Callahan, 1987, 1989).

b. Molecular Studies. Type I Niemann-Pick disease occus at least 10 times more frequently among persons of Ashkenazi Jewish ancestory than in the general population (Spence and Callahan, 1989). Studies of the acid SMase cDNA produced from RNA isolated from cultured fibroblasts or lymphoblasts has demonstrated two mutations (Ferlinz *et al.*, 1991; Levran *et al.*, 1991a,b). Both are in exon 6 of the SMase gene, suggesting this may be the catalytic site. One is a G to T transversion at the CpG dinucleotide and predicts an arginine substitution for leucine (R496L). The mutation is present in 32% of Jewish patients but only 5.6% of non-Jewish patients with the acute form of the disease. Subjects homoallelic for the mutation are affected; carriers are heteroallelic. The mutation was not seen in 180 nonaffected Jewish subjects.

A second mutation is a 3-bp deletion which predicts the removal of an arginine residue at position 608 of the SMase polypeptide (ΔR608), and which correlates with the subacute form of the disease. This mutation has been described in over 50% of the Jewish patients with subacute disease, but only 6.7% of non-Jewish patients. The genotype/phenotype correlations are instructive: R496L/R496L fibroblasts have 0.7% of normal acid SMase activity and acute disease; R496L/ΔR608 fibroblasts have 4.8% and subacute disease; ΔR608/ΔR608 fibroblasts have 12.8% of normal activity and subacute disease. Because good monospecific antibodies to the acid SMase are not available at present, it is not known whether these mutations affect the activity or stability of the enzyme polypeptide. Correction of the metabolic defect in cultured fibroblasts from type I patients by retroviral-mediated transfer of acid SMase cDNA has been reported (Suchi *et al.*, 1992).

B. Neutral, Mg^{2+}-Stimulated Sphingomyelinase

1. Tissue Distribution and Function

A SMase with a neutral pH optimum that is inhibited by EDTA and stimulated by Mg^{2+} has been described in several mammalian tissues. Although particularly enriched in brain (Rao and Spence, 1976; Gatt, 1976; Spence *et al.*, 1979), enzyme activities with similar properties have been described in liver (Hostetler and Yazaki, 1979), adrenal tissues (Bartolf and Franson, 1986), kidney (Gosh and Chatterjee, 1987), and human urine (Chatterjee and Ghosh, 1989). In cultured neuroblastoma cells and in liver, activity is concentrated in the plasma membrane and microsomes (Hostetler and Yazaki, 1979; Spence *et al.*, 1982), and, in neuroblastoma cells, most of the plasma membrane activity is oriented externally (Mohan Das *et al.*, 1984). Further evidence supporting a plasma membrane localization in cultured renal proximal tubule cells is presented elsewhere in this volume.

Highest activity levels are found in adult mammalian brain, and these levels increase from birth in parallel with neuronal maturation (Spence *et al.*, 1979, 1981). Within brain, activity is highest in gray matter, particularly in the nuclei of the basal ganglia (Spence *et al.*, 1979). Although this nucleus receives extensive afferent input from the dopaminergic system, acute treatments which alter brain dopamine and dopaminergic neuron numbers do not alter neutral SMase activity (Sperker and Spence, 1983). Levels rise in the rat neurohypophysis during dehydration, suggesting a possible role in neurosecretion (Guy *et al.*, 1983). In calf brain synaptosomes (Pellkofer and Sandhoff, 1980) and in cultured neuroblastoma cells (Mooibroek *et al.*, 1985), activity is stimulated by volatile anesthetics; the stimulation is positively correlated with anesthetic dose and potency.

In 3T3-L1 cells incubated for 4 hours with 10^{-7} *M* dexamethasone, neutral SMase activity is stimulated 4-fold (Nelson and Murray, 1982; Ramachandran *et al.*, 1990). In HL-60 human myelocytic cells, 1α,25-dihydroxyvitamin D_3, tumor necrosis factor α, and interferon cause an increase in sphingomyelin hydrolysis, presumably by the activation of a neutral SMase (Okazaki *et al.*, 1990; Kim *et al.*, 1991). These findings, and other described in more detail in other parts of this volume, have implicated the neutral SMase in the initiation of a sphingolipid cycle that is part of the effector mechanism for a variety of biologically active molecules.

2. *Purification and Proteins*

The enzyme is a lipophilic, membrane protein (Spence *et al.*, 1981) and can be solubilized by mild neutral detergents (Rao and Spence, 1976; Gatt, 1982). Activity is generally measured in the presence of detergents such as Cutscum (Gosh and Chatterjee, 1987) or Triton X-100 (Rao and Spence, 1976); studies of activity with varying substrate–Triton mole ratios show complex kinetics (Gatt, 1982). Phosphatidylcholine (Petkova *et al.*, 1986) and phosphatidylserine (Tamiya-Koizumi and Koima, 1986) have been shown to act as lipid activators, and some of the complex effects of detergents may be due to displacement of these lipids from the active site of the enzyme (Gatt, 1982). Using radiation inactivation, the molecular mass has been determined at 165 kDa in samples from three human brains (Levade *et al.*, 1985) and 740 kDa in a fourth brain (Levade *et al.*, 1986). Other studies of the brain enzyme have shown a size of <150 kDa by sucrose density gradient centrifugation and >600 kDa by gel filtration (Spence *et al.*, 1981). All of these studies suggest relatively extensive hydrophobic areas in the enzyme with lipid or detergent binding, and ready formation of aggregates.

The enzyme has been partially purified from brain (Maruyama and Arima, 1989) and purified from human urine (Chatterjee and Ghosh, 1991). Maruyama and Arima (1989) reported the copurification of acid and neutral SMases from a subcellular fraction of rat brain. The neutral enzyme was purified 254-fold with a specific activity of 48 μmol/hr/mg protein. The acid enzyme was purified 668-fold with a specific activity of 25 μmol/hr/mg protein. Although the purified enzymes had distinct pH optima and ion requirements, a similar single band with an apparent molecular mass of 67 kDa was obtained for both on SDS-PAGE. The p*I* values and amino acid compositions were very similar. Both reacted to antibody raised against the neutral enzyme. The authors suggest that the two SMases may have common polypeptides and hence are immunologically cross-reactive.

The human urine enzyme was purified 440-fold from a urine concentrate by a combination of Sephadex column chromatography, preparative isoelectric focusing, and sphingosylphosphocholine-CH Sepharose column chromatography (Chatterjee and Ghosh, 1989, 1991). The purified enzyme had a specific activity of 220 μmol/hr/mg protein but was reported to be 5 to 15-fold higher when protease inhibitors were included during urine collection and enzyme purification. The purified enzyme was a single polypeptide of 92,000 kDa and a p*I* of 6.5. It hydrolyzed sphingomyelin and had low phospholipase A_1 and A_2 activity with phosphatidylcholine as substrate. Monospecific polyclonal antibodies to this enzyme immunoprecipitated the enzyme from urine, but were without effect on acid SMase activity.

The acid and neutral SMases clearly differ in pH optima, ion requirements, tissue location and developmental profiles, inhibition by -SH reagents and by AMP, and in heat stability (Rao and Spence, 1976; Spence *et al.,* 1979). Where monospecific antibodies have been prepared there has been cross-reactivity between the acid and neutral SMase from brain (Maruyama and Arima, 1989), but not from urine (Chatterjee and Ghosh, 1991). Acid SMase cDNA in COS-1 cells expresses acid SMase but not neutral SMase activity (Schuchman *et al.,* 1991). Taken together, this evidence indicates that the acid and neutral SMases of mammalian tissues are separate and distinct entities.

C. Serum Sphingomyelinase

A SMase activity has been described in fetal bovine serum (Spence *et al.,* 1989). Of the sera tested, activity was highest in fetal bovine, followed by newborn bovine > horse > human, with more than 95% of the activity in the lipoprotein-free infranatant ($d > 1.21$). The activity is Zn^{2+} stimulatable and EDTA-inhibited; Mg^{2+} was without effect. Activity can be demonstrated with liposomal sphingomyelin suspensions, but is stimulated 15-fold by Triton X-100. The activity resembles the acid, lysosomal SMase in pH optimum and inhibition by AMP, but differs in heat lability at 55°C, inhibition by EDTA, and metal ion requirement. It resembles the neutral SMase in inhibition by EDTA and heat lability, but differs in pH optima and in specific metal ion requirement. The tissue(s) of origin and biological role(s) of this SMase are unknown.

D. Other Mammalian Sphingomyelinases

Nilsson (1969) described a SMase that is present in the crude protein fraction of human duodenal contents and in meconium, and is enriched in brush-border preparations of rat intestinal mucosa. Activity was not found in pancreatic homogenates or secretions. The activity was optimal at alkaline pH (pH 8.9 to 9.2) and was stimulated by taurocholate. The effects of metal ions and EDTA were not specifically examined. The enzymes may be concerned with the degradation of sphingomyelin in ingested foods.

Yamaguchi and Suzuki (1978) reported a SMase in purified myelin fractions from young adult rats with a neutral pH optimum (7.0) and no requirement for divalent cations. Subsequent studies from this laboratory have not reproduced this observation and the existence of this activity is doubtful (Yamanaka *et al.*, 1981).

III. Avian Sphingomyelinase

Hirshfield and Loyter (1975) described a SMase in chicken erythrocytes with properties that are very similar to those of the mammalian neutral SMases. The pH optimum is 7.0 to 9.0, and the activity is inhibited by EDTA and stimulated by Mn^{2+} and Mg^{2+}. Activity can be demonstrated with the endogenous sphingomyelin of the chicken red cell membranes, and with exogenous sphingomyelin–Triton X-100 or sphingomyelin–cholate micelles. The sphingomyelin in human red cell membranes, which have no endogenous SMase activity, can also be hydrolyzed by the SMase of the chicken erythrocyte membrane preparations (Record *et al.*, 1980). The enzyme is normally latent, and is activated by hypotonic lysis and inhibited by clumping of cells. The role(s) of the avian enzyme remains to be determined.

IV. Insect Sphingomyelinase

Spates *et al.*, (1990) have described a SMase activity in the midgut homogenates of the stable fly, *Stomoxys calcitrans* (L). The enzyme has a pH optimum of 7.6; activity is reduced by dialysis, particularly against imidazole buffers, and is not restored by additions of Mg^{2+} and Mn^{2+}.

V. Bacterial Sphingomyelinases

Several bacteria produce proteins called hemolysins, so named because their presence was first detected by their ability to promote lysis of red blood cells. At least some of these hemolysins are phospholipases and, of these, some appear to be relatively specific SMases. A SMase that hydrolyzes both micellar and membrane-bound sphingomyelin has been purified from *Staphylococcus aureus* (Wadström and Mollby, 1971a,b). The enzyme has a pH otpimum of 7.4 and activity is stimulated by Mg^{2+}. *Clostridium perfringens* secretes two phospholipase C activities resolvable by gel chromatography (Pastan *et al.*, 1968). The first activity is Ca^{2+} dependent and hydrolyzes phosphatidylcholine more rapidly than sphingomyelin. The second activity is specific for sphingomyelin, requires Mg^{2+}, and is inhibited by Ca^{2+}.

Bacillus cereus produces three extracellular phopholipase C's which hydrolyze phosphatidylcholine, sphingomyelin, and phosphatidylinositol (Tomita *et al.*, 1982; Yamada *et al.*, 1988; Gilmore *et al.*, 1989). The *B. cereus* SMase differs from the *S. aureus* in several respects, such as amino acid composition and isoelectric point. Its molecular mass has been reported at 23 and 34 kDa. The enzyme may have a loosely associated Mg^{2+} which accelerates activity. Ca^{2+} may be required for liposomal and membrane-bound substrates to enhance absorption of the enzyme to the substrate surface. The phosphatidylcholine and SMase hydrolyzing activities form a gene cluster. The function of this SMase is not known; however, its production is repressed by inorganic phosphate (P_i) in the growth medium. Guddal *et al.*, (1989) have speculated that, because *B. cereus* is a soil bacterium, and because P_i is an important limiting factor for growth in soil, the enzyme may be part of an efficient P_i retrieval system.

Leptospira interogans also contains SMase activities (Segers *et al.*, 1990). The genetic analysis predicts a protein of 63 kDa. At both the DNA and protein level, the middle part of the Leptospira SMase shares a high degree of similarity with the *S. aureus* and *B. cereus* enzymes, and may be the catalytic portion of the molecule. The amino terminals differ, possibly because the *Leptospira* SMase seems cell bound, whereas the activity of the other two bacterial species is extracelluar.

Acknowledgment

The author's research was supported by a Career Investigatorship and a program grant (PG-16) from the Medical Research Council of Canada.

References

Ahmad, T. Y., Beaudet, A. L., Sparrow, J. T., and Morrisett, J. D. (1986). *Biochemistry* **25,** 4415–4420.

Albouz, S., Le Saux, F., Wenger,D., Hauw, J. J., and Bauman, N. (1986). *Life Sci.* **38,** 357–363.

Allan, D., and Quinn, P. (1988). *Biochem. J.* **254,** 765–771.

Alpert, A., and Beaudet, A. L. (1981). *J. Clin. Invest.* **68,** 1592–1596.

Aubert-Tulkens, G., Van Hoof, F., and Tulkens, P. (1979). *Lab. Invest.* **40,** 481–491.

Barenholz, Y., and Gatt, S. (1982). *In* "Phospholipids" (J. N. Hawthorne and G. B. Ansell, eds.), pp. 129–177. Elsevier, Amsterdam.

Barenholz, Y., and Thompson, T. E. (1980). *Biochim. Biophys. Acta* **604,** 129–158.

Barenholz, Y., Roitman, A., and Gatt, S. (1966). *J. Biol. Chem.* **241,** 3731–3737.

Bartolf, M., and Franson, R. C. (1986). *J. Lipid Res.* **26,** 57–63.

Berheimer, A. W., Linder, R., and Avigad, L. S. (1980). *Infect. Immun.* **29,** 123–131.

Boegheim, J. P. J., Jr., Van Linde, M., Op den Kamp, J. A. F., and Roelofsen, B. (1983). *Biochim. Biophys. Acta* **735,** 438–442.

Brady, R. O., and Koval, G. J. (1958). *J. Biol. Chem.* **233,** 26–31.

Brady, R. O., Bradley, R. M., Young, O. M., and Kaller, H. (1965). *J. Biol. Chem.* **240,** 3693–3694.

Byers, D. M., Rastogi, S. R., Cook, H. W., Palmer, F. B. St. C., and Spence, M. W. (1989). *Biochem. J.* **262,** 713–719.

Byers, D. M., Morgan, M. W., Cook, H. W., Palmer, F. B., and Spence, M. W. (1992). *Biochim. Biophys. Acta* **1138,** 20–26.

Chakravarthy, B., Spence, M. W., CLarke, J. T. R., and Cook, H. W. (1985). *Biochim. Biophys. Acta* **812,** 223–233.

Chatterjee, S., and Ghosh, N. (1989). *J. Biol. Chem.* **264,** 12554–12561.

Chatterjee, S., and Ghosh, N. (1991) *In* "Methods in Enzymology" (E. Dennis, ed.), vol. 197, pp. 540–547. Academic Press, San Diego.

Cohen, R., and Barenholz, Y. (1984). *Biochim. Biophys. Acta* **778,** 94–104.

Cook, H. W., Palmer, F. B. St. C., Byers, D. M., and Spence, M. W. (1988). *Anal. Biochem.* **174,** 552–560.

Dawson, R. M. C., Hemington, N., and Irvine, R. F. (1985). *Biochem. J.* **230,** 61–68.

Dean, R. T. (1977). *Biochem. J.* **168,** 603–605.
de Duve, C. (1983). *FEBS Lett.* **137,** 391–397.
Demel, R. A., Jansen, J. W. C. M., Van Dijuk, P. W. M., and Van Deenen, L. L. M. (1977). *Biochim. Biophys. Acta* **465,** 1–101.
Dennis, E. A., Rhee, S. G., Billah, M. M., and Hannun, Y. A. (1991). *FASEB J.* **5,** 2068–2077.
Diringer, H., Marggraf, W. D., Koch, M. A., and Anderer, F. A. (1972). *Biochem. Biophys. Res. Commun.* **47,** 1345–1352.
Duncan, R., and Pratten, M. K. (1977). *J. Theor. Biol.* **66,** 727–735.
Epand, R. M., Boni, L. T., and Hui, S. W. (1982). *Biochim. Biophys. Acta* **692,** 330–338.
Eppler, C. M., Malewicz, B., Jenkin, H. M., and Baumann, W. J. (1987). *Lipids* **22,** 351–357.
Esumi, M., and Fujiki, H. (1983). *Biochem. Biophys. Res. Commun.* **112,** 709–716.
Ferlinz, K., Hurwitz, R., and Sandhoff, K. (1991). *Biochem. Biophys. Res. Commun.* **179,** 1187–1191.
Fowler, S. (1969). *Biochim. Biophys. Acta* **191,** 481–484.
Fujino, Y., Nakano, M., Negishi, T., and Ito, S. (1968). *J. Biol. Chem.* **243,** 4650–4651.
Futerman, A. H., Stieger, B., Hubbard, A. L., and Pagano, R. E. (1990). *J. Biol. Chem.* **265,** 8650–8657.
Gal, A. E., Brady, R. O., Barranger, J. A., and Pentchev, P. G. (1980). *Clin. Chim. Acta* **104,** 129–132.
Gatt, S. (1976). *Biochem. Biophys. Res. Commun.* **68,** 235–241.
Gatt, S. (1982). *In* "Phospholipids in the Nervous System" (L. A. Horrocks, G. B. Ansell, and G. Porcellati, eds.), pp. 181–197. Raven Press, New York.
Gatt, S., Herzl, A., and Barenholz, Y. (1973). *FEBS Lett.* **30,** 281–285.
Gatt, S., Barenholz, Y., Goldberg, R., Dinur, T., Besley, G., Leibovitz-Ben Gershon, Z., Rosenthal, J., Desnick, R. J., Devine, E. A., Shafit-Zagardo, B., and Tsuruki, F. (1981). *In* "Methods in Enzymology" (J. Lowenstein, ed.), Vol. 72, pp. 351–375. Academic Press, New York.
Gilmore, M. S., Cruz-Rodz, A. L., Leimeister-Wachter, M., Kreft, J., and Goebel, W. (1989). *J. Bacteriol.* **171,** 744–753.
Gosh, P., and Chatterjee, S. (1987). *J. Biol. Chem.* **262,** 12550–12556.
Gronberg, L., and Slotte, J. P. (1990). *Biochemistry* **29,** 3173–3178.
Guddal, P. H., Johansen, T., Schulstad, K., and Little, C. (1989). *J. Bacteriol.* **171,** 5702–5706.
Gupta, A. K., and Rudney, H. (1991). *J. Lipid Res.* **32,** 125–136.
Guy, N. C., Clarke, J. T. R., Spence, M. W., and Cook, H. W. (1983). *Brain Res. Bull.* **10,** 603–606.
Hannun, Y. A., and Bell, R. M. (1989). *Science* **243,** 500–507.
Heller, M. (1978). *Adv. Lipid Res.* **16,** 267–326.
Hirshfield, D., and Loyter, A. (1975). *Arch. Biochem. Biophys.* **167,** 186–192.
Hostetler, K. Y., and Yazaki, P. J. (1979). *J. Lipid Res.* **20,** 456–463.
Jobb, E. A., and Callahan, J. W. (1987). *J. Inherited Metab. Dis.* **10,** 326–328.
Jobb, E. A., and Callahan, J. W. (1989). *Biochem. Cell Biol.* **67,** 801–807.
Jones, C. S., Shankaran, P., and Callahan, J. W. (1981). *Biochem. J.* **195,** 373–382.
Jones, C. S., Davidson, D. J., and Callahan, J. W. (1982). *Biochim. Biophys. Acta* **701,** 261–268.
Kanfer, J. N. (1980). *Can. J. Biochem.* **58,** 1370–1380.
Kanfer, J. N., Young, O. M., Shapiro, D., and Brady, R. O. (1966). *J. Biol. Chem.* **240,** 1081–1084.
Kim, M. Y., Linardič, C., Obeid, L., and Hannun, Y. (1991). *J. Biol. Chem.* **266,** 484–489.
Kishimoto, Y., and Kawamura, N. (1979). *Mol. Cell. Biochem.* **23,** 17–25.

Kobayashi, T., Pimplikar, S. W., Parton, R. G., Bhakdi, S., and Simons, K. (1992). *FEBS Lett.* **300,** 227–231.

Kolesnick, R. N. (1991). *Prog. Lipid Res.* **30,** 1–38.

Koval, M., and Pagano, R. E. (1991). *Biochim. Biophys. Acta* **1082,** 113–125.

Kretz, K. A., Carson, G. S., Morimoto, S., Kishimoto, Y., Fluharty, A. L., and O'Brien, J. S. (1990). *Proc. Natl. Acad. Sci. U. S. A.* **87,** 2541–2544.

Lange, Y., Swaisgood, M. H., Ramos, B. V., and Steck, T. L. (1989). *J. Biol. Chem.* **264,** 3786–3793.

Levade, T., and Gatt, S. (1987). *Biochim. Biophys. Acta* **918,** 250–259.

Levade, T., Potier, M., Salvayre, R., and Douste-Blazy, L. (1985). *J. Neurochem.* **45,** 630–632.

Levade, T., Salvayre, R., Potier, M., and Douste-Blazy, L. (1986). *Neurochem. Res.* **11,** 1131–1138.

Levin, I. W., Thompson, T. E., Barenholz, Y., and Huang, C. (1985). *Biochemistry* **24,** 6282–6286.

Levran, O., Desnick, R. J., and Schuchman, E. H. (1991a). *J. Clin. Invest.* **88,** 806–810.

Levran, O., Desnick, R. J., and Schuchman, E. H. (1991b). *Proc. Natl. Acad. Sci. U. S. A.* **88,** 3748–3752.

Linder, R., Bernheimer, A. W., and Kim, K. (1977). *Biochim. Biophys. Acta* **467,** 290–300.

Liscum, L., and Faust, J. R. (1987). *J. Biol. Chem.* **262,** 17002–17008.

Marggraf, W. D., Anderer, F. A., and Kanfer, J. N. (1981). *Biochim. Biophys. Acta* **664,** 61–73.

Maruyama, E. N., and Arima, M. (1989). *J. Neurochem.* **52,** 611–618.

Masson, M., Albouz, S., Boutry, J. M, Spezzatti, B., Castagna, M., and Baumann, N. (1989). *J. Neurochem.* **52,** 1645–1647.

Mazière, J. C., Mazière, C., Mora, L., Bereziat, G., and Polonovski, J. (1981). *Biochem. Biophys. Res. Commun.* **100,** 1299–1304.

Mazière, J. C., Mazière, C., Mora, L., Gallie, F., and Polonovski, J. (1983). *Biochem. Biophys. Res. Commun.* **112,** 860–865.

Mazière, J. C., Mazière, C., Mora, L., and Polonovski, J. (1984). *FEBS Lett.* **173,** 159–163.

McIntosh, T. J., Simon, S. A., Needham, D., and Huang, C. H. (1992). *Biochemistry* **31,** 2020–2024.

Merrill, A. H., Jr., and Jones, D. D. (1990). *Biochim. Biophys. Acta* **1044,** 1–12.

Mohan Das, D. V., Cook, H. W., and Spence, M. W. (1984). *Biochim. Biophys. Acta* **777,** 339–342.

Mooibroek, M. J., Cook, H. W., Clarke, J. T., and Spence, M. W. (1985). *J. Neurochem.* **44,** 1551–1558.

Morell, P., and Radin, N. S. (1970). *J. Biol. Chem.* **245,** 342–350.

Morimoto, S., Martin, B. M., Kishimoto, V., and O'Brien, J. S. (1988). *Biochem. Biophys. Res. Commun.* **156,** 403–410.

Myher, J. J., Kuksis, A., Shepherd, J., Packard, C. J., Morrisett, J. D., Taunton, O. D., and Golto, A. (1981). *Biochim. Biophys. Acta* **666,** 110–119.

Narimatsu, S., Soeda, S., Tanaka, T., and Kishimoto, Y. (1986). *Biochim. Biophys. Acta* **877,** 334–341.

Nelson, D. H., and Murray, D. K. (1982). *Proc. Natl. Acad. Sci. U. S. A.* **79,** 6690–6692.

Nelson, D. H., Murray, D. K., and Brady, R. O. (1982). *J. Clin. Endocrinol. Metab.* **54,** 292–295.

Nilsson, A. (1969). *Biochim. Biophys. Acta* **176,** 339–347.

Noda, S., Hayasaka, S., and Setogawa, T. (1989). *Ophthalmic Res.* **21,** 200–205.

O'Brien, J. S., and Kishimoto, Y. (1991). *FASEB J.* **5,** 301–308.

O'Brien, J. S., and Sampson, E. L. (1965). *J. Lipid Res.* **6,** 545–569.
O'Brien, J. S., Kretz, K. A., Dewji, N., Wenger, D. A., Esch, F., and Fluharty, A. L. (1988). *Science* **241,** 1098–1101.
Okazaki, T., Bell, R. M., and Hannun, Y. A. (1989). *J. Biol. Chem.* **264,** 19076–19080.
Okazaki, T., Bielawaska, A., Bell, R. M., and Hannun, Y. A. (1990). *J. Biol. Chem.* **265,** 15823–15831.
Ong, D. E., and Brady, R. N. (1973). *J. Biol. Chem.* **248,** 3884–3888.
Pastan, I., Macchia, V., and Katzen, R. (1968). *J. Biol. Chem.* **243,** 3750–3755.
Pellkofer, R., and Sandhoff, K. (1980). *J. Neurochem.* **34,** 988–992.
Pentchev, P. G., Brady, R. O., Gal, A. E., and Hibbert, S. R. (1977). *Biochim. Biophys. Acta* **488,** 312–321.
Pentchev, P. G., Comly, M. E., Kruth, H. S., Vanier, M. T., Wenger, D. A., Patel, S., and Brady, R. O. (1985). *Proc. Natl. Acad. Sci. U. S. A.* **82,** 8247–8251.
Pereira, L., Desnick, R. J., Alder, D., Disteche, C. M., and Schuchman, E. H. (1991). *Genomics* **9,** 8531–8539.
Petkova, D. H., Momchilova, A. B., and Koumanov, K. S. (1986). *Biochimie* **68,** 1195–1200.
Poulos, A., Ranieri, E., Shankaran, P., and Callahan, J. W. (1984). *Pediatr. Res.* **18,** 1088–1093.
Quintern, L. E., and Sandhoff, K. (1991). *In* "Methods in Emzymology" (E. Dennis, ed.), Vol. 197, pp. 536–540. Academic Press, San Diego.
Quintern, L. E., Weitz, G., Nehrkorn, H., Tager, J. M., Schram, A. W., and Sandhoff, K. (1987). *Biochim. Biophys. Acta* **922,** 323–336.
Quintern, L. E., Zenk, T. S., and Sandhoff, K. (1989a). *Biochim. Biophys. Acta* **1003,** 121–124.
Quintern, L. E., Schuchman, E. H., Levran, O., Suchi, M., Ferlinz, K., Reinke, H., Sandhoff, K., and Desnick, R. J. (1989b). *EMBO. J.* **8,** 2469–2473.
Ramachandran, C. K., Murray, D. K., and Nelson, D. H. (1990). *Biochem. Biophys. Res. Commun.* **167,** 607–613.
Rao, B. G., and Spence, M. W. (1976). *J. Lipid Res.* **17,** 506–515.
Record, M., Loyter, A., and Gatt, S. (1980). *Biochem. J.* **187,** 115–121.
Robinson, B. S., Johnson, D. W., and Poulos, A. (1992). *J. Biol. Chem.* **267,** 1746–1751.
Rouser, G., Berry, J. F., Marinetti, G. V., and Stolz, E. (1953). *J. Am. Chem. Soc.* **75,** 310–320.
Rouser, G., Nelson, G. J., Fleischer, S., and Simon, G. (1968). *In* "Biological Membranes" (D. Chapman, ed.), pp. 5–69. Academic Press, London.
Rouser, G., Kitchevsky, G., Yamamoto, A., and Baxter, C. F. (1972). *Adv. Lipid Res.* **10,** 261–360.
Sakuragawa, N. (1982). *J. Biochem. (Tokyo)* **92,** 637–646.
Sankaram, M. B., and Thompson, T. E. (1990). *Biochemistry* **29,** 10670–10675.
Sato, M., Yoshida, Y., Sakuragawa, N., and Arima, M. (1988). *Biochim. Biophys. Acta* **962,** 59–65.
Schneider, P. B., and Kennedy, E. P. (1967). *J. Lipid Res.* **8,** 202–209.
Schneider, P. B., and Kennedy, E. P. (1968). *J. Lipid Res.* **9,** 58–64.
Schuchman, E. H., Suchi, M., Takahaski, T., Sandhoff, K., and Desnick, R. J. (1991). *J. Biol. Chem.* **66,** 8531–8539.
Schuchman, E. H., Levran, O., Pereira, L. V., and Desnick, R. J. (1992). *Genomics* **12,** 197–205.
Segers, R. P., van der, D. Rift A., de Nijs, A., Corcione, P., van der, Z. eijst BA., and Gaastra, W. (1990). *Infect. Immun.* **58,** 2177–2185.
Shapiro, D., and Flowers, H. M. (1962). *J. Am. Chem. Soc.* **84,** 1047–1050.

Sidhu, H. S., Rastogi, S. A. R., Byers, D. M., Cook, H. W., Palmer, F. B. St. C., and Spence, M. W. (1992). *Biochim. Biophys. Acta* **1124,** 29–35.

Slotte, J. P., and Bierman, E. L. (1988). *Biochem. J.* **250,** 653–658.

Sood, C. K., Sweet, C., and Zull, J. E. (1972). *Biochim. Biophys. Acta* **282,** 429–434.

Spates, G. E., Bull, D. L., and Chen, A. C. (1990). *Arch. Insect Biochem. Physiol.* **14,** 1–12.

Spence, M. W., and Burgess, J. K. (1978). *J. Neurochem.* **30,** 917–919.

Spence, M. W., and Callahan, J. W. (1989). *In* "The Metabolic Basis of Inherited Disease" (C. R. Scriver, A. L. Beaudet, W. S. Sly, and D. Valle, eds.), 6th ed., pp. 1655–1676. McGraw-Hill, New York.

Spence, M. W., Burgess, J. K., and Sperker, E. R. (1979). *Brain Res.* **168,** 543–551.

Spence, M. W., Burgess, J. K., Sperker, E. R., Hamed, L., and Murphy, M. G. (1981). *In* "Lysosomes and Lysosomal Storage Diseases" (J. W. Callahan and J. A. Lowden, eds.), pp. 219–228. Raven Press, New York.

Spence, M. W., Wakkary, J., Clarke, J. T. R., and Cook, H. W. (1982). *Biochim. Biophys. Acta* **719,** 162–164.

Spence, M. W., Clarke, J. T. R., and Cook, H. W. (1983). *J. Biol. Chem.* **258,** 8595–8600.

Spence, M. W., Byers, D. M., Palmer, F. B., and Cook, H. W. (1989). *J. Biol. Chem.* **264,** 5358–5363.

Sperker, E. R., and Spence, M. W. (1983). *J. Neurochem.* **40,** 1182–1184.

Scribney, M. (1966). *Biochim. Biophys. Acta* **125,** 542–547.

Scribney, M., and Kennedy, E. P. (1958). *J. Biol. Chem.* **233,** 1315–1322.

Stoffel, W. (1970). *Chem. Phys. Lipids* **5,** 139–158.

Stoffel, W., and Melzner, I. (1980). *Hoppe- Seyler's Z. Physiol. Chem.* **361,** 755–771.

Stoffel, W., LeKim, D., and Sticht, G. (1968). *Hoppe-Seyler's Z. Physiol. Chem.* **349,** 1637–1644.

Stoffel, W., LeKim, D., and Tschung, T. S. (1971). *Hoppe-Seyler's Z. Physiol. Chem.* **352,** 1058–1064.

Stoffel, W., Zierenberg, O., Tunggal, B., and Schreiber, E. (1974). *Proc. Natl. Acad. Sci. U. S. A.* **71,** 3696–3700.

Stoffel, W., Kruger, E., and Melzner, I. (1980). *Hoppe-Seyler's Z. Physiol. Chem.* **361,** 773–779.

Strasberg, P. M. S., and Callahan, J. W. (1988). *In* "Lipid Storage Disorders: Biological and Medical Aspects" (R. L. Salvayre, L. Douste-Blazy, and S. Gatt, eds.), pp. 601–606. Plenum, New York.

Suchi, M., Dinur, T., Desnick, R. J., Gatt, S., Pereira, L., Gilboa, E., and Schuchman, E. H. (1992). *Proc. Natl. Acad. Sci. U. S. A.* **89,** 3227–3231.

Suzuki, K., and Suzuki, Y. (1982). *In* "The Metabolic Basis of Inherited Disease" (J. B. Stanbury, J. B. Wyngaarden, D. S. Fredrickson, J. L. Goldstein, and M. S. Brown, eds.), 5th ed., pp. 857–880. McGraw-Hill, New York.

Tamiya-Koizumi, K., and Koima, K. (1986). *J. Biochem. (Tokyo)* **99,** 1803–1806.

Thomas, G. H., Tuck-Muller, C. M., Miller, C. S., and Reynolds, L. W. (1989). *J. Inherited Metab. Dis.* **12,** 139–151.

Thudicum, J. L. W. (1962). "A Treatise on the Chemical Constitution of the Brain" (a facsimile edition of the original). Baillière, London.

Tomita, M., Taguchi, R., and Ikezawa, H. (1982). *Biochim. Biophys. Acta* **704,** 90–99.

Ullman, M. D., and Radin, N. S. (1972). *Arch. Biochem. Biophys.* **152,** 767–777.

Ullman, M. D., and Radin, N. S. (1974). *J. Biol. Chem.* **249,** 1506–1512.

van den Hill, A., van Heusden, P. H., and Wirtz, K. W. A. (1985). *Biochim. Biophys. Acta* **833,** 354–357.

Vanha-Perttula, T. (1988). *FEBS Lett.* **233,** 263–267.

Vanier, M. T., Pentchev, P., Rodriguez-Lafrasse, C., and Rousson, R. (1991). *J. Inherited Metab. Dis.* **14,** 580–595.

Voelker, D. R., and Kennedy, E. P. (1982). *Biochemistry* **21,** 2753–2759.

Wadström, T., and Mollby, R. (1971a). *Biochim. Biophys. Acta* **242,** 288–307.

Wadström, T., and Mollby, R. (1971b). *Biochim. Biophys. Acta* **242,** 308.

Watanabe, K., Sakuragawa, N., Arima, M., and Satoyoshi, E. (1983). *J. Lipid Res.* **24,** 596–603.

Weinreb, N. J., Brady, R. O., and Tappel, A. L. (1968). *Biochim. Biophys. Acta* **159,** 141–146.

White, D. A. (1974). *In* "Form and Function of Phospholipids" (G. B. Ansell, J. N. Hawthorne, and R. M. C. Dawson, eds.), p. 441. Elsevier, Amsterdam.

Widnell, C. C., and Unkeless, J. C. (1968). *Proc. Natl. Acad. Sci. U. S. A.* **61,** 1050–1057.

Williams, R. D., Wang, E., and Merrill, A. H. (1984). *Arch. Biochem. Biophys.* **228,** 282–291.

Xu, X. X., and Tabas, I. (1991). *J. Biol. Chem.* **266,** 24849–24858.

Yamada, A., Tsukagoshi, N., Udaka, S., Sasaki, T., Makino, S., Nakamura, S., Little, C., Tomita, M., and Ikezawa, H. (1988). *Eur. J. Biochem.* **175,** 213–220.

Yamaguchi, S., and Suzuki, K. (1978). *J. Biol. Chem.* **253,** 4090–4092.

Yamamura, T., and Tezuka, T. (1990). *J. Dermatol. Sci.* **1,** 79–83.

Yamanaka, T., Hanada, E., and Suzuki, K. (1981). *J. Biol. Chem.* **256,** 3884–3889.

Yavin, E., and Gatt, S. (1969). *Biochemistry* **6,** 1692–1698.

Yeagle, P. L., and Young, J. E. (1986). *J. Biol. Chem.* **261,** 8175–8181.

Yoshida, Y., Arimoto, K., Sato, M., Sakuragawa, N., Arima, M., and Satoyoshi, E. (1985). *J. Biochem. (Tokyo)* **98,** 1669–1679.

Neutral Sphingomyelinase

SUBROTO CHATTERJEE

Department of Pediatrics
Lipid Research Arteriosclerosis Division
Johns Hopkins University School of Medicine
Baltimore, Maryland 21205

I. Introduction

A. Overview

Sphingomyelinases (E.C. 3.1.4.12) are a group of phospholipases that catalyze the hydrolytic cleavage of sphingomyelin via the reaction (Spence and Callahan, 1989) below.

$$\text{Sphingomyelin} \rightarrow \text{Ceramide} + \text{Phosphocholine}$$

For several decades, investigators (Bartolf and Franson, 1986) have associated the lack of sphingomyelinase, the consequences of which range from mild to severe lipid storage disorder to death in man. Recently, however, there has been renewed interest in sphingomyelinase and sphingomyelin metabolism because of the finding that the backbone of sphingomyelin (i.e., sphingosine) might serve as an inhibitor of protein kinase C. The availability of enzymes involved in sphingomyelin metabolism and corresponding antibodies and the involvement of ceramide in cell differentiation have also contributed to a resurgence of interest in this field. In addition to the possible biological roles of sphingosine and ceramide in signal transduction pathways involving cell regulation, recent evidence from several laboratories suggests that sphingomyelinase may be involved in the mobilization of cell surface cholesterol, in cholesteryl ester synthesis, and in inducing low-density lipoprotein (LDL) receptor activity. More importantly, sphingomyelinase may serve as a signal for various exogenous effectors such as antibiotics, drugs, and growth factors, which influence the normal physiology of cells.

This article aims to serve as a comprehensive review of sphingomyelinase, particularly neutral sphingomyelinase (N-SMase), and its role as a biomodulator (traffic controller) in signal transduction of reactions in various biological systems. Our emphasis here is on the isolation, characterization, and localization of N-SMase and on its role in coordinating the regulation of lipid metabolism. We also look at how exogenous effectors modulate the function of this enzyme and thereby affect the physiology of cells.

B. Sphingomyelinase, Sphingomyelin Structure, Localization, Tissue Distribution, and Cholesterol

At least three kinds of sphingomyelinases have been reported so far (Spence and Callahan, 1989). The classical acid sphingomyelinase (A-SMase) cleaves sphingomyelin at an acidic pH optimum (usually around pH 4.0). This enzyme is predominantly localized in the lysosomes. It digests LDL-associated sphingomyelin that is endocytosed via the LDL receptor pathway and delivered to the lysosomes. A lack or a deficiency of the A-SMase leads to the storage of sphingomyelin in Niemann-Pick disease. Another heat-stable, Zn^{2+} and detergent-stimulatable acid sphingomyelinase (pH optimum 5.4) has been reported to occur in the serum of fetal calf. The biological role of this enzyme has not been shown; however, the enzyme may contribute to some of the A-SMase activity present in cultured cells and thus may confound the interpretation of data. The classical neutral sphingomyelinase (N-SMase) from human urine has a neutral pH optimum, whereas the bacterial enzyme has a trimodal pH optima (see below). Direct evidence based on the use of purified N-SMase from our laboratory indicates that this enzyme is involved in the catabolism of cell surface sphingomyelin and the mobilization of cholesterol.

Sphingomyelin has three components: a long-chain sphingoid base backbone, a fatty acid, and a polar head group phosphocholine (Barenholz and Gatt, 1982). The sphingoid base is attached to the fatty acid at the carbon 2 position to form ceramide (Fig. 1). Although palmitic acid (16 : 0) is the most common fatty acid constituent of sphingosine, the fatty acid component may vary widely from tissue to tissue and may vary according to age and diet (Frederickson and Sloan, 1972). Earlier experiments by Bretscher (1973) revealed that sphingomyelin is localized on the outer leaflet of the plasma membrane of many cells. While most if not all of the cellular sphingomyelin is concentrated with the plasma membrane, some was found in microsomes, less in the nucleus, and still less in mitochondria. Most probably, sphingomyelin in human plasma is predominantly associated with the lipoproteins. Thus very-low-density lipoproteins, LDL, and high-density lipoproteins represent approximately 23, 25, and 13%, respectively, of the total phospholipid associated with such lipoproteins (Chapman, 1980).

Like sphingomyelin, most if not all of the unesterified cholesterol is located on the cell surface (Steck, 1988). Because of the selective affinity of cholesterol toward sphingomyelin, sphingomyelin may serve as a trap for cholesterol in various tissues. Accordingly, because of a strong association between sphingomyelin and unesterified cholesterol, the Niemann-Pick group of diseases have been appropriately referred to as "sphingomyelin-cholesterol lipidosis" (Spence and Callahan, 1989).

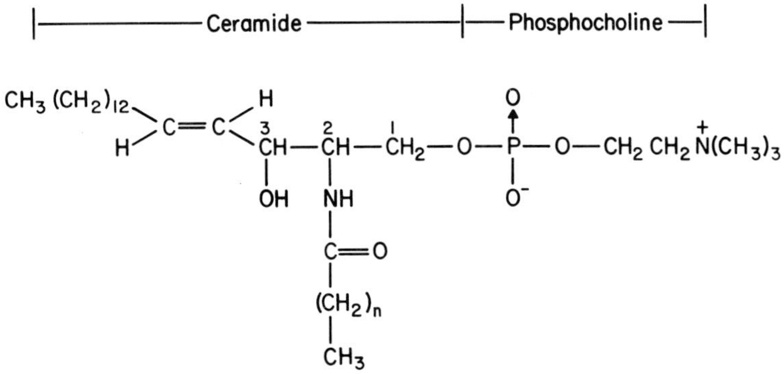

|——— Ceramide ———|— Phosphocholine —|

$CH_3(CH_2)_{12}CH{=}CH{-}\overset{3}{C}H(OH){-}\overset{2}{C}H(NH{-}C({=}O){-}(CH_2)_n{-}CH_3){-}\overset{1}{C}H_2{-}O{-}P(\rightarrow O)(O^-){-}O{-}CH_2CH_2\overset{+}{N}(CH_3)_3$

FIG. 1. Structure of sphingomyelin.

II. The Isolation and Properties of Neutral Sphingomyelinase and Its Modulation by Various Compounds

The finding in our laboratory that human urine contains N-SMase activity was purely serendipitous. It stemmed from our earlier studies that showed that patients taking gentamicin (an amino glycoside antibiotic) have lower N-SMase activity in their urinary proximal tubular cells than do untreated normolipidemic volunteers (see below). Consequently, we rationalized that the enzyme may be present in the urine in a soluble form.

A. Isolation and Purification from Human Urine of N-SMase and Its Storage

Our standard approach to isolate N-SMase from human urine has been described in considerable detail elsewhere (Chatterjee and Ghosh, 1989, 1991). In brief, we collect at least 10 liters of fresh urine from normolipidemic adult males. Sodium azide (0.001%) and phenylmethylsulfonyl fluoride (100 m*M*) are essential preservatives to avoid proteolytic cleavage of N-SMase. The urine sample is immediately centrifuged at 4°C at high speed (16,270 *g*, 30 minutes). The supernatant is subjected to ultrafiltration by use of a Millipore filter (PT-GC-OMT) with a nominal cut of M_r 10,000. This procedure, which is not labor intensive, has a dual purpose. First, it removes salts and a large number of low-molecular-weight proteins from the urine supernatant. Second, it concentrates urine from 10 liters down to 200 ml in a few hours. After concentration, the sample is subject to a purification process that employs Sephadex column chromatography, preparative isoelectric focusing, and affinity column chromatography on sphingosylphosphocholine CH-Sepharose columns. The purified enzyme can be stored at 4°C for up to 6 weeks without loss of activity. However, prolonged storage (6 months or more) at 4°C results in loss of activity.

B. Isolation of N-SMase from Tissues

We have been able to isolate and solubilize a membrane-bound N-SMase from cultured human kidney proximal tubular (PT) cells. Our approach has been to isolate plasma membranes from such cells by employing classical subcellular fractionation procedures (Chatterjee and Ghosh, 1989, 1991). The enzyme is solubilized from the plasma membrane preparations through extraction with 0.2 *M* Tris glycine buffer, pH 7.4, containing 0.01% cutscum. The solubilized preparation is next subjected to the purification steps described above. This protocol may also be adopted to isolate N-SMase from solid tissues such as human kidney.

C. Characterization and Determination of Properties

Upon polyacrylamide gel electrophoresis, the urinary N-SMase resolves as a single protein band having an apparent M_r of 92,500. The apparent M_r of the antibody-recognizable proteins in the urinary N-SMase and the membrane-bound N-SMase was approximately 92,000 (Chatterjee and Ghosh, 1989).

The properties of N-SMase from bacteria, human urine, and human brain (Levade *et al.,*) 1986 are compared in Table I. Using a natural substrate (^{14}C) sphingomyelin, we found that a commercially available sphingomyelinase from *Streptomyces* species has trimodal pH optima of 4.2, 6.5, and 8.0. In contrast, the urinary N-SMase has a pH optimum of 7.4. N-SMase clearly differs from A-SMase (described elsewhere in this monograph; also see Slife *et al.,* 1989) in several ways. For instance, N-SMase is inactive toward sphingomyelinase at acid pH and it is denatured at 55°C. In addition, its requirement for metal ions (e.g., Mg^{2+}) and its inactivation by EDTA are other physical or chemical properties that clearly distinguish this enzyme from A-SMase. Most importantly, antibody against N-SMase does not immunoprecipitate A-SMase and vice versa.

Sphingolipid activator proteins (SAPs) or saposins have been shown to stimulate the catabolism of sphingomyelin via A-SMase (Ho and O'Brien, 1971; Morimoto *et al.,* 1990). At least four kinds of saposins (A, B, C,

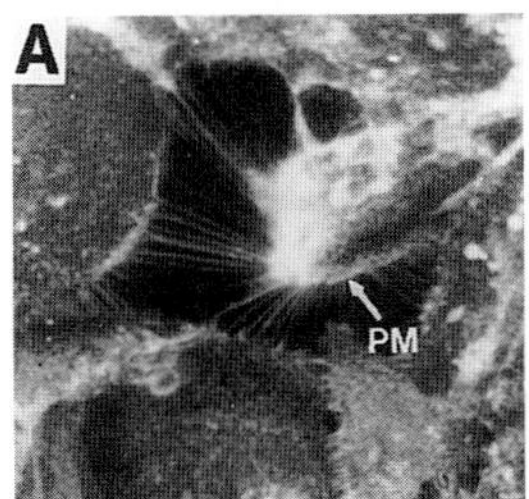

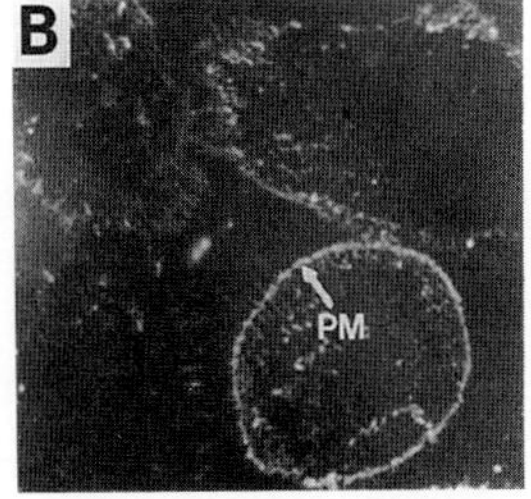

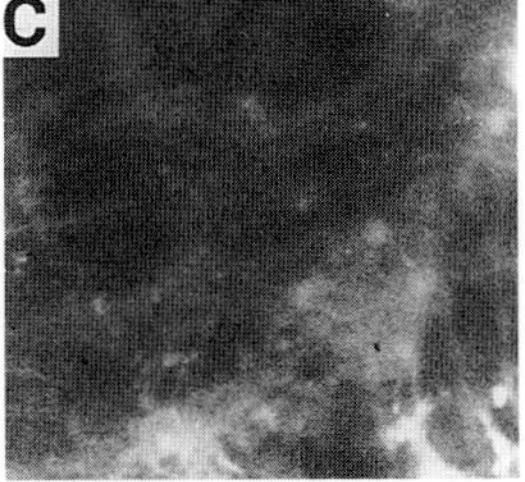

FIG. 2. Fluorescence micrographs of cultured human kidney proximal tubular (PT) cells after incubation with FITC-conjugated antibody against neutral sphingomyelinase (N-SMase). Human PT cells ($\times 10^5$) were seeded on glass coverslips in 60 × 16 mm^2 plastic petri dishes and grown for 7 days in medium containing 10% fetal calf serum and no antibiotics. Next, the dishes were washed with ice-cold phosphate-buffered saline and incubated at 4°C with immunopurified antibody against N-SMase (dilution 1 ÷ 10 in saline) for 1 hour. The cells were washed and examined under a fluorescent light microscope with epi-illumination and photographed (×400). (A) cells incubated with FITC-labeled antibody against N-SMase; (B) cells permeabilized with 0.1% Triton X-100 for 1 minute at room temperature followed by immunostaining; (C) cells incubated with FITC-labeled preimmune rabbit IgG. PM, Plasma membrane.

Table I
PROPERTIES OF NEUTRAL SPHINGOMYELINASE FROM VARIOUS SOURCES

	Bacteria	Human urine	Human brain[a]
I. Physical/chemical properties			
Molecular weight	N/A	92,500	167,000
pH optima	4.2–8.0	7.4	7.0–7.5
p*I* value	N/A	6.5	N/A
Heat stability (50–55°C)	N/A	Unstable	Unstable
Effect of Mg^{2+}	Activatable	Activatable	Activatable
Detergent requirement	Triton X-100, Brij	Cutscum	Triton X-100
Effect of 5′-adenosine monophosphate	N/A	Ineffective	Ineffective
Effects of trifluoroacetic acid	N/A	Inactive enzyme	N/A
II. Substrate specificity			
Sphingomyelin	Yes	Yes	Yes
Trinitrophenylamino-lauryl sphingomyelin	Yes	Yes	Yes
Pyrenedecanoylsphingomyelin	N/A	N/A	Yes
2 *N*-Hexadecanoylamide-nitro-phenylphosphorylcholine	N/A	N/A	No
Phosphatidylcholine	High activity	Low activity	No
Phosphatidylglycerol	N/A	N/A	
Saposin D	N/A	Ineffective	N/A

[a] Taken from Levade *et al.* (1986) with the permission of the authors.

and D) have been isolated. Their apparent M_r is approximately 8,000–13,000 and all are derived from the lysosomes. Since saposin D has been shown to act specifically on sphingomyelinase without complexing with the substrate (i.e., sphingomyelin), we were interested to assess whether this protein would have an effect on our purified urinary N-SMase. Accordingly, N-SMase activity was measured in the presence of saposin D (a gift from Dr. Yasuo Kishimoto, University of California, San Diego) and with and without detergents and metal ions and bovine serum albumin. The activity of A-SMase (bovine spleen) in both the presence and the absence of these compounds was also measured simultaneously to serve as a positive control. We found that saposin D did not alter N-SMase activity. Thus, reaction to saposin D may be considered another criterion on which to distinguish N-SMase from A-SMase. Most probably, N-SMase and A-SMase have diverse primary structures but share the same catalytic domain required to hydrolyze sphingomyelin.

D. Tissue Distribution and Localization

Nilsson (1968) was the first to report on a sphingomyelinase of intestinal origin that cleaved sphingomyelin at neutral or alkaline pH. Gatt (1976) characterized N-SMase in rat livers, and Spence and associates (Rao and Spence, 1976; Mooibrock *et al.,* 1985) characterized it in human brain and in cultured murine neuroblastoma cells. Gray matter appears to have highest N-SMase activity (Rao and Spence, 1976), as do dopamine-enriched areas in brain. We have shown that human kidney, cultured human PT cells, and skin fibroblasts have N-SMase activity. The presence of N-SMase in human urine is presumed to derive from the exfoliation of cells from human kidney and the urinary tract (Chatterjee and Ghosh, 1989). Most probably, N-SMase is a ubiquitous enzyme in mammalian tissues.

Localization studies employing subcellular fractionation (Leelavathi *et al.,* 1970) and measurement of enzymatic activity revealed that N-SMase is localized predominantly in the plasma membrane and to a lesser extent in microsomes in rat liver (Hostetler and Yakazaki, 1979; Slife *et al.,* 1989), bovine adrenal glands (Bartolf and Franson, 1986), human kidney, and cultured human PT cells (Ghosh and Chatterjee, 1987, see Table II). However, other reports reveal that N-SMase is exclusively localized and externally oriented in the plasma membranes of neuroblastoma cells (Das *et al.,* 1984). In none of these studies has N-SMase been associated with the lysosomes.

We have taken the localization studies on N-SMase a step further in that we have employed two diverse approaches to study localization of this enzyme in human PT cells (Chatterjee, 1990). In the first, we prepared fluorescein isothiocyanate (FITC)-conjugated N-SMase antibody and used the preparation to localize N-SMase in cultured human PT cells. In addition, we used C_6-NBD-sphingomyelin (a gift from Dr. Richard Pagano) to study its hydrolysis in our PT cells. Further details are given in the figure legends of Figs. 2 and 3. In the second approach, we carried out cell surface labeling of protein with ^{125}I. This labeling was followed by immunoprecipitation of N-SMase with corresponding antibody. Next, the immunoprecipitate was solubilized and subjected to gel electrophoresis and autoradiography.

Table II
SPHINGOMYELINASE ACTIVITY IN VARIOUS SUBCELLULAR FRACTIONS OF CULTURED HUMAN PT CELLS INCUBATED ± GENTAMICIN[a]

Subcellular fraction	Control		Gentamicin	
	A-SMase	N-SMase	A-SMase	N-SMase
Homogenate	106.075	137.48	75.245	55.75
Lysosomes	2012.385 (99.5)	ND	1556.780 (99.3)	ND
Mitochondria	6.370 (0.09)	ND	4.450 (0.10)	ND
Microsomes	1.540 (0.05)	475.335 (31.2)	1.050 (0.04)	288.625 (29.4)
Golgi apparatus	16.800 (0.36)	14.910 (0.67)	16.310 (0.52)	14.645 (1.20)
Plasma membrane	ND	1343.525 (68.14)	ND	772.050 (69.3)

[a] Cultured human PT cells incubated for 21 days ± gentamicin were subjected to subcellular fractionation. The activity of A-SMase and N-SMase was measured in suitable aliquots of cell homogenates and various subcellular fractions. The data in parentheses represent the percentage of distribution of total activity (Ghosh and Chatterjee, 1987). ND, Not determined.

Immunofluorescence detection of glutaraldehyde fixed cultured human PT cells revealed strong immunofluorescence on the plasma membrane (Fig. 2A). The distribution of the N-SMase on the plasma membrane of these cells appeared nonhomogeneous from cell to cell. For example, some cells (with large membranous area) were strongly stained; in contrast, small rounded cells (presumably mitotic cells) were poorly stained or not stained at all.

On permeabilization of cells with Triton X-100, we found very little intracellular staining as compared to background. By contrast, very strong and striking staining of the plasma membrane occurred under these conditions (Fig. 2B). In control experiments, incubation of PT cells with FITC-labeled preimmune rabbit serum IgG revealed nonspecific staining (Fig. 2C). These findings suggest that the antigen (N-SMase) is predominantly localized on the surface of PT cells. Preincubation of cells with 1,10-phenanthroline, an inhibitor of phospholipase D, did not impair the surface staining of cells with FITC-conjugated antibody against N-SMase (data not shown). These findings indicate the specific nature of the interaction between N-SMase antibody and N-SMase.

Previously, we and others have reported that Mg^{2+} stimulates and ethylenediaminetetraacetic acid (EDTA) inhibits N-SMase activity (Ghosh and Chatterjee, 1987; Rao and Spence, 1976). When cells were incubated in the presence of Mg^{2+} and C_6-NBD sphingomyelin, strong cell surface staining was observed (Fig. 3A). By contrast, preincubation with EDTA decreased the surface staining of cells (Fig. 3B). Following their incubation with C_6-NBD-sphingomyelin, the cells were extracted with organic solvents. The total lipid extracts were separated by thin-layer chromatography (TLC) (Chatterjee, 1987). Gel areas corresponding to C_6-NBD-ceramide, C_6-NBD-sphingomyelin, and C_6-NBD-fatty acid were scraped and eluted, and the absorbance values were measured at 465 nm. Such studies revealed that ceramide was the major product of catabolism of sphingomyelin. For example, the absorbance values obtained for cell-

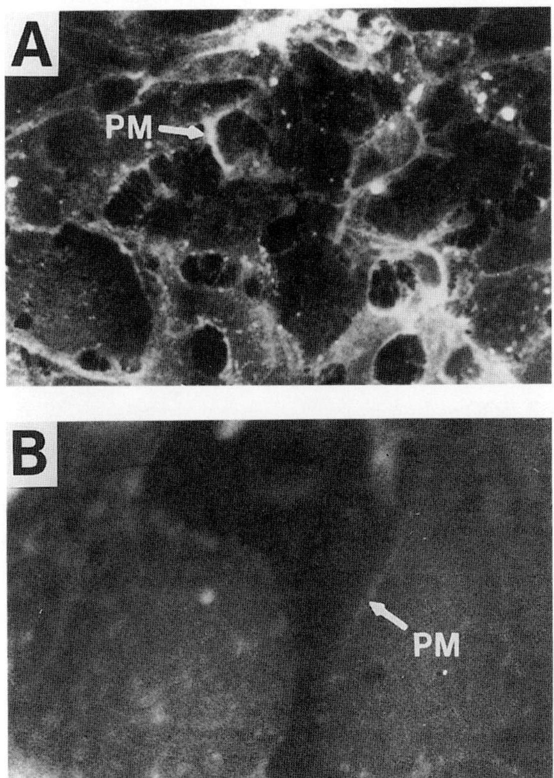

FIG. 3. Fluorescence micrographs of PT cells after incubation with C_6-NBD-sphingomyelin. Monolayer cultures were washed twice with ice-cold phosphate-buffered saline (PBS) and next incubated with C_6-NBD-sphingomyelin in medium containing lipoprotein-deficient serum (1 mg protein/ml) for 1 hour at 4°C. The cells were then washed five times with medium containing lipoprotein-deficient serum and observed under a microscope. Cells were incubated with C_6-NBD-SM in the presence of MG^{2+} (0.5 m*M*) alone (A) ($\times 200$) and (EDTA) (0.5 m*M*) alone (B) ($\times 400$). PM, plasma membrane.

associated ceramide were 0.090, 0.010, and 0.0 when cells were incubated with C_6-NBD-sphingomyelin plus Mg^{2+}, C_6-NBD-sphingomyelin plus EDTA, and C_6-NBD-sphingomyelin, respectively. These findings are consistent with the hypothesis that sphingomyelin may be hydrolyzed to ceramide at the surface of cells predominantly by the plasma membrane-bound neutral sphingomyelinase.

When cells were labeled at the surface with ^{125}I and a membrane-rich fraction was subjected to immunoprecipitation and gel electrophoresis, most of the radioactivity was associated with a protein of apparent M_r of 92,000 (Fig. 4B, lane 2). This radiolabeled protein had an electrophoretic migration similar to pure N-SMase (Fig. 4A, lane 2). Employing Western immunoblot assay, we have previously shown that the N-SMase antibody recognizes (immunoprecipitates) a protein with an apparent M_r of 92,000 from partially purified N-SMase derived from PT cell plasma membrane. Taken together, our present studies reveal that, on immunoprecipitation, the antibody recognized the same N-SMase protein as did the Western

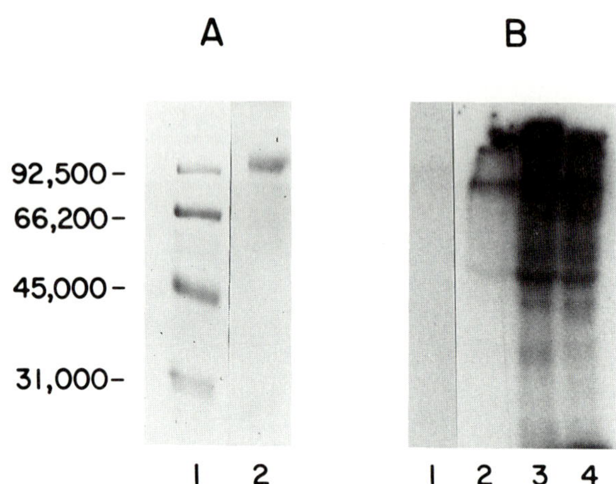

FIG. 4. Cell surface labeling of N-SMase with ^{125}I. Cultured human PT cells were labeled on the cell surface with ^{125}I using iodogen. Cells were homogenized, and plasma membranes were isolated, solubilized, and immunoprecipitated with antibody against N-SMase. The immunoprecipitate was solubilized, subjected to polyacrylamide gel electrophoresis, dried, and subjected to autoradiography. (A) Polyacrylamide gel electorphoretogram, lane 1, molecular weight standards; lane 2, purified N-SMase from human urine, stained with Coomassie Blue, (B) Autoradiogram, lane 1, ^{125}I-labeled plasma membrane immunoprecipitated with rabbit IgG; lane 2, ^{125}I-labeled cell immunoprecipitated with antibody against N-SMase; lane 3, ^{125}I-labeled plasma membrane; lane 4, ^{125}I-labeled whole-cell homogenate.

blots and that this protein is oriented on the surface of PT cells. Interestingly, a phosphatidylinositol-specific phospholipase C involved in regulation of cell growth has also been localized to the external surface of Swiss 3T3 cells (Ting and Pagano, 1990).

As a whole, the evidence is strong for the localization of N-SMase at the external surface of mammalian cells. Such a topology may provide N-SMase direct access to extracellular sphingomyelin associated with lipoproteins, and perhaps to the cell membrane sphingomyelin (on perturbation of the lipid bilayer).

E. Modulators of N-SMase Activity in Mammalian Systems

The cell surface topology of N-SMase makes it an easily accessible target for modulation via a wide variety of exogenous compounds (see Table III). Perhaps this may be nature's way of utilizing N-SMase as a signal transduction molecule to conduct a variety of physiological reactions *in vitro*.

The differentiation of monocytes to macrophages or to granulocytes is an important event in human biology. Earlier studies showed that a human leukemic cell line (HL-60), comprising promyelocytes under basal conditions, undergoes differentiation into macrophages on incubation with phorbol esters (e.g., TPA) (Rivera *et al.*, 1979a,b). This event is marked by growth inhibition, morphological changes, and the development of specific cell surface markers and enzyme markers such as acid phosphatase and α-naphthylacetate esterase. On the other hand, incubation of HL-60 cells with dimethyl sulfoxide differentiates them into granulocytes. Several laboratories have taken advantage of the HL-60 cell system to study various aspects of cellular differentiation (Collins *et al.*, 1978; Dressler and Kolesnick, 1990; Dressler *et al.*, 1992; Kolesnick, 1989; Mathais *et al.*, 1991; Okazaki *et al.*, 1990). Studies by Okazaki *et al.* (1990) provided perhaps the first evidence that the action of a N-SMase may be involved in 1α,25-dihydroxyvitamin D_3-mediated differentiation of HL-60 cells to macrophages. Dexamethasone increases N-SMase activity in 3T3-LI mouse fibroblasts and increases the cellular level of sphingomyelin. Whether such changes can explain the physiological effects of this glucocortoid horomone on glucose transport and lipolysis (Murray *et al.*, 1979) needs to be investigated further.

That one of the products of N-SMase reaction, ceramide, may be internalized and alter cell differentiation has been shown in HL-60 cells (Okazaki *et al.*, 1990). Corroborative evidence that this process may be due

Table III
MODULATORS OF NEUTRAL SPHINGOMYELINASE IN MAMMALIAN SYSTEMS

Modulator	Mammalian system	Effects	Reference
Hormones, steroids			
1,25-Dihydroxy-vitamin D_3	Human myelocytic leukemic HL-60 cells	Activation of N-SMase and differentiation of monocytes to macrophages	Okazaki *et al.* (1990)
Dexamethasone	3T3-LI mouse fibroblasts	Activation of enzyme	Murray *et al.* (1979)
Drugs, antibiotics			
Gentamicin	Human kidney proximal tubular (PT) cells	Inactivates enzyme	Ghosh and Chatterjee (1987)
Gentamicin	Urinary PT cells from patients *in vivo*	Inactivates enzyme	Chatterjee and Bose (1988)
Cyclosporin	Urinary PT cells from patients *in vivo*	No effect	Chatterjee and Bose (1988)
Growth factors			
Tumor necrosis factor	Human myelocytic leukemic HL-60 cells	Activation of enzyme and differentiation of monocytes to macrophages	Young-Kim *et al.* (1991); Dressler *et al.* (1992)
γ-Interferon	Human myelocytic leukemic HL-60 cells	Activation of enzyme and differentiation of monocytes to macrophages	Young-Kim *et al.* (1991)

to sphingomyelinase action has also appeared (Okazaki *et al.*, 1990). Furthermore, on degradation, ceramide is known to release sphingosine via conversion to *N,N'*-dimethylsphingosine (Igarashi *et al.*, 1990), which in turn could modulate protein kinase C and regulate cell growth and a variety of other physiologically important reactions. However, recent studies have shown that ceramide itself can stimulate epidermal growth factor receptor phosphorylation in A431 human epidermal carcinoma cells and that sphingosine may act in part by conversion to ceramide (Goldhorn *et al.*, 1991).

Hanun and co-workers have recently shown that tumor necrosis factor

α (TNF-α) and interferon-α (IFN-α) stimulated N-SMase activity as an early event in monocytic differentiation of HL-60 cells (Young-Kim *et al.*, 1991). Synthetic ceramide (C2-ceramide) further enhanced the action of these agents (Young-Kim *et al.*, 1991). Further studies by Kolesnick and co-workers revealed that ceramide may be phosphorylated to ceramide-1-phosphate, which in turn may be involved in mediating TNF-α induced cellular differentiation in HL-60 cells (Dressler and Kolesnick, 1990; Kolesnick, 1989; Mathais *et al.*, 1991).

III. The Effects of Neutral Sphingomyelinase on Lipoprotein and Sterol Metabolism

As mentioned, both unesterified cholesterol and sphingomyelin are predominantly localized within the plasma membrane. Accordingly, it is not surprising to observe a coordinated regulation of these molecules by LDL (see below).

Cholesterol homeostasis in mammalian systems is controlled by human plasma low-density lipoprotein (Goldstein and Brown, 1989). LDL are initially bound to a specific high-affinity receptor located within the coated pit on the cell surface. The receptor-bound LDL are internalized by absorptive endocytosis. Subsequently, LDL are transferred to the lysosomes where the apoB is degraded and cholesteryl esters are hydrolyzed. Thus, the exogenously derived cholesterol or LDL suppresses endogenous synthesis of cholesterol by suppressing the rate-limiting enzyme hydroxymethylglutaryl-CoA (HMG-CoA) reductase (EC 1.1.1.88). In addition, excess exogenous cholesterol is utilized for the endogenous synthesis of cellular membranes and cholesteryl esters by use of acyl-CoA: cholesterol acyltransferase, (EC 2.3.1.26) (ACAT). The exogenous cholesterol may also suppress LDL receptor, thereby completing the cycle involved in cholesterol homeostasis in the cell (Goldstein and Brown, 1989).

Our laboratory has shown in cultured normal human kidney PT cells that downregulation of cholesterol metabolism by exogenous LDL also suppresses the incorporation of [^{14}C]serine into ceramide and sphingomyelin synthesis (Fig. 5A,B). Conversely, in homozygous form of familial hypercholesterolemia (patients lacking LDL receptor) the lack of regulation of cholesterol synthesis was accompanied by the lack of suppression of [^{14}C]serine incorporation into ceramide and sphingomyelin (Chatterjee *et al.*, 1986). This lack of regulation of sphingomyelin synthesis may account for the increased plasma levels of sphingomyelin in the plasma of patients with familial hypercholesterolemia.

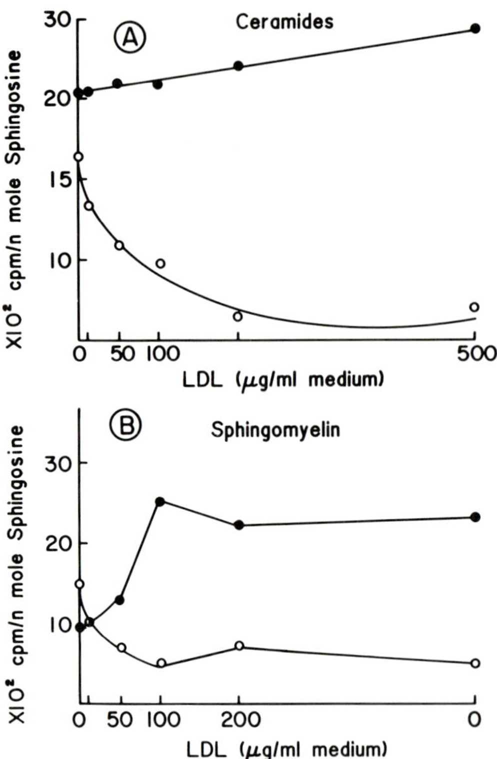

FIG. 5. Effects of LDL on the incorporation of [^{3}H]serine into ceramide and sphingomyelin in normal and familial hypercholesterolemic homozygous proximal tubular cells. Cells were incubated for 24 hours in medium containing lipoprotein-deficient serum (1 mg protein/ml) plus [^{3}H]serine. Ceramide and sphingomyelin were purified from cell extracts and subjected to acid-catalyzed methanolysis. The long-chain bases (sphingosines) were eluted by extracting the methanolysates with ethyl ether. Suitable aliquots of the ether extracts were subjected to radioactivity and sphingosine measurements. ○, Normal cells; ●, FH cell. The incorporation of [^{3}H]serine into ceramide (A) and sphingomyelin (B) is plotted against increasing concentrations of LDL (Chatterjee *et al.*, 1986).

The mobilization of cholesterol from the cell surface or of exogenous LDL-derived cholesterol to the interior of the cells allows access to ACAT, a microsomal enzyme. This enzyme converts cholesterol to cholesteryl esters (the storage form of cholesterol). The following discussion is aimed at providing evidence to support the notion that alterations in the cellular levels of sphingomyelin can modulate sterol and lipoprotein metabolism in cultured mammalian cells. The effects of exogenous N-SMase on other important biological events are summarized in Table IV.

A. Effects on Mobilization of Cell Surface Cholesterol and Endogenous Cholesteryl Ester Synthesis

In studies of the effects of sphingomyelinase on the synthesis of cholesteryl esters, human fibroblasts were incubated with a bacterial sphingo-

Table IV
Effects of Exogenous N-SMase on Cell Differentiation, Lipid and Lipoprotein Metabolism

Sphingomyelinase source	Mammalian system	Effects	Reference
Staphylococcus aureus	GH_3 rat pitutary cells	Inactivation of protein kinase C	Mathais *et al.* (1991)
S. aureus	HL-60 human leukemic monocytes	Inhibition of differentiation to macrophages	Kolesnick (1989)
S. aureus	Adherent macrophages	Detachment of macrophages	Kolesnick (1989)
S. aureus + vitamin D	HL-60	Induction of differentiation to macrophages	Okazaki *et al.* (1990)
S. aureus + TNF-α	HL-60	Induction of differentiation to macrophages	Young-Kim *et al.* (1991)
Human urine	Human skin fibroblasts	Increase in LDL binding, internalization, and degradation	Chatterjee (1993a)
S. aureus	Human skin fibroblasts	Increase in cholesteryl ester synthesis	Slotte and Bierman (1988)
Human urine	Human skin fibroblasts	Increase in cholesteryl ester synthesis	Chatterjee (1993a)
S. aureus	Human fibroblasts Human hepatic blastoma cells (HEP G-2) Rat intestinal cells (IEC-6)	Decrease in HMG-CoA reductase activity	Gupta and Rudney (1991)
LDL + *S. aureus*	J-774 macrophages	Increase binding of LDL and cholesteryl ester synthesis	Xu and Tabas (1991)

myelinase. This resulted in a marked reduction (30–60% less than control) in the cellular levels of sphingomyelin and plasma membrane-bound cholesterol (Chatterjee, 1993a; Slotte and Bierman, 1988). This reduction was accompanied by a time-dependent and N-SMase concentration-dependent increase in the synthesis of cholesteryl esters (see Table IV). When human fibroblasts were incubated with human N-SMase and corresponding antibody, the stimulatory effects of N-SMase on the synthesis of cholesteryl esters were compromised (Chatterjee, 1993a).

Although the mechanism of action of N-SMase on the increased synthesis of cholesteryl esters has not been established, it most probably involves providing more substrate (i.e., cholesterol), which in turn increases the activity of ACAT (Slotte and Bierman, 1988). Indirect evidence demonstrating that the ACAT inhibitor compromised the stimulatory effects of N-SMase on synthesis of cholesteryl esters further supports this notion (Chatterjee, 1993b; Slotte and Bierman, 1988).

B. Effects on HMG-CoA Reductase Activity

The consequence of depleting cellular sphingomyelin by employing the action of bacterial sphingomyelinase on HMG-CoA reductase has been extensively studied by Gupta and Rudney (1991). They showed that sphingomyelinase exerted a concentration-dependent decrease in HMG-CoA reductase activity in rat intestinal epithelial cells (IEC-6), human hepatic blastoma cells (Hep-G2), and human skin fibroblasts (GM-43). Ceramide and phosphocholine had no effects on HMG-CoA reductase activity, but sphingosine stimulated the activity of this enzyme. Such results reveal that N-SMase-mediated inhibition of HMG-CoA reductase activity occurs via a different mechanism than does sphingosine-mediated activation of HMG-CoA reductase activity. Although the bacterial SMase has both A-SMase and N-SMase activities, the following additional observations revealed that the effects on HMG-CoA reductase were indeed due to N-SMase action in IEC-6 cells. (1) Incubation of IEC-6 cells with tricyclic antidepressants (e.g., chlorpromazine and desipramine) and calmodulin antagonists such as W-7 (inhibitors of endogenous A-SMase activity) increased the activity of HMG-CoA reductase two- to threefold. By contrast, these compounds were unable to compromise the inhibitory effects of exogenous SMase on HMG-CoA reductase activity. (2) When cells were incubated with ketoconazole (an inhibitor of oxysterol formation) it compromised the inhibitory effect of N-SMase on HMG-CoA reductase. Gupta and Rudney (1991) concluded that the membrane-derived cholesterol may be converted to an oxysterol that, in turn, inhibits the activity of HMG-CoA reductase in IEC-6 cells.

C. Effects on the Binding, Internalization, and Degradation of LDL

Because LDL and the LDL receptor are intimately involved in the regulation of cholesterol biosynthesis, we were interested in the implications of altering cellular levels of sphingomyelin by employing the action of N-SMase on the binding, internalization, and degradation of ^{125}I-labeled LDL (Chatterjee, 1993a). When cells were incubated with human N-SMase, we found a substantial decrease (55% less than control) in cellular sphingomyelin content; this was accompanied by a N-SMase concentration-dependent increase in ^{125}I-labeled LDL binding, internalization, and degradation in cultured human fibroblasts (Chatterjee, 1993a). Antibody against N-SMase and heat inactivation of N-SMase compromised the stimulatory effects of N-SMase on ^{125}I-labeled LDL metabolism. Incubation of cells with phospholipase D and phospholipase C did not alter ^{125}I-labeled LDL binding, internalization, and degradation. These observations suggest that the stimulatory effect of N-SMase on LDL metabolism in fibroblasts is specific. Additional studies revealed that unlabeled LDL competitively displaced ^{125}I-labeled LDL from binding to N-SMase-treated cells.

To assess the mechanism of action of N-SMase on LDL receptor activity in fibroblasts, we incubated cells with ceramide, phosphocholine, sphingophosphocholine, and sphingosine. These precursors of sphingomyelin failed to mimic the stimulatory effects of N-SMase on ^{125}I-labeled LDL metabolism. In fact, sphingosine and C_2-ceramide (a water-soluble ceramide) decreased ^{125}I-labeled LDL receptor activity in human fibroblasts. It is possible that these precursors of sphingomyelin may be used by the cells to synthesize sphingomyelin, which in turn decreases LDL receptor activity in fibroblasts. Alternatively, sphingomyelin may be downregulating the activity of N-SMase in human fibroblasts and thus indirectly reducing LDL receptor activity. The effects of N-SMase may be at the intracellular or nuclear level, which may contribute to an increase in a recirculating pool of LDL receptors or newly synthesized LDL receptors. Clearly, further studies will be required to understand the N-SMase-mediated increase in LDL receptor activity in fibroblasts at the biochemical level and molecular level. The preceding observations were made in fibroblasts pretreated with N-SMase. Subsequently, cells were washed and shown to have similar levels of N-SMase before (23 pg/mg protein) and after (21.6 pg/mg protein) treatment. Moreover, following incubation for 4 hours with LDL, the levels of phosphatidylcholine and sphingomyelin in LDL incubated with cells previously incubated ± N-SMase did not change. Taken together, these studies suggest that the increase in LDL receptor activity in the fibroblasts study was due to a cellular decrease in sphingomyelin levels, not LDL. In another study, LDL modified on incubation with a bacterial sphingomyelinase increased the binding, internalization, degradation, and synthesis of cholesteryl esters in a macrophage cell line, J774 (Xu and Tabas, 1991). Since both LDL and sphingomyelinase may be present in atherosclerotic lesions, it is possible that sphingomyelinase-treated LDL may serve as an atherogenic lipoprotein. However, whether human atherosclerotic plaques contain sphingomyelinase-modified LDL remains to be established.

IV. The Effects of Antibiotics on Neutral Sphingomyelinase Activity and LDL Receptor Activity

In this section, I review our work on the effects of various compounds on N-SMase activity *in vitro* in cultured human kidney PT cells and *in vivo* in patients taking gentamicin and cyclosporin for various ailments.

A. Effects of Gentamicin and Cyclosporin *in Vitro* and *in Vivo* on Sphingomyelinase Activity in Man

Gentamicin is an aminoglycoside antibiotic that combats gram-negative bacterial infections in man. It is nephrotoxic to humans and experimental animals (Josephovitz *et al.*, 1985; Kaloyanides and Pastoriza-Munoz, 1980; Humes *et al.*, 1982). Previous studies have shown that renal biopsies from individuals exhibiting gentamicin-induced nephrotoxicity show pathological changes in the PT cells, changes ranging from mild subcellular damage in the form of "myeloid bodies" to overt necrosis (Whelton and Solez, 1982). Studies with cultured porcine PT cells (Schwertz *et al.*, 1986) and human PT cells incubated with gentamicin (Chatterjee *et al.*, 1987) have also shown the appearance of numerous "myeloid bodies." Such morphological changes induced by gentamicin are accompanied by gross biochemical changes, including the accumulation of phosphatidylcholine and sphingomyelin in human PT cells (Fig. 6).

To assess biochemical mechanisms underlying gentamicin-mediated accumulation of sphingomyelin in PT cells, we pursued detailed studies of the sphingomyelinases (Ghosh and Chatterjee, 1987). We found that cultured normal human PT cells exhibit biomodal pH optima for sphingomyelinase activity. Maximum sphingomyelinase activity was observed at pH 5.6 (K_m 0.07×10^{-7} M) and at pH 7.4 (K_m 1.8×10^{-7} M) (Ghosh and Chatterjee, 1987). In contrast to a moderate decrease in A-SMase activity, the activity of N-SMase in cells incubated with gentamicin was decreased 2.7-fold compared to control.

Subcellular fractionation (Table II) studies revealed that the distribution of N-SMase activity in microsomes, Golgi apparatus, and plasma membrane in cells incubated with gentamicin was similar to the corresponding

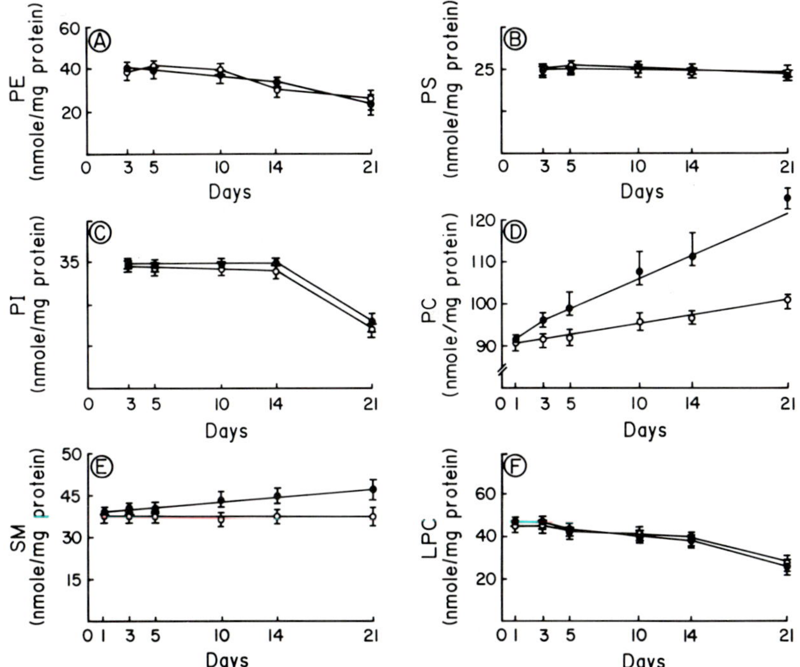

FIG. 6. Effects of time of incubation of PT cells with and without gentamicin on the cellular levels of individual phospholipids. Cells were grown and incubated with and without gentamicin (0.03 m*M*) for 2 days. At the indicated time intervals, cells were harvested, washed, and centrifuged. Cell pellets were then homogenized in distilled water and suitable aliquots withdrawn for the measurement of protein content. Total lipids were extracted from cell homogenates. Phospholipids were separated on TLC plates and their phosphorus content measured. The phosphorous content was multiplied by 25 to calculate the amount of individual phospholipid. The results were derived from four separate experiments analyzed in duplicate. (A) ○, phosphatidylethanolamine (PE) levels in control cells; ●, PE levels in cells incubated with gentamicin. (B) ○, Phosphotidylserine (PS) levels in control cells; ●, PS levels in cells incubated with gentamicin. (C) ○, Phosphatidylinositol (PI) levels in control cells; ●, PI levels in cells incubated with gentamicin. (D) ○, Phosphatidylcholine (PC) levels in control cells; ●, PC levels in cells incubated with gentamicin. (E) ○, Sphingomyelin (SM) levels in control cells; ●, SM levels in cells incubated with gentamicin. (F) ○, Lysophosphatidylcholine (LPC) levels in control cells; ●, LPC levels in cells incubated with gentamicin (Chatterjee, 1987).

control subcellular fractions (Ghosh and Chatterjee, 1987). Interestingly, the specific activity of N-SMase in the microsomal and plasma membrane fraction was lower by approximately twofold in cells incubated with gentamicin than in control cells. These studies suggest but do not prove that gentamicin may not have altered the processing of N-SMase. Further studies employing immunoblot assays in our laboratory have revealed that the N-SMase mass in cells incubated with gentamicin is lower compared to control cells. Our preliminary data indicate that the decrease in N-SMase activity in cells incubated with gentamicin was not due to the accumulation of this antibiotic which, in turn, may have altered cellular pH, but rather was due to a decrease in cellular levels of the enzyme. Further studies on the molecular cloning of N-SMase are warranted to understand the biochemical mechanism involved in gentamicin-induced accumulation of sphingomyelin in man.

B. Effects of Gentamicin on LDL Receptor Activity in PT Cells

In contrast to the finding that incubation with exogenous N-SMase raised LDL receptor activity in human fibroblasts, we found that a decrease in N-SMase activity in PT cells incubated with gentamicin led to lowered LDL receptor activity and decreased cholesteryl ester synthesis (Chatterjee *et al.,* 1987).

C. Effects of Gentamicin and Cyclosporin A on N-SMase Activity

To complement our *in vitro* studies, we pursued *in vivo* studies on the morphological and biochemical effects of gentamicin on urinary cell phospholipids and phospholipases in man (Chatterjee and Bose, 1988). Compared to normolipidemic volunteers, patients receiving gentamicin shed higher levels of phosphatidylcholine, (78%), phosphatidylethanolamine (38%), and sphingomyelin (30%). This was accompanied by a 10-fold decrease in N-SMase activity in patients receiving gentamicin as compared to control. On the other hand, the levels of sphingomyelin and N-SMase were similar to control in patients receiving cyclosporin A (CsA). Taken together, our *in vivo* studies reveal that deficient N-SMase activity precedes phospholipid overloading and gross pathological changes in patients receiving gentamicin but not in patients receiving CsA.

V. Future Research Direction and Clinical Potential of Neutral Sphingomyelinase

Our preliminary data indicate that TNF-α increases the activity of N-SMase and cholesteryl ester synthesis in cultured human fibroblasts (Chatterjee, 1993b). Interestingly, when cells were incubated with TNF-α plus antibody against N-SMase, the stimulatory effects of TNF-α on cholesteryl ester synthesis were compromised. Since in cells incubated with an ACAT inhibitor (Sandoz) plus TNF-α the utilization of cell membrane-derived (3H) cholesterol for cholesteryl ester synthesis was compromised, it is suggested that the enzyme ACAT was involved in the conversion of cell membrane cholesterol to cholesteryl esters in cells incubated with TNF-α alone. Considered together, our preliminary studies reveal the possibility that N-SMase action may initiate signal transduction of TNF-α and stimulate cholesteryl ester synthesis. In another recent report, TNF-α was found to induce LDL receptors in human vascular endothelial cells (Hamanaka *et al.,* 1992). In addition, we have found that in human fibroblast, interleukin-1 can mimic the stimulatory effects on cholesteryl ester synthesis in a fashion similar to TNF-α. Such findings indicate that cholesterol metabolism can be regulated independently of LDL. Further studies are required to assess whether advantage can be taken of such observations to develop and implement novel drugs for therapy of patients with hypercholesterolemia.

The studies reported in this review may represent the proverbial tip of the iceberg insofar as the biochemical mechanisms involved in the interaction of various biomolecules with N-SMase and the signal transduction processes are unknown. Of particular interest is whether G proteins are involved. Molecular cloning of N-SMase will open up a novel approach to studying the regulation and the regulatory elements involved in mammalian systems for this enzyme. Since sphingomyelin metabolism and cholesterol metabolism are tightly linked, investigation is warranted into the roles of this enzyme in atherosclerosis in general, in foam cell formation, and in cell proliferation in patients with coronary artery disease or cancer.

VI. Summary

Although we have accumulated few data that would definitely designate the role of N-SMase in modern cell biology, current evidence indicates that N-SMase may play a central role in signal transduction (Fig. 7). Biomodulators such as hormones, antibiotics, drugs, growth factors, and lipoproteins may interact with N-SMase either directly or in combination with other components (receptors) on the cell surface. The cell surface topology would provide easy access for such interactions.

According to our hypothetical model shown in Fig. 7, activation of N-SMase via biomodulators, e.g., TNF-α, leads to the mobilization of cell surface cholesterol to the interior of the cell. We speculate that this process may be facilitated by sterol carrier proteins. This phenomenon may alter

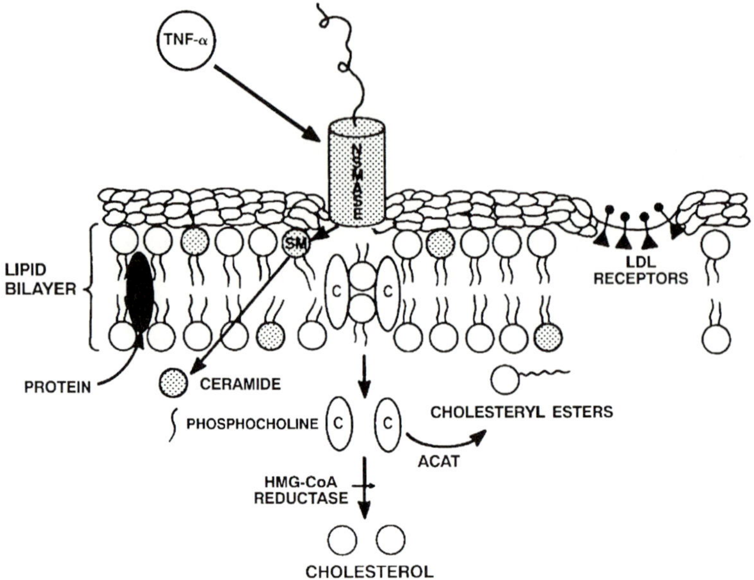

FIG. 7. Hypothetical model depicting the role of neutral sphingomyelinase in signal transduction of tumor necrosis factor, α, and in the stimulation of LDL receptor activity, synthesis of cholesteryl esters, and inhibition of cholesterol synthesis in human fibroblasts. TFN-α, Tumor necrosis factor α; SM, sphingomyelin; N-SMase, neutral sphingomyelinase; Cer, ceramide; ACAT, fatty acyl-CoA:cholesterol acyltransferase; HMG-CoA reductase, hydroxymethylglutaryl-coenzyme A reductase.

the lipid bilayer and make N-SMase accessible to its substrate, sphingomyelin. Ceramide released as a consequence of N-SMase reaction would then carry out various biological phenomena, such as cell proliferation and differentiation. Cholesterol on the other hand, may undergo oxidation and inhibit HMG-CoA reductase activity; it may also be utilized by ACAT to form cholesteryl esters. A decrease in cell membrane cholesterol levels may stimulate LDL receptor activity and LDL receptor recycling and/or synthesis. In contrast, inactivation or decreased activity of N-SMase, as in the case of gentamicin-treated cells, may have the opposite effect, i.e., decreased LDL receptor activity and decreased cholesteryl ester synthesis. Thus, N-SMase may indirectly modulate cholesterol metabolism in cells independent of LDL. It can also modify LDL, allowing its rapid uptake and foam cell formation in macrophages. We can safely conclude that N-SMase action may be required to carry out a myriad of important cellular functions either directly or by cholesterol mobilization, or generation of ceramide, ceramide-1-phosphate, and sphingoid bases.

The present data support the view that N-SMase action may initiate signal transduction for such biomodulators as growth factors, lipoprotein, and hormones.

References

Barenholz, Y., and Gatt, S. (1982). *In* "Phospholipids" (J. N. Hawthorne and G. D. Ansel, eds.), pp. 129–177. Elsevier, Amsterdam.

Bartolf, M., and Franson, R. C. (1986). *J. Lipid Res.* **26,** 57–63.

Bretscher, M. S. (1973). *Science* **181,** 622–662.

Chapman, M. J. (1980). *In* "Methods in Enzymology" (J. Segrest and J. Albers, eds.), Vol. 128, pp. 70–143. Academic Press, New York.

Chatterjee, S. (1987). *J. Biochem. Toxicol* **2,** 181–201.

Chatterjee, S. (1990). *FASEB J.* **4,** 783.

Chatterjee, S. (1993a). *J. Biol. Chem.* **268,** 3400–3406.

Chatterjee, S. (1993b). *J. Biol. Chem.* Submitted for publication.

Chatterjee, S., and Bose, S. (1988). *J. Biochem. Toxicol.* **3,** 47–58.

Chatterjee, S., and Ghosh, N. (1989). *J. Biol. Chem.* **246,** 12554–12561.

Chatterjee, S., and Ghosh, N. (1991). *In* "Methods in Enzymology" (E. Dennis, ed.), Vol. 197, pp. 540–554. Academic Press, San Diego.

Chatterjee, S., Clarke, K. S., and Kwiterovich, P. O. (1986). *J. Biol. Chem.* **261,** 13474–13479.

Chatterjee, S., Trifillis, A. L., and Regec, A. L. (1987). *Can. J. Biochem. Cell Biol.* **65,** 1044–1056.

Collins, S. J., Ruscetti, F. W., Gallagher, R. E., and Gallo, R. C. (1978). *Proc. Natl. Acad. Sci. U.S.A.* **75,** 2458–2462.

Das, D. V. M., Cook, M. W., and Spence, M. W. (1984). *Biochim. Biophys. Acta* **771,** 339–342.

Dressler, K. A., and Kolesnick, R. N. (1990). *J. Biol. Chem.* **265,** 14917–14922.

Dressler, K. A., Mathais, S., and Kolesnick, R. N. (1992). *Science* **255,** 1715–1718.

Frederickson, D. S., and Sloan, H. (1972). *In* "The Metabolic Basis of Inherited Disease" (J. B. Stanley, J. B. Wyngaarden, and D. F. Frederickson, eds.), 3rd ed., p. 783. McGraw-Hill, New York.

Gatt, S. (1976). *Biochem. Biophys. Res. Commun.* **68,** 235–241.

Ghosh, P., and Chatterjee, S. (1987). *J. Biol. Chem.* **262,** 12550–12556.

Goldhorn, T., Dressler, K. A., Muimhi, J., Radin, N. S., Mendelson, J., Menaldino, D., Liotta, D., and Kolesnick, R. N. (1991). *J. Biol. Chem.* **266,** 16902–16097.

Goldstein, J. L., and Brown, M. S. (1989). *In* "The Metabolic Basis of Inherited Disease" (C. R. Scriver, A. L. Beaudet, W. S. Sly, and D. Vallee, eds.), 6th ed., pp. 1215–1250. McGraw-Hill, New York.

Gupta, A. K., and Rudney, H. (1991). *J. Lipid Res.* **32,** 125–136.

Hamanaka, R., Kohnot, K., Seguchi, T., Okamura, K., Morimoto, A., Ono, M., Ogata, J., and Kuwano, M. (1992). *J. Biol. Chem.* **267,** 13160–13165.

Ho, M. W., and O'Brien, J. S. (1971). *Proc. Natl. Acad. Sci. U.S.A.* **68,** 2810–2813.

Hostetler, K. Y., and Yakazaki, P. J. (1979). *J. Lipid Res.* **20,** 456–463.

Humes, H. D., Weinberg, J. M., and Knauss, T. C. (1982). *Kidney Dis.* **2,** 529.

Igarashi, Y., Kitamura, K., Toyokumi, F., Dean, B., Fenderson, B., Ogawa, T., and Hakomori, S.-I. (1990). *J. Biol. Chem.* **265,** 5385–5389.

Josephovitz, C., Farruggella, T., Levine, R., Lane, B., and Kalyonides, G. J. (1985). *J. Pharmacol. Exp. Ther.* **235,** 810–819.

Kaloyanides, G. J., and Pastoriza-Munoz, E. (1980). *Kidney Int.* **18,** 571–582.
Kolesnick, R. N. (1989). *J. Biol. Chem.* **264,** 7617–7624.
Leelavathi, D. E., Estes, L. W., Feingold, D. S., and Lombardi, B. (1970). *Biochim. Biophys. Acta* **211,** 124–138.
Levade, T., Salvayre, R., and Douste-Blazy, L. (1986). *J. Clin. Chem. Biochem.* **24,** 205–220.
Mathais, S., Dressler, K. A., and Kolesnick, R. N. (1991). *Proc. Natl. Acad. Sci. U.S.A.* **88,** 10009.
Mooibrock, M. J., Cook, H. W., Clarke, J. T. R., and Spence, M. W. (1985). *J. Neurochem.* **49,** 1551–1558.
Morimoto, S., Kishimoto, Y., Tonuch, J., Weiler, S., Ohashi, T., Barranger, J. A., Kretz, K. A., and O'Brien, J. S. (1990). *J. Biol. Chem.* **265,** 1933–1937.
Murray, D. K., Rohmann-Wennhold, A., and Nelson, D. H. (1979). *Endocrinology (Baltimore)* **105,** 774–777.
Nilsson, A. (1968). *Biochim. Biophys. Acta* **176,** 339–347.
Okazaki, T., Bielawski, A., Bill, R. M., and Hunnun, Y. A. (1990). *J. Biol. Chem.* **265,** 15823–15831.
Rao, B. G., and Spence, M. W. (1976). *J. Lipid Res.* **17,** 506–515.
Rivera, G., O'Brien, T. G., and Diamond, L. (1979a). *Science* **204,** 868–870.
Rivera, G., Santoli, D., and Damsky, C. (1979b). *Proc. Natl. Acad. Sci. U.S.A.* **76,** 2779–2783.
Schwertz, D. W., Kreissberg, J. I., and Venkatachalam, M. A. (1986). *J. Pharmacol. Exp. Ther.* **236,** 254–262.
Slife, C. W., Wang, E., Hunter, R., Wang, S., Burgess, C., Liotta, D. C., and Merrill, A. M. (1989). *J. Biol. Chem.* **264,** 10371–10377.
Slotte, J. P., and Bierman, E. L. (1988). *Biochem. J.* **250,** 653–658.
Spence, M. W., and Callahan, J. N. (1989). *In* "The Metabolic Basis of Inherited Disease" (C. R. Schriver, A. C. Beaudet, W. J. Sly, and D. Valle, eds.), 6th ed., pp. 1655–1676. McGraw-Hill, New York.
Steck, T. (1988). *J. Biol. Chem.* **263,** 13023–13031.
Ting, A. E., and Pagano, R. (1990). *J. Biol. Chem.* **265,** 5357–5340.
Whelton, A., and Solez, K. (1982). *J. Lab. Clin. Med.* **99,** 148–155.
Xu, X. X., and Tabas, I. (1991). *J. Biol. Chem.* **266,** 24849–24858.
Young-Kim, M., Linardič, C., Obeid, L., and Hannun, Y. (1991). *J. Biol. Chem.* **266,** 484–489.

ADVANCES IN LIPID RESEARCH, VOL. 26

Ceramidases: Enzymology and Metabolic Roles

DANIEL F. HASSLER AND ROBERT M. BELL

Department of Biochemistry and
Section of Cell Growth, Regulation, and Oncogenesis
Duke University Medical Center
Durham, North Carolina 27710

I. Introduction

The enzymes responsible for ceramide hydrolysis to sphingosine and fatty acid are known as ceramidases (EC 3.5.1.23). *In vitro* activity measurements and metabolic studies suggest that ceramide degradation constitutes their principal biochemical role.

Ceramidases are believed to exist as isoenzymes which differ with respect to catalytic pH optimum, substrate specificity, subcellular location, tissue distribution, and possibly metabolic function. The isoenzymes with acidic pH optima reside in lysosomes and are thought to function primarily to degrade ceramide for further catabolism or recycling of the sphingosine moiety. Ceramidases with activity at neutral to alkaline pH have been proposed to function in signal transduction pathways to produce sphingosine for regulatory purposes.

II. Enzymology

A. *In Vitro* Activity

Ceramidases were first identified and characterized by their *in vitro* catalytic activity (1). The activity is universally associated with membrane-containing fractions and requires detergent for extraction from membranes and enzymatic activity *in vitro*. An interesting aspect of ceramidases is that they catalyze not only the hydrolysis of ceramides but also their synthesis from free fatty acids and sphingoid bases (1–3). This "reverse ceramidase reaction," in which fatty acids unlinked to coenzyme A are used for the formation of an amide bond, is unusual, although not unprecedented (4–6). Ceramides synthesized with radiolabel in the sphingosine or fatty acid moieties serve as substrate for the hydrolysis reaction. The reverse reaction, with radiolabel in either sphingosine or fatty acid, permits testing of a variety of fatty acids as substrate. Assay methods (7) involve solubilization of substrate(s) in a mixture of Triton X-100 and bile salts with or without Tween 20. Very little activity is observed without added detergent, and the bile salt component seems to be least dispensable.

B. pH Optima

The primary attribute used to distinguish and characterize the ceramidases is their catalytic pH optimum. Most sources of the activity have an optimum at or below pH 5 (2,8,9). Ceramidase activities with optima at neutral to alkaline pH have also been described (10–12). Some tissues contain optima at both acidic and alkaline pH (8,13–15). Difficulties in determining pH optima arise from the use of bile salts, which precipitate at acidic pH, in activity assays. Assays lacking bile salts have measureably less activity.

Evidence that multiple pH optima are due to distinct isoenzymes comes from assays performed on fibroblasts and cerebellum from patients with Farber's lipogranulomatosis. The lysosomes of patients with this genetic disorder accumulate ceramides (16,17), a condition which is eventually fatal. The disease is characterized by a deficiency of acid ceramidase activity (18). Ceramidase activity at alkaline pH, however, is comparable to that found in control subjects (8,13).

C. Tissue Distribution

By assaying the catalytic activity at various pH values, the tissue and subcellular distributions of the ceramidase isoenzymes have been determined. Activity has been assayed *in vitro* using crude cellular or tissue homogenates and subcellular fractions from a variety of sources. Acid ceramidase has been detected in rat brain (1,14), kidney, liver, and spleen (14), cultured human fibroblasts (19), human kidney and cerebellum (8,18), leukocytes (15,20), spleen (9), and plasma (20). Neutral to alkaline activity has been found in pig intestinal mucosa (10), rat organs (14), human cerebellum (8) and fibroblasts (13), pig lens epithelium (11), human leukocytes (15), and porcine epidermis (12). Some discrepancy has been found between human and rat kidney. Human kidney only contains activity at acidic pH (8), whereas rat kidney contains activity at both acidic and alkaline pH (14).

D. Subcellular Distribution

Attempts to fractionate and characterize ceramidases have focused on the enzymes with acidic pH optimum. Studies employing fluorescent ceramide analogs or radiolabeled ceramides have implicated lysosomes as the site of acid ceramidase activity (16,17). Membrane fractions containing lysosomes were used as the enzyme source in most studies.

While it is clear that the acid ceramidase is located within lysosomes, the subcellular location of the neutral to alkaline activities is indeterminate. There are few reports of ceramidase activity from subcellular fractions largely free of lysosomes. A microsomal ceramidase has been described (21) and a plasma membrane activity has been postulated (22). Both reports were descriptive in nature, and no attempts to characterize the activity further were described.

E. Substrate Specificity

Ceramides most commonly used to assay ceramidase activity usually contain either palmitic or oleic acid. The same fatty acids and D-*erythro*-sphingosine are utilized in the reverse reaction. Ceramides containing lauric, myristic, palmitic, stearic, oleic, linoleic, and linolenic acids were hydrolyzed to varying degrees by fibroblast homogenates (13). In the reverse reaction, hexanoic, octanoic, decanoic, lauric, myristic, palmitic, and stearic acids but not acetic, butyric, or lignoceric acids were utilized by the enzyme from rat brain (2). Dihydrosphingosine and dihydroceramides also served as substrates.

Different substrate specificities have been reported for the ceramidase isoenzymes. Momoi *et al.* (13) reported that the alkaline ceramidase activity from human fibroblasts could hydrolyze only ceramides containing unsaturated fatty acids, whereas the acid ceramidase utilized ceramides containing both saturated and unsaturated fatty acids. Unlike the human fibroblast enzyme, alkaline ceramidase from rat organs hydrolyzed *N*-palmitoylsphingosine quite readily (14).

The lysosomal acid ceramidase from human fibroblasts hydrolyzes *N*-lauroylsphingosine much better than does the neutral/alkaline activity, whereas both hydrolyze *N*-oleoylsphingosine quite readily. This differential substrate utilization has been used as a diagnostic test for deficiency of the acid ceramidase (20).

Analysis of the ceramides from tissues of patients with Farber's disease shows not only increased levels of this lipid but marked differences with respect to the composition of the fatty acid moiety. Ceramides from Farber's patients contain a high percentage of hydroxy fatty acids, whereas normal subjects' ceramides contain little or none of this type (23). These findings suggest that the lysosomal acid ceramidase may be the isoenzyme used to hydrolyze ceramides containing hydroxy fatty acids. These types of fatty acid are found primarily in cerebrosides and sulfatides (24), sphingolipids whose headgroups are removed by lysosomal enzymes.

F. Kinetic and Mechanistic Characterization

Kinetic and mechanistic studies on ceramidases have been performed on partially purified activities. The hydrolytic activity from rat brain at pH 4.8 had a K_m of 300 μM for *N*-oleoylsphingosine (2). The activity from pig lens epithelium at pH 10 had a K_m of 100–300 μM for *N*-palmitoylsphingosine and, in the reverse reaction at pH 8, had K_ms of 82 and 245 μM, respectively, for palmitic acid and sphingosine (11).

The hydrolysis reaction was inhibited by both free fatty acids and sphingosine (2). A synthetic ceramide analog, *N*-oleoylethanolamine, was found to inhibit hydrolysis of *N*-oleoylsphingosine, as did cerebrosides (8). Inhibition by *N*-oleoylethanolamine is especially interesting in light of the finding that it can be synthesized, in analogy to the reverse ceramidase reaction, from free fatty acids and ethanolamine (5,6).

Studies to determine the equilibrium constant for hydrolysis or synthesis were performed using a partially purified preparation from rat brain (3). Discrepancies in the values obtained by hydrolysis or synthesis were observed, being much higher for the synthesis reaction. The discordance was explained by possible local concentration differences for the different substrates, including water, in the micellar assay system employed.

Dependence of ceramide formation on exogenous free fatty acid requires an enzyme source free of fatty acids and other lipids. This complete dependence on added fatty acid by a lipid-depleted enzyme sample is crucial in providing evidence for the reaction mechanism. The absence of endogenous lipid is essential to rule out the possibility of transacylation of free fatty acids into endogenous lipid and the subsequent transfer of this "activated" fatty acid to sphingosine.

G. Purification

A number of enzymological questions regarding ceramidases remain unresolved due to the lack of studies on highly purified enzymes. Purification is necessary to obtain accurate vital data such as occurrence of isoenzymes, molecular masses, pH optima and profiles of individual isoenzymes, substrate specificities, sensitivity to inhibitors, and mechanistic details of the hydrolytic and reverse reactions.

Few reports have been published regarding the purification of ceramidases. Ceramidase activity from rat brain has been purified more than 100-fold relative to a homogenate to a specific activity of 12.5 nmol/min/mg protein following detergent solubilization (3). The fraction of a rat brain homogenate sedimenting after 30 minutes at 27,000 *g* following a brief period of sonication was used for further purification. The sediment was extracted with sodium cholate, centrifuged at high speed, and the supernatant dialyzed to remove detergent. The dialysate was clarified by centrifugation, fractionated with ammonium sulfate, and redialyzed. Following treatment with trypsin and chymotrypsin to degrade most proteins, the sample was passed over a column of Sephadex G-150 and activity eluted near the void volume. Ceramidase activity was not appreciably impaired by the proteolysis, although a slight mobility shift of activity on Bio-Gel P-150 was noticed after proteolysis.

The partially purified preparation was subjected to a variety of physical and chemical treatments in an attempt to separate the hydrolytic from the synthetic activity. None of the treatments accomplished separation of the activities, prompting the assertion that both were due to the same enzyme.

It is uncertain whether this partially purified preparation contained multiple activities or a single isoenzyme. Although the lysosomal ceramidases from rat brain have pH optima between 4.5 and 5.0, assays of the enzymatic activity were performed at pH 7.4 to facilitate solubility of the bile salts necessary for activity *in vitro*.

There are apparently no reports on purification of ceramidases from sources free of lysosomes.

H. Molecular Biological Studies

Since no ceramidase isoenzyme has been purified from any source, practically nothing is known about the primary structure of these enzymes.

Purification of the polypeptides would allow for attempts at obtaining peptide sequence data, production of antibodies, and an entrance into molecular biological studies.

III. Metabolic Roles

A. Activity in Cultured Cells

The role of ceramidases in cultured cells can be studied using exogenous radiolabeled sphingolipids or sphingolipid precursors. One study has shown that, when exogenous ceramides containing radiolabeled *N*-acyl chains are added to SV40-transformed Schwann cells, the radiolabel appears in glycerolipids (25). This suggests that deacylation or transacylation of ceramide occurs in these cells. Deacylation of ceramide was also demonstrated in guinea pig epidermal cells labeled with [^{3}H]palmitate (26). Chasing of prelabeled cells with albumin or serum resulted in a marked release of label from ceramides as free palmitate in the medium. Studies employing ceramides labeled in the fatty acid moiety demonstrated that cultured human hepatoma cells and macrophages converted much of the radiolabel into their glycerolipids (27).

Radiolabeled ceramides injected into the brains of young rats were shown to undergo fatty acid replacement (28). It was suggested that ceramidases could be responsible both for the deacylation and reacylation of these ceramides, and that the enzyme could be a primary determinant of the fatty acid composition of a cell's ceramides.

Another study, using cultured fibroblasts from normal and Farber patients, showed that cells can take up exogenous ceramide by at least two routes (29). When added as a complex with apolipoprotein, ceramides are routed to the lysosomes. When added as a component of liposomes, the ceramide is metabolized extralysosomally.

These studies suggest that *N*-deacylation of ceramide via ceramidases is a common metabolic event in many types of cells and tissues.

It would be of great interest to demonstrate clearly a role for ceramidases in the production of sphingosine for some regulatory purpose in cultured cells. Since hydrolysis of ceramides via ceramidase is the only established route for production of sphingosine in cells, some measure of the ceramidase activity could be obtained by quantifying changes in free sphingosine. Changes in sphingosine levels have been measured in human neutrophils treated with a variety of agonists (30). There remains a dearth of systems, however, in which sphingosine levels have been shown to be modulated by normal cellular effector agonists.

Metabolism of sphingosine, either by reacylation to ceramide, phosphorylation, or other possible metabolic fates, makes difficult the assessment of the role ceramidases play in regulating sphingosine levels. Inhibitors of ceramide biosynthesis (31), sphingosine phosphorylation (32), or ceramidase (8) could be useful tools in these studies.

B. Production of Endogenous Sphingosine

Exogenous sphingosine has interesting regulatory effects on a number of enzymes and cellular processes (33,34). Consequently, it has been proposed that cells may utilize ceramidases as a means to produce endogenous sphingosine. Metabolic studies have demonstrated that sphingosine is produced only via the degradation of more complex sphingolipids which contain it (35), presumably via the ceramidase reaction. Its original incorporation into more complex sphingolipids occurs via dehydrogenation of dihydroceramides which contain a fully saturated long-chain base, dihydrosphingosine. In addition to being a component of sphingolipids, sphingosine has been shown to exist as the free base in a variety of cell types and tissues (36,37). Consistent with its possible role as an important regulatory molecule, free sphingosine levels are maintained on the order of 0.01 to 0.04% of the total cellular lipid in most tissues and cells examined (37)

A few systems have been identified in which measurable changes in free sphingosine occur. Freshly isolated human neutrophils were found to contain free sphingosine (30). Incubation of neutrophils resulted in a time- and temperature-dependent increase in sphingosine. Serum, plasma, and plasma lipoproteins stimulated sphingosine production by neutrophils. Several neutrophil agonists blunted the increases in free sphingosine, presumably by stimulating activity of the acyl-CoA-dependent ceramide synthase. Another sphingosine-forming system of note is the plasma membrane of rat liver (22). Time- and temperature-dependent increases in free sphingosine from endogenous substrates occurred on incubation of a plasma membrane fraction in the presence of divalent cations at neutral to alkaline pH. Addition of sphingomyelinase, sphingomyelin, ceramide, or deoxycholate to the plasma membranes resulted in much greater formation of free sphingosine. These results suggest the concerted action of both sphingomyelinase and ceramidase in the production of free sphingosine in plasma membranes.

More experimental data are needed to support this functional role for ceramidases. Despite the lack of conclusive evidence that sphingosine produced endogenously via ceramide hydrolysis has a regulatory role, it remains that ceramide hydrolysis via ceramidases is the only established pathway for production of sphingosine in cells and that exogenous sphingosine affects cells in many interesting ways.

IV. Conclusion

The precise metabolic and biological roles of ceramidases are not entirely clear. *In vitro* catalytic activity and metabolic studies suggest that ceramide degradation is their primary biochemical role. The possibility that ceramidases function to synthesize ceramide *in vivo* as well as *in vitro* cannot be excluded.

From analysis of Farber's disease patients and tissues, the lysosomal activity seems to be necessary to prevent ceramides, including those containing hydroxy fatty acids, from accumulating excessively. Both lysosomal and non-lysosomal activities may function to recycle the sphingosine moiety of a cell's sphingolipids. The lysosomal enzyme would serve to recycle sphingolipids degraded by other lysosomal hydrolases, whereas a non-lysosomal activity could be used to salvage unused ceramides accrued during sphingomyelin or cerebroside synthesis. Conjecturally, ceramidases may also perform the task of regulating the cellular levels of sphingosine or ceramide, with the effect that molecular targets of these molecules are appropriately supplied with their effectors. Finally, ceramidases may function to regulate the types and levels of particular fatty acids incorporated into a cell's ceramides, and, consequently, its overall complement of sphingolipids.

The potential importance of the ceramidases in producing sphingosine for a regulatory purpose within cells makes them prime candidates for study. Some primary objectives are to characterize the isoenzymes with respect to their catalytic activity, subcellular location, molecular biology, and biological roles.

References

1. Gatt, S. (1963). *J. Biol. Chem.* **238,** PC3131–3133.
2. Gatt, S. (1966). *J. Biol. Chem.* **241,** 3724–3730.
3. Yavin, E., and Gatt, S. (1969). *Biochemistry* **8,** 1692–1698.
4. Fukui, T., and Axelrod, B. (1961). *J. Biol. Chem.* **236,** 811–816.
5. Colodzin, M., Bachur, N. R., Weissbach, H., and Udenfriend, S. (1963). *Biochem. Biophys. Res. Commun.* **10,** 165–170.
6. Bachur, N. R., Masek, K., Melmon, K. L, and Udenfriend, S. (1965). *J. Biol. Chem.* **240,** 1019–1024.
7. Gatt, S., and Yavin, E. (1969). *In* "Methods in Enzymology" (J. Lowenstein, ed.), Vol. 14, pp. 139–144. Academic Press, New York.

8. Sugita, M. Williams, M., Dulaney, J. T., and Moser, H. W. (1975). *Biochim. Biophys. Acta* **398,** 125–131.
9. Al, B. J. M., Tiffany, C. W., Gomes de Mesquita, D. S., Moser, H. W., Tager, J. M., and Schram, A. W. (1989). *Biochim. Biophys. Acta* **1004,** 245–251.
10. Nilsson, A. (1969). *Biochim. Biophys. Acta* **176,** 339–347.
11. Shen, Y.-W., and Tao, R. V. P. (1982). *Invest. Ophthalmol. Visual Sci.* **22,** 734–743.
12. Wertz, P. W., and Downing, D. T., (1990). *FEBS Lett.* **268,** 110–112.
13. Momoi, T., Ben-Yoseph, Y., and Nadler, H. L. (1982). *Biochem. J.* **205,** 419–425.
14. Spence, M. W., Beed, S., and Cook, H. W. (1986). *Biochem. Cell Biol.* **64,** 400–404.
15. Momoi, T., Mikawa, H., Ben-Yoseph, Y., and Nadler, H. L. (1988). *Ann. Paediatr. Jpn.* **34,** 22–26.
16. Chen, W. W., Moser, A. B., and Moser, H. W. (1981). *Arch. Biochem. Biophys.* **208,** 444–455.
17. Chen, W. W., and Decker, G. L. (1982). *Biochim. Biophys. Acta* **718,** 185–192.
18. Sugita, M., Dulaney, J. T., and Moser, H. W. (1972). *Science* **178,** 1100–1102.
19. Dulaney, J. T., Milunsky, A., Sidbury, J. B., Hobolth, N., and Moser, H. W. (1976). *J. Pediatr.* **89,** 59–61.
20. Ben-Yoseph, Y., Gagne, R., Parvathy, M. R., Mitchell, D. A., and Momoi, T. (1989). *Clin. Genet.* **36,** 38–42.
21. Stoffel, W., and Melzner, I. (1980). *Hoppe-Seyler's Z. Physiol. Chem.* **361,** 755–771.
22. Slife, C. W., Wang, E., Hunter, R., Wang, S., Burgess, C., Liotta, D. C., and Merrill, A. H., Jr. (1989). *J. Biol. Chem.* **264,** 10371–10377.
23. Sugita, M., Connolly, P., Dulaney, J. T., and Moser, H. W. (1973). *Lipids* **8,** 401–406.
24. O'Brien, J. S., and Rouser, G. (1964). *J. Lipid Res.* **5,** 339–342.
25. Chen, G. L., and Chen, W. W. (1986). *NATO ASI Ser., Ser. A* **116,** 615–619.
26. Kondoh, H., Kanoh, H., and Ono, T. (1983). *Biochim. Biophys. Acta* **753,** 97–106.
27. Stein, O., Oette, K., Hollander, G., Dabach, Y., Ben-Naim, M., and Stein, Y. (1989). *Biochim. Biophys. Acta* **1003,** 175–182.
28. Okabe, H., and Kishimoto, Y. (1977). *J. Biol. Chem.* **252,** 7068–7073.
29. Sutrina, S. L., and Chen, W. W. (1982). *J. Biol. Chem.* **257,** 3039–3044.
30. Wilson, E., Wang, E., Mullins, R. E., Uhlinger, D. J., Liotta, D. C., Lambeth, J. D., and Merrill A. H., Jr. (1988). *J. Biol. Chem.* **263,** 9304–9309.
31. Wang, E., Norred, W. P., Bacon, C. W., Riley, R. T., and Merrill, A. H., Jr. (1991). *J. Biol. Chem.* **266,** 14486–14490.
32. Buehrer, B., and Bell, R. M. (1992). *J. Biol. Chem.* **267,** 3154–3159.
33. Merrill, A. H., Jr., and Stevens, V. L. (1989). *Biochim. Biophys. Acta* **1010,** 131–139.
34. Merrill, A. H., Jr. (1991). *J. Bioenerg. Biomembr.* **23,** 83–104.
35. Merrill, A. H., Jr., and Wang, E. (1986). *J. Biol. Chem.* **261,** 3764–3769.
36. Merrill, A. H., Jr., Wang, E., Mullins, R. E., Jamison, W. C. L., Nimkar, S., and Liotta, D. C. (1988). *Anal. Biochem.* **171,** 373–381.
37. Van Veldhoven, P. P., Bishop, W. R., and Bell, R. M. (1989). *Anal. Biochem.* **183,** 177–189.

ADVANCES IN LIPID RESEARCH, VOL. 26

Sphingosine Kinase: Properties and Cellular Functions

BENJAMIN M. BUEHRER AND ROBERT M. BELL

Department of Biochemistry and
Section of Cell Growth, Regulation, and Oncogenesis
Duke University Medical Center
Durham, North Carolina 27710

I. Introduction

Catabolism of sphingosine requires two independent and sequential steps. Keenan and Maxam reported the production of ethanolamine phosphate and palmitaldehyde from dihydrosphingosine in a cell-free system (1). Two discernible activities were present, one requiring ATP and magnesium, the other requiring pyridoxal phosphate. Sphingosine kinase mediates the ATP- and magnesium-dependent phosphorylation of the hydroxyl on the first carbon of sphingosine. Sphingosine-1-phosphate lyase cleaves the product into ethanolamine phosphate and the corresponding aldehyde. The cleavage products of this reaction can be scavenged to synthesize other lipids such as phosphatidylethanolamine (PE) (2,3) and the ether lipids (4,5). Due to the relative inabundance of sphingosine-1-phosphate in cells, phosphorylation by sphingosine kinase is thought to be the rate-limiting step in this process (6). Sphingosine kinase was thought to only have a degradative function; however, recent evidence supports the hypothesis that sphingosine-1-phosphate may be a potent cellular regulator (7,8).

The importance of sphingosine and other sphingolipids is evident in recent findings implicating these compounds in many cellular functions (9,10). Gangliosides have long been known to be markers of differentiation, to function in contact inhibition, and to be specifically expressed in human melanomas and carcinomas. Roles for *N*-deacylated sphingolipids, lysosphingolipids, have only recently been suggested. The observation that sphingosine inhibits protein kinase C-mediated processes *in vitro* and in living cells (11–13) suggested that lysosphingolipids may act as cellular second messengers or lipid mediators. These molecules could be derived from complex sphingolipids which would serve as reservoirs for the production of bioactive metabolites. Many cellular responses have since been shown to be affected by sphingosine through a protein kinase C-dependent mechanism (10). Sphingosine can also exert effects independent of protein kinase C. Epidermal growth factor (EGF) receptor has an intrinsic, ligand-activated, tyrosine kinase activity stimulated by EGF binding. Sphingosine can activate receptor tyrosine kinase activity to the level induced by EGF (14,15). Sphingosine also induces receptor autophosphorylation, increases the number of EGF-binding sites expressed on the surface of A431 cells, and increases EGF receptor affinity (14–16). Recently, sphingosine has been shown to increase cytosolic phosphatidic acid (PA) levels and increase thymidine incorporation in Swiss 3T3 fibroblasts (17). The specificity of these effects has not been determined, and there is evidence indicating that sphingosine-1-phosphate is involved (8).

Sphingosine is cytotoxic to cells, but the toxic concentration is dependent on cell type. Long-chain bases (added in ethanol), including sphingosine, were found to be toxic at relatively low concentrations in CHO cells; greater than 5 μM dihydrosphingosine dramatically reduced cell number (18). Pittet *et al.* determined that 12 μM dihydrosphingosine or 25 μM D-sphingosine added in dimethyl sulfoxide (DMSO) induced cell permeabilization of human neutrophils (19). The permeabilization of neutrophils was found to be dependent on the method of long-chain base delivery. When sphingosine was complexed to fatty acid-free bovine serum albumin, cell integrity was maintained, whereas sphingosine added in DMSO dramatically affected permeability (20). It is not clear whether the growth inhibition caused by sphingosine addition is due to protein kinase C inhibition or is a result of other cellular actions (21).

II. Metabolism of Sphingosine

Sphingosine is thought to be generated from the degradation of complex sphingolipids, and not through *de novo* synthesis (see Fig. 1). Formation of the complex sphingolipids occurs initially through condensation of palmitoyl-CoA and serine to generate 3-ketodihydrosphingosine. This compound is rapidly reduced to dihydrosphingosine, and subsequently acylated to dihydroceramide. It is at this point where the 4-*trans* double bond is introduced into the molecule (22,23) and also where the synthetic pathways for more complex sphingolipids diverge. Glycosphingolipids are synthesized by successive transfer of carbohydrate groups linked to sugars that are attached to the 1-hydroxyl group present on ceramide. Sphingomy-

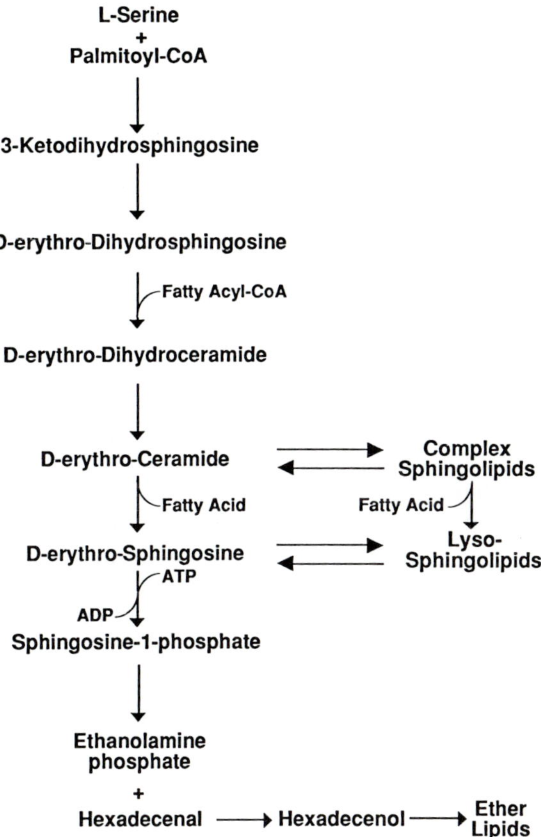

FIG. 1. Pathway of sphingosine metabolism.

elin is produced by the transfer of phosphorylcholine from phosphatidylcholine to ceramide. Sphingosine is generated through the degradation of these complex lipids by a series of glycosidases, phosphodiesterases, and ceramidases. The regulation of these processes remains to be characterized, and may be as complex as that of glycerolipids and that proposed for the "sphinogmyelin cycle" observed in HL-60 cells (24,25).

Cellular levels of free sphingosine are low (26–30) and have been shown to be modulated in at least one case by a variety of growth factors (31). Low sphingosine levels may reflect the potency of long-chain bases to modulate cellular functions and thereby require regulated production. There are multiple routes of long-chain base metabolism. In the brain, sphingosine may be condensed with galactose to form psychosine through a process mediated by UDPgalactose:sphingosine galactosyltransferase (32). More commonly, sphingosine may be acylated to ceramide by a fatty acyl-CoA:sphingosine acyltransferase (33). This process, however, generates ceramide, another bioactive sphingolipid (25,34). Another common pathway of sphingosine metabolism results in long-chain base degradation via sphingosine kinase and sphingosine-1-phosphate lyase (35).

III. Purification and Characterization of Sphingosine Kinase

Sphingosine kinase activity is present in a variety of tissues and cells, including bovine brain (36) and kidney (37), rat liver (1), human and porcine platelets (38), *Hansenula ciferrii* (39), and *Tetrahymena pyriformis* (5,40). While complete purification of the enzyme from any source has not been achieved, initial characterizations of partially purified preparations have been carried out. Sphingosine kinase is a soluble, cytoplasmic enzyme in platelets (38) and is peripherally associated with membranes in rat brain (41) and other tissues. However, in *T. pyriformis* (5,40), sphingosine kinase is more tightly associated with the mitochondrial and microsomal fractions. This localization is in agreement with that of sphingosine-1-phosphate lyase, which resides predominantly in particulate fractions (35,42,43). No sphingosine-1-phosphate lyase activity has been reported in platelets, the only reported source of completely cytoplasmic sphingosine kinase (44,45).

Sphingosine kinase requires a phosphate donor and divalent cation for activity. There have been slight differences in the nucleotide specificity reported for the different preparations of enzyme activity. In most cases, ATP is the preferred phosphate donor molecule; other nucleotides are less active. Louie *et al.* (36), determined an apparent K_m for ATP of sphingosine kinase from bovine brain to be 310 μM, whereas Stoffel *et al.* (5) found an apparent K_m of 90 μM in *T. pyriformis*. In this latter case, GTP supported higher activity than did ATP. The divalent cation requirement for sphingosine kinase from all sources is fulfilled by magnesium, manganese, and calcium. Magnesium activates sphingosine kinase to a greater extent than do manganese and calcium. Early studies determined that equimolar concentrations of divalent cation and ATP are needed for activation of the kinase (38), and that a fixed ratio of 5 : 1 (magnesium to ATP) maximally stimulates activity (36). Sphingosine kinase from all sources reported has a neutral pH optimum. Kinase activity is supported over a range of pH (pH 6–8), but maximal activity is generally achieved at pH 7.2–7.5.

Purification of sphingosine kinase activity from any source has proved to be an extremely difficult task. The most pure preparation reported produced only a 234-fold purification from starting material and a specific activity of 3.75 nmol/min/mg (36). The high-speed, cytosolic fraction of bovine brain homogenate was concentrated by ammonium sulfate precipitation. The 25–55% ammonium sulfate pellet was suspended in buffer and subjected to DEAE-cellulose chromatography. Sphingosine kinase activity eluted from the ion-exchange column in two distinct peaks. A Sephadex G-200 column fractionated enzyme activity from the second DEAE peak consistent with an apparent molecular weight of 190,000. Following G-200 chromatography, the preparation was run over a hydroxylapatite resin. This chromatography step also afforded two distinct peaks of activity. The second peak of activity contained the highest specific activity of sphingosine kinase and resulted in a 5% yield from the starting cytosol. The presence of multiple peaks of sphingosine kinase activity during two steps in the purification process has been attributed to the possible presence of multiple enzyme forms. The two peaks of activity from DEAE-cellulose eluted with the same apparent molecular weight when fractionated by Sephadex G-200. Further differentiation of the multiple forms of sphingosine kinase has not been reported, and the presence of isoenzymic forms cannot be ruled out.

Sphingosine exists in four isomeric states: D-*erythro*-(2*S*,3*R*), L-*erythro*-(2*R*,3*S*), D-*threo*-(2*R*,3*R*), and L-*threo*-(2*S*,3*S*) (see Fig. 2). These isomers have been tested for their ability to serve as substrates for sphingosine kinase. Although the results for the various isomers differ depending on the source of enzyme, one isomer, D-*erythro*-sphingosine, functions as a substrate in all. This isomer is the naturally occurring enantiomer of sphingosine and has been found to be the best substrate in all of the sources examined. Recent work, using rat brain and human platelet enzyme (41), showed that only sphingosine with the *erythro*-conformation functions as a substrate in a mixed micellar assay system (see Table I). Sphingosine kinase activity in isolated whole platelets displays a similar substrate specificity for *erythro*-sphingosine. *erythro*-Dihydrosphingosine and phytosphingosine (4-hydroxyl-dihydrosphingosine) can also be phosphorylated by rat brain sphingosine kinase in mixed micelles. Phytosphingosine is most commonly found in plants, fungi, and yeast (3,35,46), but can be found in mammalian tissues as well (mainly through ingestion) (47). Although phytosphingosine serves as a substrate for sphingosine kinase from mammalian sources, it is not as active as dihydrosphingosine or sphingosine. Dihydrosphingosine is as active a substrate as sphingosine, but has a higher K_m value than its unsaturated analog.

The chain length specificity of sphingosine kinase was investigated.

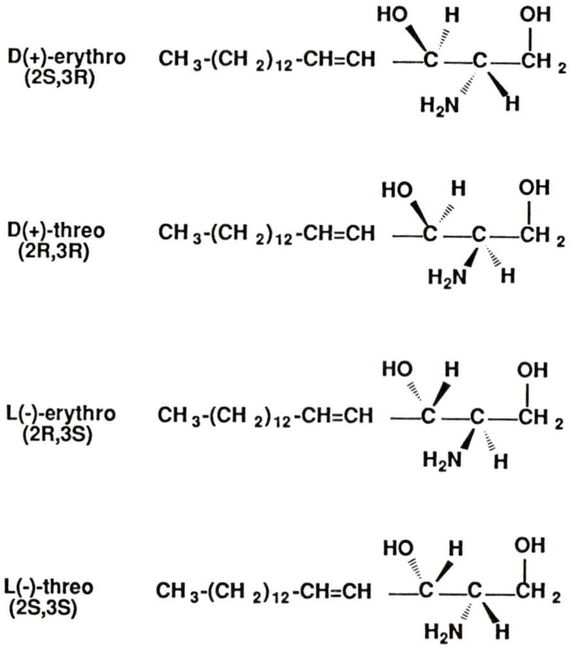

D(+)-erythro (2S,3R) CH_3-$(CH_2)_{12}$-CH=CH—C—C—CH_2 (HO, H, OH; H_2N, H)

D(+)-threo (2R,3R) CH_3-$(CH_2)_{12}$-CH=CH—C—C—CH_2 (HO, H, OH; H_2N, H)

L(-)-erythro (2R,3S) CH_3-$(CH_2)_{12}$-CH=CH—C—C—CH_2 (HO, H, OH; H_2N, H)

L(-)-threo (2S,3S) CH_3-$(CH_2)_{12}$-CH=CH—C—C—CH_2 (HO, H, OH; H_2N, H)

FIG. 2. Sphingosine isomers. Shown are the four stereoisomers of sphingosine and the orientations around carbons 2 and 3.

D-*erythro*-Dihydrosphingosines ranging from 14 to 20 carbons in length were tested *in vitro* using sphingosine kinase from human platelets (38). Sphingosine kinase activity increased with the length of the carbon chain.

Some of the sphingosine isomers have been found to have inhibitory properties (38,41). The nonphosphorylatable *threo*-sphingosine (and dihydrosphingosine) enantiomers function as competitive inhibitors of sphingosine kinase *in vitro* and in isolated platelets. Furthermore, a mixture of D- and L-*threo*-dihydrosphingosine was found to inhibit sphingosine kinase activity to an extent that influenced the metabolism of exogenous D-*erythro*-sphingosine in platelets. This inhibition was found to have effects on protein kinase C-mediated events. Sphingosine preincubation can inhibit thrombin-induced 40 (47)-kDa protein phosphorylation in isolated whole platelets (11). The length of sphingosine preincubation reduces the potency of the inhibitory effect, most likely through rapid metabolism of sphingosine (48). By perturbing one of the pathways of sphingosine metabolism, this time-dependent loss of inhibition was prevented (41).

IV. Sphingosine-1-Phosphate

Recently, sphingosine kinase has been implicated to have an important role in cellular regulation. Ghosh *et al.* found that sphingosine stimulated the release of calcium in permeablized smooth muscle cells in an ATP-dependent manner. This was subsequently attributed to the formation of sphingosine-1-phosphate (7). Others have provided evidence to support this notion of sphingosine-1-phosphate as a potent cell regulator (8,49). Responses in 3T3 fibroblasts associated with sphingosine have been shown to occur concomitantly with sphingosine-1-phosphate generation (8). Cell growth, intracellular calcium, and DNA synthesis were all increased in response to levels of sphingosine-1-phosphate lower than those of sphingosine. 3T3 cells pretreated with phorbol ester to downregulate protein kinase C remained responsive to sphingosine-1-phosphate and sphingosine. Therefore, sphingosine (as well as sphingosine-1-phosphate) may serve as a positive cellular regulator independently and in addition to its ability to inhibit protein kinase C-mediated events.

V. Concluding Remarks

The assessment of the role of sphingolipids in cells has been broadened from being stable complex lipids to being reservoirs of bioactive lipid

Table I
KINETIC VALUES DETERMINED FOR SPHINGOSINE AND DIHYDROSPHINGOSINE ISOMERS[a,b]

	Rat brain		Human platelet	
	Substrate (K_m)	Inhibition (K_i)	Substrate (K_m)	Inhibition (K_i)
D-*erythro*-Sphingosine	0.35	—	0.21	—
L-*erythro*-Sphingosine	0.38	—	0.15	—
D-*threo*-Sphingosine	—	0.46	—	0.30
L-*threo*-Sphingosine	—	0.40	—	0.23
D,L-*erythro*-Dihydrosphingosine	1.0	—	0.11	—
D,L-*threo*-Dihydrosphingosine	—	0.21	—	0.10

[a] Data from Buehrer and Bell (41) for sphingosine kinase from rat brain and human platelets.

[b] All values are expressed in terms of mole %.

second messengers. Knowledge of the pathways and functions of sphingolipid metabolites are expanding, yet in no instance has a complete and satisfying set of data been collected. Due to the emerging importance of sphingosine and sphingosine-1-phosphate in cellular regulation, it is important to investigate the structure and function of sphingosine kinase.

References

1. Keenan, R. W., and Maxam, A. (1969). *Biochim. Biophys. Acta* **176,** 348–356.
2. Henning, R., and Stoffel, W. (1969). *Hoppe-Seyler's Z. Physiol. Chem.* **350,** 827–835.
3. Stoffel, W. (1973). *Mol. Cell. Biochem.* **1,** 147–155.
4. Stoffel, W., LeKim, D., and Heyn, G. (1970). *Hoppe-Seyler's Z. Physiol. Chem.* **351,** 875–883.
5. Stoffel, W., Bauer, E., and Strahl, J. (1974). *Z. Phys. Chem.* **355,** 61–74.
6. Stoffel, W., and Bister, K. (1973). *Z. Phys. Chem.* **354,** 169–181.
7. Ghosh, T. K., Bian, J., and Gill, D. L. (1990). *Science* **248,** 1653–1656.
8. Zhang, H., Desai, N. N., Olivera, A., Seki, T., Brooker, G., and Spiegel, S. (1991). *J. Cell Biol.* **114,** 155–167.
9. Hannun, Y. A., and Bell, R. M. (1989). *Science* **243,** 500–507.
10. Merrill, A. H. J. (1991). *J. Bioenerg. Biomembr.* **23,** 83–104.
11. Hannun, Y. A., Loomis, C. R., Merrill, A. H. J., and Bell, R. M. (1986). *J. Biol. Chem.* **261,** 12604–12609.
12. Merrill, A. H. J., Sereni, A. M., Stevens, V. L., Hannun, Y. A., Bell, R. M., and Kinkade, J. M. J. (1986). *J. Biol. Chem.* **261,** 12610–12615.
13. Wilson, E., Olcott, M. C., Bell, R. M., Merrill, A. H. J., and Lambeth, J. D. (1986). *J. Biol. Chem.* **261,** 12616–12623.
14. Davis, R. J., Girones, N., and Faucher, M. (1988). *J. Biol. Chem.* **263,** 5373–5379.
15. Wedegaertner, P. B., and Gill, G. N. (1989). *J. Biol. Chem.* **264,** 11346–11353.
16. Faucher, M., Girones, N., Hannun, Y. A., Bell, R. M., and Davis, R. J. (1988). *J. Biol. Chem.* **263,** 5319–5327.
17. Zhang, H., Desai, N. N., Murphey, J. M., and Spiegel, S. (1990). *J. Biol. Chem.* **265,** 21309–21316.
18. Merrill, A. H. J. (1983). *Biochim. Biophys. Acta* **754,** 284–291.
19. Pittet, D., Krause, K.-H., Wolheim, C. B., Bruzzone, R., and Lew, D. P. (1987). *J. Biol. Chem.* **262,** 10072–10076.
20. Lambeth, J. D., Burnham, D. N., and Tyagi, S. R. (1988). *J. Biol. Chem.* **263,** 3818–3822.
21. Stevens, V. L., Nimkar, S., Jamison, W. C. L., Liotta, D. C., and Merrill, A. H. J. (1990). *Biochim. Biophys. Acta* **1051,** 37–45.
22. Ong, D. E., and Brady, R. N. (1973). *J. Biol. Chem.* **248,** 3884–3888.
23. Merrill, A. H., and Wang, E. (1986). *J. Biol. Chem.* **261,** 3764–3769.
24. Okazaki, T., Bell, R. M., and Hannun, Y. A. (1989). *J. Biol. Chem.* **264,** 19076–19080.
25. Okazaki, T., Bielawska, A., Bell, R. M., and Hannun, Y. A. (1990). *J. Biol. Chem.* **265,** 15823–15831.
26. Merrill, A. H., Wang, E., Mullins, R. E., Jamison, W. C. L., Nimkar, S., and Liotta, D. C. (1988). *Anal. Biochem.* **171,** 373–381.
27. Kobayashi, T., Mitsuo, K., and Goto, I. (1988). *Eur. J. Biochem.* **172,** 747–752.

28. Van Veldhoven, P. P., Bishop, W. R., and Bell, R. M. (1989). *Anal. Biochem.* **183,** 177–189.
29. Wertz, P. W., and Downing, D. T. (1989). *Biochim. Biophys. Acta* **1002,** 213–217.
30. Wertz, P. W., and Downing, D. T. (1990). *J. Invest. Dermatol.* **94,** 159–161.
31. Wilson, E., Wang, E., Mullins, R. E., Uhlinger, D. J., Liotta, D. C., Lambeth, J. D., and Merrill, A. H. (1988). *J. Biol. Chem.* **263,** 9304–9309.
32. Friedrich, V. L., Jr., and Hauser, G. (1973). *J. Nerochem.* **20,** 1131–1141.
33. Narimatsu, S., Soeda, S., Tanaka, T., and Kishimoto, Y. (1986). *Biochim. Biophys. Acta* **877,** 334–341.
34. Dobrowsky, R. T., and Hannun, Y. A. (1992). *J. Biol. Chem.* **267,** 5048–5051.
35. Stoffel, W., Sticht, G., and LeKim, D. (1968). *Hoppe-Seyler's Z. Physiol. Chem.* **349,** 1745–1748.
36. Louie, D. D., Kisic, A., and Schroepfer, G. J. J. (1976). *J. Biol. Chem.* **251,** 4557–4564.
37. Keenan, R. W., and Haegelin, B. (1969). *Biochem. Biophys. Res. Commun.* **37,** 888–894.
38. Stoffel, W., Hellenbroich, B., and Heimann, G. (1973). *Hoppe-Seyler's Z. Physiol. Chem.* **354,** 1311–1316.
39. Stoffel, W., Sticht, G., and LeKim, D. (1968). *Hoppe-Seyler's Z. Physiol. Chem.* **349,** 1149–1156.
40. Keenan, R. W. (1972). *Biochim. Biophys. Acta* **270,** 383–396.
41. Buehrer, B. M., and Bell, R. M. (1992). *J. Biol. Chem.* **267,** 3154–3159.
42. Stoffel, W., LeKim, D., and Sticht, G. (1969). *Hoppe-Seyler's Z. Physiol. Chem.* **350,** 1233–1241.
43. Van Veldhoven, P. P., and Mannaerts, G. P. (1991). *J. Biol. Chem.* **266,** 12502–12507.
44. Stoffel, W., Assmann, G., and Binczek, E. (1970). *Hoppe-Seyler's Z. Physiol. Chem.* **351,** 635–642.
45. Stoffel, W., Heimann, G., and Hellenbroich, B. (1973). *Hoppe-Seyler's Z. Physiol. Chem.* **354,** 562–566.
46. Stoffel, W., and Assman, G. (1972). *Hoppe-Seyler's Z. Physiol. Chem.* **353,** 965–970.
47. Assmann, G., and Stoffel, W. (1972). *Hoppe-Seyler's Z. Physiol. Chem.* **353,** 971–979.
48. Khan, W. A., Mascarella, S. W., Lewin, A. H., Wyrick, C. D., Carroll, F. I., and Hannun, Y. A. (1991). *Biochem. J.* **278,** 387–392.
49. Zhang, H., Buckley, N. E., Gibson, K., and Spiegel, S. (1990). *J. Biol. Chem.* **265,** 76–81.

ADVANCES IN LIPID RESEARCH, VOL. 26

Sphingosine-Phosphate Lyase

PAUL P. VAN VELDHOVEN AND GUY P. MANNAERTS

Katholieke Universiteit Leuven
Fakulteit Geneeskunde
Campus Gasthuisberg
Afdeling Farmakologie
B-3000 Leuven, Belgium

I. Introduction and Historical Perspective

Sphingolipids are composed of three basic constituents: a sphingoid base,[1] a fatty acid, and a polar headgroup. The polar headgroup distinguishes neutral sphingolipids like ceramide, phosphosphingolipids like sphingomyelin, and glycosphingolipids like cerebrosides and gangliosides. Variations in the chain length and the degree of unsaturation and hydroxylation of the fatty acid, combined with those occurring in the sphingoid

[1] Throughout this review, the term sphingoid base is used to describe a group of related long-chain aliphatic 2-amino-1,3-diols. Sphinganine, previously known as dihydrosphingosine, refers to D(+)-*erythro*-2-amino-1,3-octadecanediol [2D-amino-1,3D-octadecanediol; (2*S*,3*R*)-2-amino-1,3-octadecanediol]. Sphinganine containing a 4-*trans*- double bond is called sphingenine [(4*E*)-sphingenine; 4-*t*-sphingenine], formerly known as sphingosine. Sphinganine possessing an additional hydroxyl group at C-4 is referred to as 4*D*-hydroxysphinganine (formerly phytosphingosine). Unless otherwise specified, a chain length of 18 carbon atoms is assumed for these three sphingoid bases. In addition, the use of these names implies that the compounds referred to possess the D-*erythro* configuration, as found in naturally occurring sphingenine. The nomenclature of the *erythro*- and *threo*-diasteromers causes a lot of confusion, since C-2 or C-3 are arbitrarily chosen as points of references. The *S*/*R* system, although unequivocal, does not result in identical prefixes for the same atom in different sphingoid bases. Therefore, we prefer to specify the substituents at C-2, C-3, and C-4 by the prefix *D* or *L*. This prefix refers to the orientation of the substituents to the right or left, respectively, of the carbon chain drawn vertically in a Fischer projection with C-1 at the top (see Nomenclature of Lipids, 1977).

base, further add to the bewildering complexity and diversity of the sphingolipids. Despite this complexity, our understanding of sphingolipid breakdown is quite comprehensive (Kanfer, 1983a; Sweeley, 1991). The occurrence of several lysosomal storage disorders, in which different sphingolipids accumulate, has certainly been beneficial to the uncovering of the degradation pathway (Kanfer, 1983b). Through the action of different lysosomal hydrolases, sphingolipids are degraded to ceramide, which is then hydrolyzed by the lysosomal ceramidase to its components: a fatty acid and a sphingoid base. The predominant sphingoid bases found in nature are sphingenine, sphinganine, and 4*D*-hydroxysphinganine[1] (Karlsson, 1970a,b). The common structure of these sphingoid bases, 2-amino-1,3-diol-alkane, is unique in nature.

In the late 1960s, when procedures to synthesize specifically labeled sphingoid bases became available (Gatt, 1966; Stoffel and Sticht, 1967c; Keenan and Okabe, 1968; Nilsson, 1968), studies on the breakdown of sphingoid bases were initiated by the groups of Gatt, Karlsson, Stoffel, Keenan, and Nilsson. *In vivo* studies in rats, intravenously injected with labeled sphingoid bases, demonstrated that sphingoid base uptake was highest in liver. Intracellularly, the sphingoid bases were actively metabolized, either via a biosynthetic route to ceramide and other sphingolipids or via a degradative pathway, which yielded fatty acids that were recovered in an esterified form in neutral glycerolipids and phospholipids. Further work revealed that the three different sphingoid bases are degraded *in vivo* to a two-carbon unit, corresponding to carbons 1 and 2 of the sphingoid base, and a hydrocarbon fragment, two carbon atoms shorter than the original base (Barenholz and Gatt, 1967, 1968; Roitman *et al.*, 1967; Stoffel and Sticht, 1967a,b; Gatt and Barenholz, 1968; Keenan and Okabe, 1968; Nilsson, 1968). Similar findings were reported in *Hansenula ciferrii* (Karlsson *et al.*, 1967; Stoffel *et al.*, 1968a) and later in *Tetrahymena pyriformis* (Stoffel *et al.*, 1974). The fragments, both of which can undergo considerable metabolic changes, were initially thought to be ethanolamine and a fatty acid. *In vitro* experiments with rat liver homogenates showed that the degradation of sphingoid bases was dependent on ATP and that, in addition to a fatty acid, considerable amounts of fatty aldehyde were formed (Gatt and Barenholz, 1968; Stoffel *et al.*, 1968b; Keenan and Maxam, 1969). ATP was assumed to serve as a phosphate donor for the phosphorylation of the primary hydroxyl group of the sphingoid bases, since the degradation of synthetic sphinganine-1-phosphate in rat liver homogenates was not stimulated by ATP (Stoffel *et al.*, 1968b). Subsequent studies indeed showed the existence of sphingosine kinase(s)[2] in different mammalian tissues (Keenan and Haegelin,

1969; Hirschberg *et al.*, 1970; Stoffel *et al.*, 1970a, 1973a; Louie *et al.*, 1976).

Degradation of sphingoid bases thus turned out to consist of two specific enzymatic steps. Sphingoid bases are first phosphorylated at the 1-hydroxyl group, and then cleaved to a long-chain aldehyde and ethanolamine-phosphate (Stoffel, 1973a,b; Kanfer, 1983b). Sphingosine kinase, the enzyme catalyzing the phosphorylation, is treated in another article of this volume (see Buehrer and Bell, this volume), while sphingosine-1-phosphate lyase, the cleavage enzyme responsible for the ultimate step in sphingolipid catabolism, is the subject of this review.

Since the enzyme can act on different phosphorylated sphingoid bases (see Section III), we will refer to it as sphingosine-phosphate lyase.[2] Other names used in the literature include dihydrosphingosine-1-phosphate aldolase, sphinganine-1-phosphate alkanal-lyase, and sphinganine-1-phosphate lyase.

More information on sphingoid base breakdown can be found in some older reviews on sphingosine metabolism written by Stoffel (1970, 1973a,b). In these reviews, sphingoid base degradation is only briefly treated, with some cursory notes on sphingosine-phosphate lyase. To our knowledge, this is the first comprehensive review on the enzyme.

II. Preparation of Substrate; Assay Methods

Sphingosine-phosphate lyase has been studied by only a limited number of researchers. The main reason is that phosphorylated sphingoid bases are not readily available. Moreover, due to the low activity and the stereospecificity of the enzyme (see Section III), substrates should preferentially be radioactive and possess a (2*D*,3*D*) configuration, constituting another drawback for the study of this enzyme.

Since 1957, when Weiss reported on the chemical synthesis of sphinganine-phosphate[3], no further attempts to chemically prepare phosphory-

[2] The generic term sphingosine is used to allude to sphingoid bases without specification of their substituents or degree of unsaturation, but possessing the (2*D*,3*D*) configuration. In the same sense, the names sphingosine kinase and sphingosine-phosphate lyase are used, since these enzymes act not only on sphingenine (previously called sphingosine), but also on the other natural sphingoid bases.

[3] Sphinganine-phosphate, sphingenine-phosphate, sphingosine-phosphate, and 4*D*-hydroxysphinganine-phosphate refer to sphingoid bases carrying a phosphate group at the C-1 hydroxyl group.

lated sphingoid bases have been made until recently. Starting from sphingenine[4], sphinganine-phosphate was obtained by Weiss (1957) by four steps involving hydrogenation, carbobenzoxylation of the amino group, phosphorylation with diphenylphosphorylchloride, and removal of the protecting group. Although attempted, sphingenine-phosphate cannot be prepared in this way, probably due to the presence of the allylic hydroxyl group. By suitable protection of this group by acylation (of N,O-protected sphingenine) with pivaloylchloride, as recently published by Hakomori and co-workers, this problem has been solved (Sadahira *et al.,* 1992). Selective deprotection of the primary hydroxyl group with tosylchloride, followed by its phosphorylation and deblocking of the primary amino group and of the allylic hydroxyl group, results in sphingenine-phosphate possessing the natural 2*D*, 3*D*-configuration. Other alternatives to prepare sphingenine-phosphate (as well as sphinganine- or 4*D*-hydroxysphinganine-phosphate) are based on biochemical approaches. Cytosolic fractions from rat liver (Hirschberg *et al.,* 1970) or microsomal fractions from *T. pyriformis* (Keenan, 1972) have been used as a crude source of sphingosine kinase. Isolated platelets, which are apparently not able to metabolize phosphorylated sphingoid bases, can also be employed (Stoffel *et al.,* 1970a,1973a). Further purification of the produced phosphate esters is tedious, however, and hampered by their sparing solubility in polar and nonpolar solvents (Stoffel *et al.,* 1969a, 1973a; Keenan and Haegelin, 1969; Van Veldhoven *et al.,* 1989b). Another possible method relies on the phosphorylation of ceramide[4] by *Escherichia coli* diacylglycerol kinase,[4] followed by acidic hydrolysis of the generated ceramide-1-phosphate (Van Veldhoven *et al.,* 1989b; Dressler and Kolesnick, 1990). However, due to the allylic hydroxyl group, isomerization at the C-3 carbon and other transformations are likely to occur during the hydrolysis (Karlsson, 1970a,b). Recently, one of us described a relatively easy biochemical method which uses the enzymatic digestion of sphingenine-1-phosphocholine[4] with phospholipase D from *Streptomyces chromofuscus*.[4] On incubation of sphingenine-1-phosphocholine with this phospholipase, a flocculous precipitate is formed that is enriched in sphingenine-phosphate, which can easily be further purified by selective extractions and precipitations (Van Veldhoven *et al.,* 1989b) or thin-layer chromatography (Zhang *et al.,* 1991). The phospholipase D works on sphinganine-

[4] These compounds or enzymes are commercially available. Commercial sphingenine-1-phosphocholine, also called sphingosyl-1-phosphorylcholine, prepared by acidic hydrolysis of sphingomyelin, might contain up to 20% impurities.

1-phosphocholine as well, generating sphinganine-phosphate (Van Veldhoven *et al.,* 1989b).

In order to prepare a radioactive substrate, labeled at a suitable site to monitor the sphingosine-phosphate lyase reaction, one can start with [4,5-^{3}H]sphinganine, obtained by catalytic reduction of sphingenine in the presence of tritium gas, and synthesize the phosphate ester chemically or by means of sphingosine kinase (see above; Hirschberg *et al.,* 1970), or, using phospholipase D, one can cleave [4,5-^{3}H]sphinganine-1-phosphocholine obtained by acidic hydrolysis of sphingomyelin which is hydrogenated with tritiated sodium borohydride (see above; Van Veldhoven and Mannaerts, 1991). The preparation of phosphorylated sphingoid bases, ^{14}C- or ^{3}H-labeled at other specific sites, requires more chemical skills (Nilsson, 1968; Stoffel and Assmann, 1970; Stoffel *et al.,* 1969a,b,1970a).

Before we discuss the assay methods, one has to realize that phosphorylated sphingoid bases, either as free acids or alkali-salts, are barely soluble in water, although being zwitterionic in nature (Weiss, 1957; Stoffel *et al.,* 1969b; Van Veldhoven *et al.,* 1989b). Reported pK_a values for sphinganine-phosphate are 2.0, 8.1, and 10.7 (Stoffel *et al.,* 1969a). Similar solubility problems have been noticed with sphinganine-phosphonate (Stoffel and Grol, 1974) and capnine (2-amino-3-hydroxy-15-methylhexadecane-1-sulfonic acid), a sulfonated long-chain base found in gliding bacteria belonging to the genus *Capnocytophaga* (Godcheaux and Leadbetter, 1980). Presumably, the low solubility is caused by intermolecular interactions between the oppositely charged groups, combined with hydrophobic interactions of the hydrocarbon chains. In order to obtain a consistent solubilization of the substrates during *in vitro* assays, the use of Triton X-100 has been advocated (Stoffel *et al.,* 1969b). In retrospect, this choice was fortuitous since a number of other neutral and zwitterionic detergents are inhibitory to the enzyme (Van Veldhoven and Mannaerts, 1991).

With regard to the lyase measurements, the following points should be considered. The produced fatty aldehyde will undergo further metabolism in crude homogenates, postnuclear supernatants, and nuclear and heavy mitochondrial fractions, not only oxidation but also reduction (Van Veldhoven and Mannaerts, 1991; see Section VI). Thus, when [4,5-^{3}H]sphinganine-phosphate is used in homogenates (or in the above mentioned subcellular fractions), all three possible products, fatty aldehyde, fatty acid, and long-chain fatty alcohol, should be analyzed. In microsomal fractions, further processing of palmitaldehyde is negligible ($\sim$5%). In all tissues screened so far, phosphatases are present which act on the coenzyme (pyridoxal-phosphate; see Section III) and on the phosphorylated sphingoid bases (P. P. Van Veldhoven and G. P. Mannaerts, unpublished data). The pyridoxal-phosphate phosphatase can be blocked by fluoride, chelators, and high phosphate concentrations (Lumeng and Li, 1979). Stoffel *et al.* (1969b) reported that the addition of NaF to the assay mixtures suppresses the sphingosine-phosphatase(s) present in rat liver to such degree that hydrolysis of the substrate does not interfere with the lyase assay. Preliminary data indicate, however, the presence of at least two sphingosine-phosphatases, one of which is fluoride insensitive (P. P. Van Veldhoven and G. P. Mannaerts, unpublished data). In some tissues, especially those of neuronal origin, the activity of the fluoride-insensitive phosphatase under the assay conditions of sphingosine-phosphate lyase is 10- to 20-fold higher than the cleavage rate of the substrate. Hence, for reliable measurements, such tissues should be diluted appropriately so that no more than 30% of the substrate (initial concentration 40 μM) is consumed in the assay by both reactions. Finally, for optimum activity, the presence of chelators and pyridoxal-phosphate is required in the assay mixture, and concentrations of compounds containing primary amine groups should be kept low (Van Veldhoven and Mannaerts, 1991; see Section III).

III. Enzyme Characteristics; Reaction Mechanism

So far, sphingosine-phosphate lyase has not been purified, and only the enzymatic activity present in rat liver has been reasonably well characterized. For the sake of clarity, we would like to stress that throughout the further discussion the term sphingosine-phosphate lyase refers to the enzymatic activities measured with the three different phosphorylated sphingoid bases, since we assume that all three activities reside with a single protein.

The cleavage of phosphorylated sphingoid bases by the lyase proceeds optimally at a pH of 7.4 to 7.6 and is irreversible (Stoffel *et al.*, 1969b). The primary reaction products have been unequivocally identified as a fatty aldehyde and ethanolamine-phosphate by different groups (Stoffel *et al.*, 1968b, 1969b; Stoffel and Assmann, 1970, 1972; Shimojo and Schroepfer, 1976). Hence, sphingosine-phosphate lyase has been classified within the group of carbon–carbon lyases, a subclass of aldehyde-lyases (EC 4.1.2.27) (Enzyme Nomenclature 1984).

Like some other aldolases, sphingosine-phosphate lyase requires pyridoxal-phosphate as coenzyme (Keenan and Maxam, 1969; Stoffel *et al.*, 1969b; Nilsson, 1970). After dialysis of the enzyme source, stimulation by pyridoxal-phosphate is several-fold, indicating that this coenzyme is loosely associated (Keenan and Maxam, 1969; Stoffel *et al.*, 1969b). The analog deoxypyridoxine-phosphate acts as a, possibly competitive, inhibitor (Stoffel *et al.*, 1969b). The inhibition of the lyase by semicarbazide, cyanide, bisulfite, and cysteine is likely due to their complexation with the cofactor (Keenan and Maxam, 1969; Nilsson, 1970; Van Veldhoven and Mannaerts, 1991). Also, the lower activity in Tris-based systems (Nilsson, 1970; Van Veldhoven and Mannaerts, 1991) as compared to phosphate buffers might be caused by the formation of a Schiff base between the Tris buffer and the carbonyl group of the coenzyme (Matsuo, 1957). According to the postulated reaction mechanism (Stoffel, 1970, 1973a; see Figs. 1 and 2), pyridoxal-phosphate forms a Schiff base with the free amino group of the phosphorylated sphingoid base. During cleavage, the coenzyme probably remains associated with the two-carbon fragment and trapped within the catalytic site of the enzyme. This could explain the stereospecific incorporation of one atom of solvent hydrogen into the *R*- configuration of the produced ethanolamine-1-phosphate, resulting in the net retention of the configuration at C-2 (Shimojo *et al.*, 1976; Fig. 1).

The lyase activity is sensitive toward sulfhydryl reagents like *N*-ethylmaleimide, iodoacetamide, and *p*-chloromercuribenzoate (Stoffel *et al.*, 1969b; Van Veldhoven and Mannaerts, 1991). The inhibition has been defined as an alkylation of an SH- group which attacks the substrate at C-3 (Stoffel, 1970; Figs. 1 and 2).

Phosphorylated sphingoid bases, like all sphingolipids, possess two optical centers, located at the second and third carbon atoms of the sphingoid base. The naturally occurring isomer of sphinganine possesses the (2*D*,3*D*) [or D(+)-*erythro*] configuration (Karlsson, 1970a,b). In contrast to some sphingosine kinases which can act on all four possible isomers (Keenan, 1972; Stoffel and Bister, 1973; Stoffel *et al.*, 1973b), rat liver sphingosine-phosphate lyase recognizes only the (2*D*,3*D*) isomer, at least *in vivo* (Stoffel and Bister, 1973). Less specificity is shown toward the type of sphingoid base, however. Microsomal preparations of rat liver cleave phosphorylated sphinganine (Stoffel *et al.*, 1969b), sphingenine (Stoffel and Assmann, 1970), and 4*D*-hydroxysphinganine (Stoffel and Assmann, 1972). The corresponding fatty aldehydes produced are hexadecanal (palmitaldehyde), 2-*trans*-hexadecanal, and 2*D*-hydroxyhexadecanal (see also Section VI). Also, the chain length appears to be less critical, since, after intravenous administration to rats, the short-chain sphinganine analog 2-amino-1,3-heptanediol (Stoffel *et al.*, 1969a), as well as the longer C_{20}-sphinganine (Stoffel and Scheid, 1969), both give rise to ethanolamine-phosphate.

For most of these substrates, no detailed kinetic analysis is available. According to Stoffel and Assmann (1970), sphingenine- and sphinganine-phosphate are degraded at similar rates. The K_m values reported in different studies for sphinganine-phosphate agree fairly well: 16 μM (absence of Triton X-100; Stoffel and Grol, 1974) and 9 μM (assayed in the presence

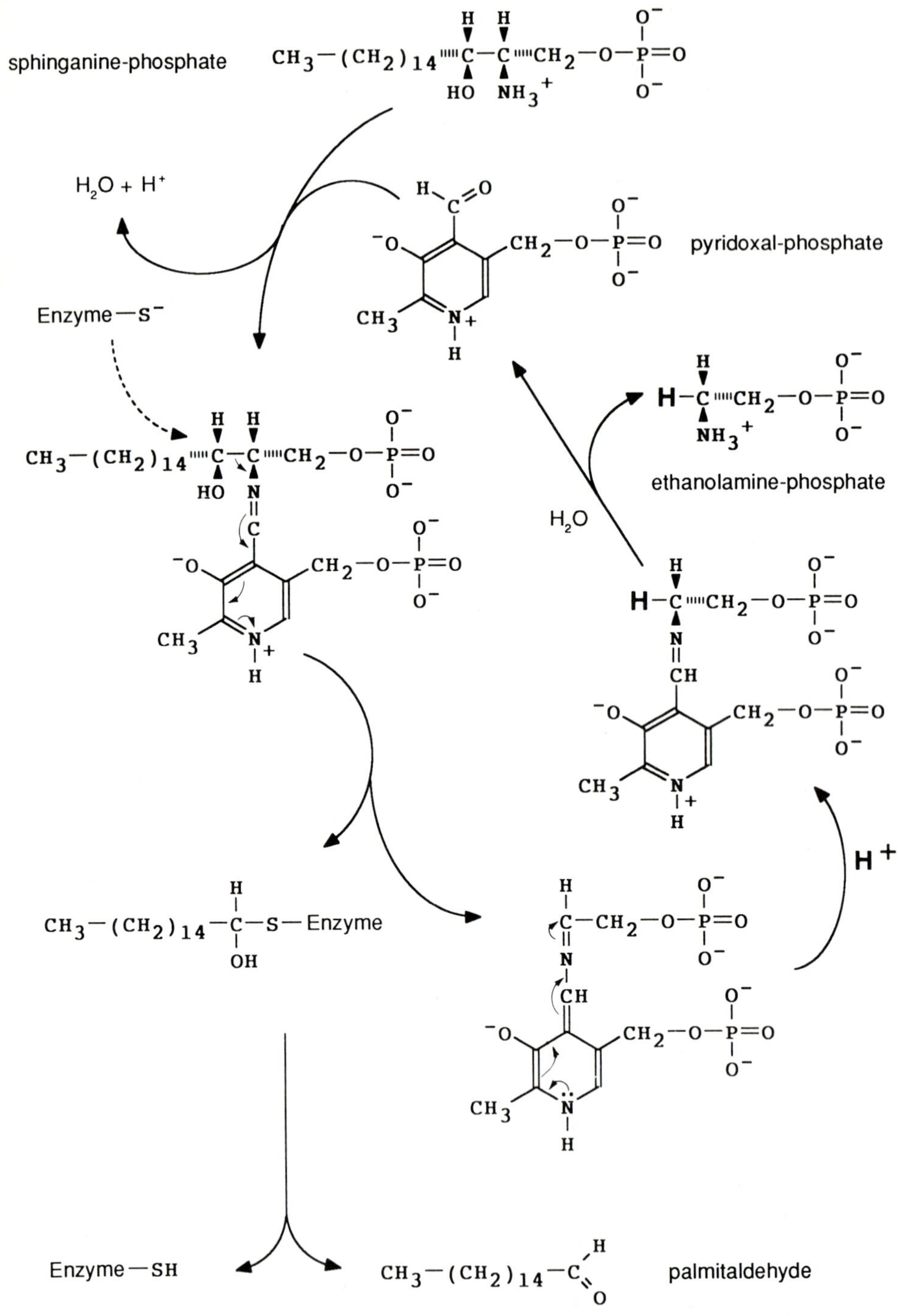

sphinganine-phosphate

$CH_3-(CH_2)_{14}-CH(OH)-CH(NH_3^+)-CH_2-O-PO_3^{2-}$

$H_2O + H^+$

pyridoxal-phosphate

Enzyme—S^-

ethanolamine-phosphate

H_2O

H^+

$CH_3-(CH_2)_{14}-CH(OH)-S-$Enzyme

Enzyme—SH

$CH_3-(CH_2)_{14}-CHO$ palmitaldehyde

of Triton X-100; Van Veldhoven and Mannaerts, 1991). No substrate inhibition was observed with sphinganine-phosphate at concentrations as high as 180 μM (P. P. Van Veldhoven and G. P. Mannaerts, unpublished data).

Only one substrate with a modified headgroup, sphinganine-1-phosphonate, has been used (Stoffel and Grol, 1974). Although the K_m values for sphinganine-phosphate and the phosphonate analog are nearly identical, cleavage of the phosphonate analog is 10-fold slower. This analog behaves as a competitive inhibitor of sphinganine-phosphate degradation (K_i 5 μM).

Substrates with modified amino groups have not been tested. Theoretically, either *N*-ac(et)ylation or *N*-(di)methylation should result in an inactive substrate, being unable to complex with pyridoxal-phosphate. This does not exclude that such compounds (e.g., *N,N*-dimethylsphingenine-phosphate) might be potent inhibitors (see Section VI).

Product inhibition by palmitaldehyde and ethanolamine-phosphate has not been observed, and serine and threonine, which bear some analogy with the 2-amino-1,3-diol structure, did not behave as competitive inhibitors (Stoffel *et al.,* 1969b). Hexadecanol or palmitic acid did not affect the enzymatic activity either (Van Veldhoven and Mannaerts, 1991).

Optimum activity requires the presence of chelators since the enzyme is sensitive toward heavy metal ions, especially Ca^{2+} and Zn^{2+}. Ni^{2+} on the other hand does not affect the enzyme (Van Veldhoven and Mannaerts, 1991).

Finally, with the exception of Triton X-100, the lyase is sensitive toward detergents. Various nonionic detergents belonging to differents classes, like octylglucoside, lauryldimethylammonium oxide, and octanoyl-*N*-methylglucamide, suppressed the activity even at concentrations below their critical micellar concentration (Van Veldhoven and Mannaerts, 1991). Triton X-100, however, was tolerated over a wide range [0 to 1.0% (w/v)] (Stoffel *et al.,* 1969b; Van Veldhoven and Mannaerts, 1991), while the zwitterionic detergent CHAPS inhibited the enzyme at concentrations

---

FIG. 1. Possible reaction mechanism for sphingosine-phosphate lyase. The mechanism, illustrated here with sphinganine-phosphate, is based on the role of pyridoxal-phosphate as coenzyme, an essential thiol group on the enzyme, the specificity of the enzyme for the 2*D*,3*D* configuration, and a stereospecific uptake of a proton in the produced ethanolamine-phosphate (redrawn from Shimojo *et al.,* 1976). After formation of a Schiff base between sphinganine-phosphate and the coenzyme, if C-2 has the correct 2*D* configuration, a nucleophilic attack occurs at C-3 by a sulfhydryl anion. The latter reaction requires a 3*D* configuration. The formed thiohemiacetal decomposes into palmitaldehyde and the sulfhydryl group. The Schiff-based two-carbon unit stereospecifically incorporates one solvent proton (**H**) at C-2 before being hydrolyzed to ethanolamine-phosphate and pyridoxal-phosphate.

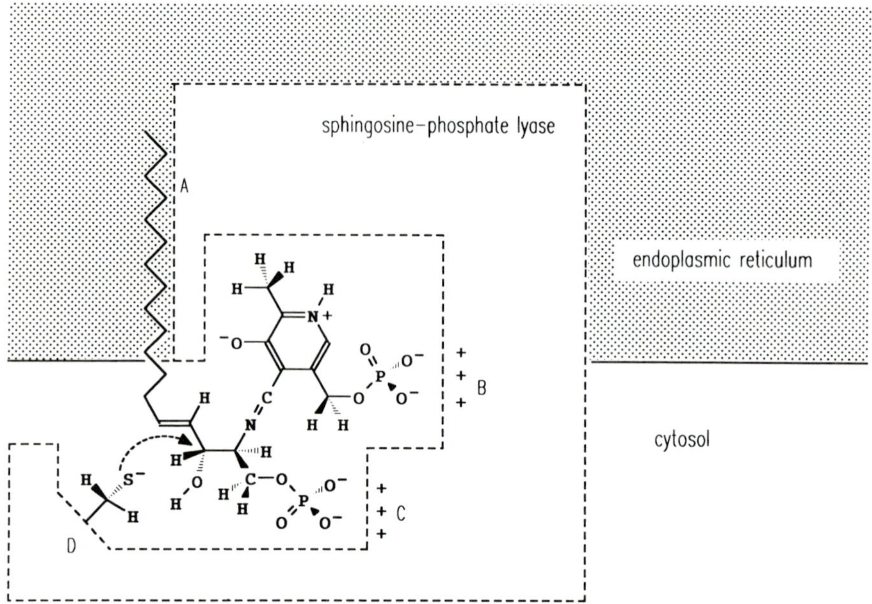

FIG. 2. Hypothetical two-dimensional presentation of sphingosine-phosphate lyase and its catalytic site. Sphingosine-phosphate lyase is shown as an integral membrane protein, embedded in the membrane of the endoplasmic reticulum with its catalytic site exposed to the cytosol (see Section III). At least four sites (A to D) seem to be involved in the interaction with the substrate and the coenzyme [sites B to D have been previously proposed by Stoffel (1973a)]. Binding of the substrate at site C through its phosphate group presents the molecule in the correct orientation to allow the coupling with the coenzyme (which is bound at site B via its phosphate group) and a nucleophilic attack by the sulfhydryl anion (site D) at C-3, provided the amino group at C-2 and the hydroxyl group at C-3 have the right orientation. The substrate is drawn as a Schiff base with pyridoxal-phosphate since, under our assay conditions, the major part of the substrate is complexed with this coenzyme (P. P. Van Veldhoven and G. P. Mannaerts, unpublished data). The competition by the (2*D*,3*L*) isomer can be explained by its occupation, when complexed with the coenzyme, of the two phosphate recognition sites, while the attack at C-3 is sterically hindered. Possibly, the (2*L*,3*L*) isomer, which has no effect on the enzyme, has no access to the catalytic site (see Section III). The sulfhydryl group, belonging to cysteine, is readily accessible to water-soluble reagents, an indication that is located in an aqueous environment. Finally, the strong inhibition of the lyase by monomeric detergents with a flexible alkyl chain seems to point to the involvement of a hydrophobic recognition site (site A). As substrate, sphingenine-phosphate is shown. Due to the 4-*trans* double bond, this substrate is more constrained than the other possible substrates. The hydrocarbon chain of all substrates [having the natural (2*D*,3*D*) configuration and fixed into the catalytic site], can extend similarly into the outer leaflet of the membrane, making contact with the hydrophobic recognition site A. The structure of inhibitory detergents is such that their alkyl chain can also interact with this site, while their polar moiety has access to the catalytic site.

above 0.2% (w/v) (Van Veldhoven and Mannaerts, 1991). The latter concentration is close to the critical micellar concentration. Indirect evidence indicates that taurocholate may also have little effect on the enzyme (Gatt and Barenholz, 1968).

Although no deleterious effects are observed when Triton X-100 is present in the assay (see above), partial inactivation of the lyase occurred when it was solubilized by means of this detergent (Van Veldhoven and Mannaerts, 1991). Acetone or butanol extractions also appear to cause a loss in activity (Gatt and Barenholz, 1968). Embedded in its lipid environment, the rat liver enzyme is quite stable. Only a modest decrease in activity (10–15%) is noted when liver samples are kept at −80°C for up to 2 months. Stored at −20°C for the same period of time, a decrease of about 25% was found (P. P. Van Veldhoven and G. P. Mannaerts, unpublished data). Lyophilized microsomal preparations apparently retain their full activity if stored at −20°C (Gatt and Barenholz, 1968).

Besides in rat liver, sphingosine-phosphate lyase has also been studied in some detail in *T. pyriformis* (Stoffel *et al.*, 1974). In most respects, the enzyme resembles the rat liver enzyme (pH optimum, pyridoxal-phosphate as coenzyme, sensitivity toward sulfhydryl reagents). A rather high K_m value of 110 μM for sphinganine-phosphate was reported. The role of the substrate configuration has been studied more thoroughly with this enzyme. The (2*L*,3*L*) stereoisomer of sphinganine-phosphate was not inhibitory, while the (2*D*,3*L*) optical isomer, which has the same chirality at C-2 as the natural substrate, behaved as a competitive inhibitor (K_i 9.7 μM) (Stoffel *et al.*, 1974). According to Keenan (1972), the lyase of this protozoan is labile and destroyed by lyophilization. Other authors claimed that the activity is not influenced by lyophilization (Stoffel *et al.*, 1974).

IV. Subcellular Localization and Topology

Well-documented subcellular localization studies have been performed only with rat liver. The pioneer experiments by Stoffel and co-workers indicated that the sphingosine-phosphate lyase activity is membrane-bound. The major portion of the activity was recovered in the microsomal fraction, although substantial activity was also found in the mitochondrial fraction and claimed to be associated with the inner mitochondrial membrane (Stoffel *et al.*, 1969b). By means of more refined cell fractionation techniques, the subcellular distribution of the lyase was reinvestigated recently in rat liver (Van Veldhoven and Mannaerts, 1991). The highest specific activity was present in the microsomal fraction. Some activity was found in nuclear and heavy mitochondrial fractions, but it could be totally accounted for by contamination with microsomes. Since the vesicles, present in microsomal fractions, are derived from different subcellular structures, the microsomal fractions were further separated by sucrose density gradient centrifugation. The gradient distribution of the lyase coincided with that of glucose-6-phosphatase, one of the marker enzymes for vesicles derived from the endoplasmic reticulum (designated as group c by Beaufay *et al.,* 1974). A small fraction of the activity was also associated with another class of vesicles derived from the endoplasmic reticulum (group b) and possessing NADPH–cytochrome *c* reductase as marker enzyme. No significant association with the mitochondrial membranes, or with the Golgi apparatus, or with the plasma membrane could be demonstrated (Van Veldhoven and Mannaerts, 1991).

Also, in other tissues or organisms the lyase seems to be localized in the endoplasmic reticulum. Highest specific activities were found in microsomal fractions prepared from the yeast *H. ciferrii* (Stoffel *et al.,* 1969b), *T. pyriformis* (Stoffel *et al.,* 1974), and intestinal mucosa of the guinea pig (Nilsson, 1970).

The membrane topology of the lyase was studied in rat liver microsomes by Van Veldhoven and Mannaerts (1991). Stripping of the microsomal membrane by high-salt treatment could not extract any activity, but the lyase could be solubilized by the addition of Triton X-100. Although partial inactivation of the lyase occurred during solubilization, these data indicate that the lyase is an integral membrane protein. Since Triton X-100 did not stimulate the activity in rat liver homogenates or microsomal fractions (Stoffel *et al.,* 1969b; Van Veldhoven and Mannaerts, 1991), it was unlikely that the enzyme is latent. In order to exclude the remote possibility that the detergent-induced inactivation masked a concomitant stimulation, latency was further studied under isotonic conditions in isolated intact microsomes. Concentrations of Triton X-100 or CHAPS, sufficient to abolish the latency of mannose-6-phosphatase, did not enhance the lyase activity, indicating that it is indeed not latent. As mentioned earlier (see Section III), other neutral detergents, like octylglucoside, lauryldimethylammonium oxide, and octanoyl-*N*-methylglucamide, could not be employed in these latency studies because they were inhibitory to the lyase at concentrations that did not affect the membrane integrity (Van Veldhoven and Mannaerts, 1991).

Finally, proteolytic treatments under isotonic conditions with trypsin, thermolysin, chymotrypsin, and subtilisin destroyed the activity, while mannose-6-phosphatase latency remained high (Van Veldhoven and Mannaerts, 1991).

From these data, one can conclude that the lyase is an integral membrane protein, the catalytic sites or essential domains of which face the cytosol in the cell. This implies that the substrate does not have to cross the membrane. The subcellular site where phosphorylation of sphingoid bases occurs is probably the cytosol (Keenan and Haegelin, 1969; Hirschberg *et al.*, 1970; Stoffel *et al.*, 1973a; Louie *et al.*, 1976), although membrane-associated sphingosine kinase activities have been found (Keenan and Haegelin, 1969; Ghosh *et al.*, 1990) (see Buehrer and Bell, this volume). In *T. pyriformis* a microsomal sphingosine kinase exists (Stoffel *et al.*, 1974). A second conclusion is that the reaction products of the lyase are released into the cytosol. The initial metabolic conversions of the products, which are discussed in Section VI, occur mainly in the cytosol or at the cytosolic face of the endoplasmic reticulum (see Vance, 1985).

V. Tissue Distribution

The sphingosine-phosphate lyase activities, presented in Table I, were measured under optimized conditions (see Section II). The specific activities in liver, spleen, and lung are about 10-fold higher than those reported by Stoffel *et al.* (1969b) [probably as a result of the presence of ethylenediaminetetraacetic acid (EDTA)]. A marked interorgan variation in specific activities is found. The lowest activity is present in skeletal muscle, the highest is intestinal mucosa. Uptake of sphingoid bases from dietary sources [free or liberated in the lumen from more complex sphingolipids (Nilsson, 1969)] by the mucosa is high (Nilsson, 1968; Assmann and Stoffel, 1972), and the lyase is likely to assist in their degradation. Interestingly, a high lyase activity is noted in Harderian gland, a sebaceous gland containing high levels of ether lipids (see Rock, 1977). This could indicate that, in some tissues, the enzyme plays an indirect role in the synthesis of ether lipids (see Section VI).

From direct enzymatic determinations, or from indirect evidence from *in vivo* sphingoid base breakdown, the enzyme appears also to be present in liver and kidney of pigs, in beef liver (Stoffel *et al.*, 1969b), in intestinal mucosa of guinea pig (Nilsson, 1970), and in lower eukaryotes like *H. ciferrii* (Stoffel *et al.*, 1968a, 1969b) and *T. pyriformis* (Keenan, 1972; Stoffel *et al.*, 1974).

In light of the recent findings that sphingenine-phosphate might trigger the release of intracellular Ca^{2+} (see Section VII), sphingosine-phosphate lyase activity has also been studied in a number of cell lines, some of which are commonly used for studies on signal transduction (see Table

Table I
DISTRIBUTION OF SPHINGOSINE-PHOSPHATE LYASE IN RAT TISSUES[a]

	Sphingosine-phosphate lyase activity	
Tissue	nmol/min. g of wet tissue	pmol/min.mg of protein
Intestinal mucosa	34.4	311
Harderian gland	13.2	112
Liver (4)	18.2 ± 0.9	90.2 ± 6.0
Spleen	12.3	77.9
Pancreas (1)	8.45	61.1
Spinal cord (1)	4.83	54.4
Testis	4.50	52.8
Cerebellum	4.45	41.8
Cerebrum	3.89	33.2
Kidney	3.97	31.4
Lung	2.86	28.9
Heart	0.86	11.2
Skeletal muscle	0.51	8.2

[a] Rat tissues were removed and homogenized in 0.25 *M* sucrose, 1m*M* EDTA, pH 7.2, 1 m*M* dithiothreitol, 0.1% (v/v) ethanol. For pancreas, the medium contained in addition 5 μg of trypsin inhibitor and 0.1 U of α_2-macroglobulin/ml. Homogenates were assayed for sphingosine-phosphate lyase activity, essentially as described by Van Veldhoven and Mannaerts (1991), except that NaF was increased to 50 m*M*. Activities represent the average of two separate experiments, unless otherwise indicated. For liver, values are means ± S.E.M.

II). The highest specific activity was seen in 3T3-L_1 cells. Activities in cell lines derived from liver or kidney are comparable to those found in these rat organs. Lowest activities were seen in human skin fibroblasts. Interestingly, the lyase activity in muscle-derived cell lines is about 10-fold higher than in skeletal muscle from rat.

From these data it seems that sphingosine-phosphate lyase is ubiquitously distributed, being present in yeast, protozoa, and mammals. One notable exception are platelets. Platelets from man, rabbit, and pig are devoid of sphingosine-phosphate lyase (Stoffel *et al.*, 1970a, 1973a).

VI. Fate of Reaction Products

As mentioned in Section III, the primary breakdown products of phosphorylated sphingoid bases are ethanolamine-phosphate and fatty aldehydes. Both products are very actively metabolized *in vivo* as summarized in Figs. 3 and 4. The depicted pathways are based on the conversions observed in rat liver or rat brain after intravenous, respectively intracerebral injection of sphingoid bases labeled at specific positions (see Stoffel, 1973a,b), but they can be generalized to other rat tissues and other species like yeast and protozoa (Karlsson *et al.*, 1967; Nilsson, 1968, 1970; Stoffel *et al.*, 1968a, 1974; Assmann and Stoffel, 1972).

Figure 3 illustrates the metabolic conversions of the cleavage products of sphinganine and sphingenine, while those of 4*D*-hydroxysphinganine are presented in Fig. 4. For sphinganine, the pathways are based on reports by Roitman *et al.* (1967), Stoffel and Sticht (1967a,b), Barenholz and Gatt (1968), and Stoffel *et al.* (1969a, 1970b); for sphingenine, on reports by Stoffel and Sticht (1967a,b), Barenholz and Gatt (1968), and Stoffel and Henning (1968); and, for 4*D*-hydroxysphinganine, on reports by Barenholz and Gatt (1967,1968), Roitman *et al.* (1967), Gatt and Barenholz (1968), and Assmann and Stoffel (1972). In Figs. 3 and 4 the sphingoid bases are considered to be C_{18} bases, but the shorter or longer homologs are metabolized in a similar way (Stoffel and Scheid, 1969; Stoffel *et al.*, 1969a).

Table II

SPHINGOSINE-PHOSPHATE LYASE ACTIVITY IN CULTURED CELLS[a]

Cells	ATCC number	Origin	Sphingosine-phosphate lyase (pmol/min.mg of protein)
Hep-G2	HB 8065	Human hepatoma	86.8
BW1		Mouse hepatoma	70.4
MDCK	CCL 34	Dog kidney	55.8
cos-7	CRL 1651	Monkey kidney	38.9
LLC-PK_1	CRL 1392	Pig kidney	19.6
A_7r_5	CRL 1444	Rat aorta	100
C_2C_{12}	CRL 1772	Mouse skeletal muscle	83.8
BC_3H1	CRL 1443	Mouse brain tumor (smooth muscle like)	85.8
3T3-L_1	CL 173	Mouse embryo	145
Swiss 3T3	CCL92	Swiss mouse embryo	69.5
Fibroblasts		Human skin	24.9
HeLa	CCL2	Human cervical carcinoma	31.6

[a] Confluent cell cultures were harvested by trypsinization, pelleted, and washed twice with phosphate-buffered saline. After homogenization in the medium described in Table I, sphingosine-phosphate lyase activity and protein content were determined. Values represent the average of two separate experiments.

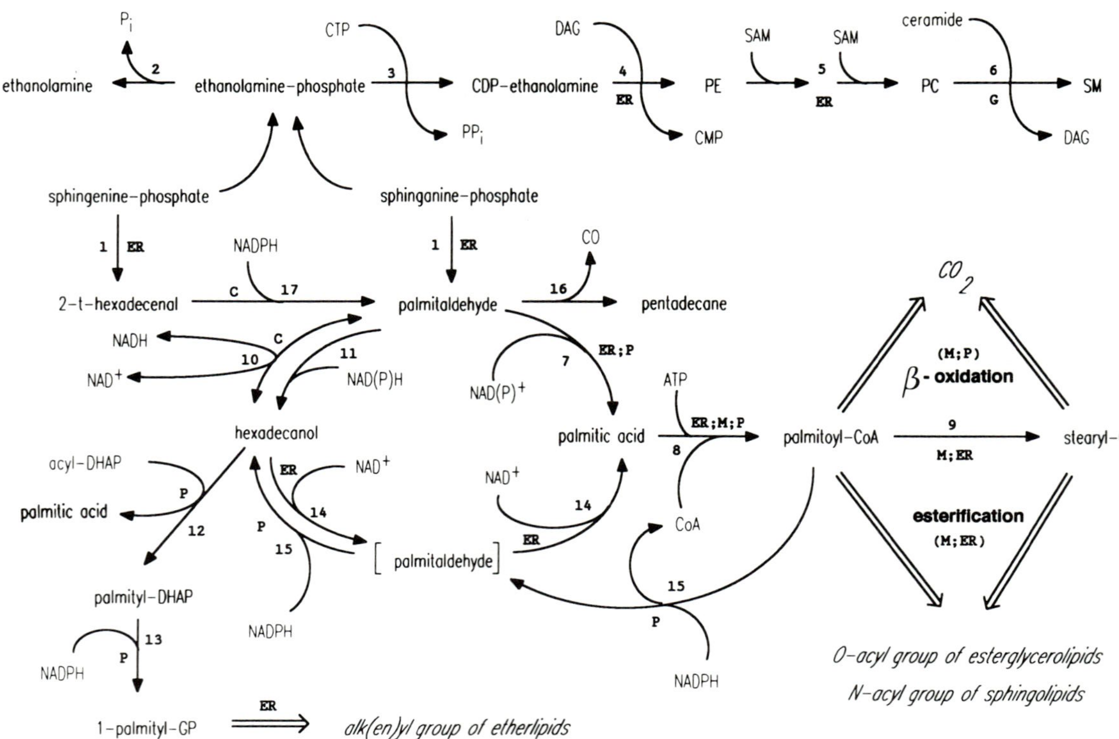

ethanolamine
P_i
2
ethanolamine−phosphate
CTP
3
PP_i
CDP−ethanolamine
DAG
4
ER
CMP
PE
SAM
5
ER
SAM
PC
ceramide
6
G
SM
DAG
sphingenine−phosphate
sphinganine−phosphate
1
ER
1
ER
2−t−hexadecenal
NADPH
C
17
palmitaldehyde
CO
16
pentadecane
NADH
C
10
NAD^+
11
NAD(P)H
7
ER;P
$NAD(P)^+$
hexadecanol
palmitic acid
ATP
ER;M;P
8
CoA
palmitoyl−CoA
9
M;ER
stearyl−CoA
CO_2
(M;P)
β- oxidation
esterification
(M;ER)
acyl−DHAP
P
palmitic acid
12
palmityl−DHAP
ER
NAD^+
14
P
15
NADPH
[palmitaldehyde]
NAD^+
14
ER
15
P
NADPH
13
P
NADPH
1−palmityl−GP
ER
alk(en)yl group of etherlipids
O−acyl group of esterglycerolipids
N−acyl group of sphingolipids

If the label is originally present in C-1 or C-2 of the sphingoid base, radioactive ethanolamine-phosphate is generated. The latter can be dephosphorylated to ethanolamine or, after conversion to CDP-ethanolamine, incorporated in phosphatidylethanolamine. Subsequent methylation of this phospholipid (a major pathway only in liver), followed by a transfer of the phosphocholine group to ceramide, explains the label found in phosphatidylcholine and sphingomyelin, respectively (see Fig. 3). Small amounts of label can be found in phosphocholine, likely resulting from the hydrolysis of phosphatidylcholine and sphingomyelin by phospholipase C and sphingomyelinase, respectively [see Vance (1991) for an overview on phospholipid metabolism].

The fate of the label, if present in C-3 to C-*n* of the alkyl fragment, depends on the nature of the sphingoid base. Sphinganine gives rise to palmitaldehyde, the major portion of which is oxidized to palmitic acid in rat liver homogenates. To a lesser extent hexadecanol is formed. In the intact organ, where oxidation also predominates (Stoffel *et al.*, 1970b), these compounds are further metabolized. When [3-^{3}H]sphinganine is given, the label present in generated palmitaldehyde will be lost during oxidation. Presently, it is not known which enzymes catalyze the initial conversions of the generated palmitaldehyde (see Fig. 3) [see Turner and Whittle (1981) and Riendeau and Meighem (1985) for further reading on fatty aldehyde metabolism].

In heavy mitochondrial fractions, palmitic acid is the major metabolite (Van Veldhoven and Mannaerts, 1991), indicating that the enzyme oxidiz-

---

FIG. 3. Metabolic conversions *in vivo* of the cleavage fragments of sphinganine-phosphate and sphingenine-phosphate. This figure demonstrates the many metabolic conversions of the reaction products formed from sphinganine- and sphingenine-phosphate. Involved enzymes are indicated by numerals, while their subcellular sites, if identified, are referred to by ER (endoplasmic reticulum), C (cytosol), P (peroxisomes), M (mitochondria), and G (Golgi apparatus). Both phosphorylated sphingoid bases are cleaved to ethanolamine-phosphate, the metabolism of which is depicted in the upper part of the figure. The processing of the (un)saturated fatty aldehyde is drawn in the lower part. 1, Sphingosine-phosphate lyase; 2, ethanolamine-phosphatase; 3, CTP-phosphoethanolamine cytidyltransferase; 4, CDP-ethanolamine:1,2-diacylglycerol phosphoethanolamine transferase; 5, phosphatidylethanolamine-*N*-methyltransferase; 6, phosphatidylcholine:ceramide cholinephosphotransferase; 7, aldehyde dehydrogenases; 8, acyl-CoA synthetase; 9, fatty acid elongase; 10, alcohol dehydrogenase; 11, aldehyde reductases; 12, alkyl-DHAP synthase; 13, acyl/alkyl-DHAP reductase; 14, long-chain alcohol dehydrogenase; 15, acyl-CoA reductase; 16, aldehyde decarbonylase; 17, 2-*trans*-alkenal reductase. CDP, Cytidine diphosphate; CMP, cytidine monophosphate; CTP, cytidine triphosphate; DAG, *sn*-1,2-diacylglycerol; DHAP, dihydroxyacetone-phosphate; GP, glycerol-3-phosphate; PE, phosphatidylethanolamine; PC, phosphatidylcholine; P_i, inorganic phosphate; PP_i, inorganic pyrophosphate; SAM, *S*-adenosylmethionine; SM, sphingomyelin.

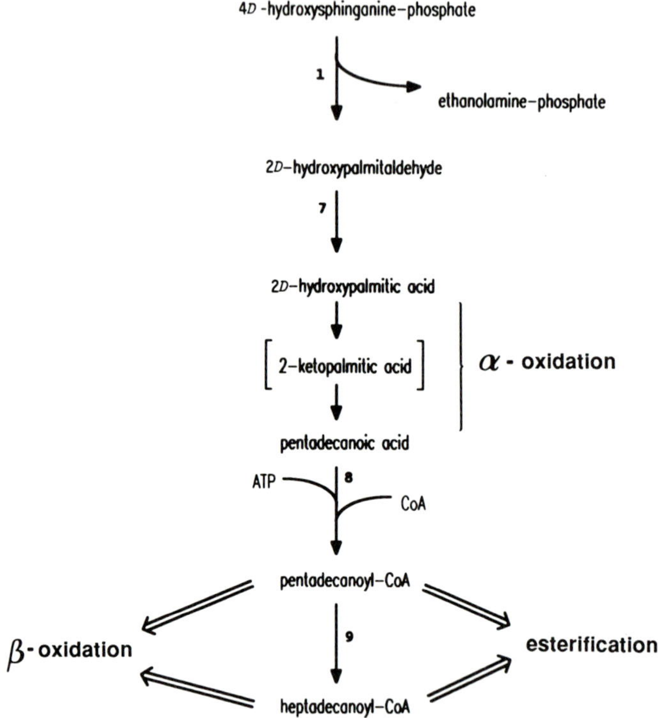

FIG. 4. Metabolic conversions of the cleavage fragments of 4*D*-hydroxysphinganine-phosphate. Ethanolamine-phosphate, produced by the action of sphingosine-phosphate lyase on 4*D*-hydroxysphinganine-phosphate, is metabolized as shown in Fig. 3. The other fragment, 2*D*-hydroxypalmitaldehyde, is oxidized and subsequently decarboxylated to pentadecanoic acid, probably via an intermediary 2-keto fatty acid. Perhaps the involved enzymes are similar to those responsible for the decarboxylation of 3-methyl branched fatty acids (this process is called α-oxidation). After activation, the acyl-CoA is either degraded via β-oxidation or esterified in glycerolipids. Involved enzymes are numbered identically to those in Fig. 3.

ing palmitaldehyde is present in the mitochondria or, more likely, that the mitochondrial $NAD(P)^+$ pool (Mannaerts *et al.*, 1982) becomes available for the reaction catalyzed by an enzyme present in contaminating organelles. Possible candidates are $NAD(P)^+$-dependent (fatty) aldehyde dehydrogenases, present in endoplasmic reticulum and peroxisomes (Nakayasu *et al.*, 1978; Antonenkov *et al.*, 1985; Panchenko *et al.*, 1985). The participation of long-chain alcohol dehydrogenase, which converts fatty alcohols to fatty acids through an intermediary fatty aldehyde, in palmitaldehyde oxidation is less likely, due to its low affinity for free fatty aldehydes (Lee, 1979).

After activation by one of the different cellular acyl-CoA synthetases (Mannaerts *et al.,* 1982), the fate of palmitoyl-CoA is influenced by the nutritional state of the animal (see Mannaerts and Debeer, 1979). In the fed state, the major part will be *O*-esterified in neutral lipids, (mostly triacylglycerols) and the main phospholipids (e.g., phosphatidylcholine, phosphatidylethanolamine, and sphingomyelin) and *N*-acylated into sphingolipids. The amount of label recovered in esterified form is about twice as much as label present in an amide linkage. During starvation (which was mostly the case in the reported studies), the CoA-esters are preferentially degraded via β-oxidation, either in mitochondria or peroxisomes (see Mannaerts and Van Veldhoven, 1991), to acetyl-CoA, which can be further oxidized to carbon dioxide. When C-3 is ^{14}C-labeled, the formation of CO_2 is substantial (after 10 hours about 25% of the injected label) (Stoffel and Sticht, 1967a,b).

Elongation to stearyl-CoA, occurring in the mitochondria or the endoplasmic reticulum (see Cook, 1991), which then follows the same routes as palmitoyl-CoA, has also been observed.

Conversion of palmitaldehyde to hexadecanol is most likely carried out by the cytosolic alcohol dehydrogenases (Stoffel and Assmann, 1970; Stoffel *et al.,* 1970b). Other enzymes able to reduce fatty aldehydes are NAD(P)H-dependent aldehyde reductases and acyl-CoA reductase. The latter enzyme is responsible for the NADPH-dependent reduction of acyl-CoAs to fatty alcohols (a reaction which proceeds via an intermediary fatty aldehyde), but can also act on free fatty aldehydes (see below). The further metabolism of hexadecanol is treated later in this article.

Decarbonylation of fatty aldehydes to alkanes has only been described so far in plants (Cheesbrough and Kolattukudy, 1984) and in avian uropygial gland (a sebaceous gland) (Cheesbrough and Kolattukudy, 1988). Given the high sphingosine-phosphate lyase activity in Harderian gland (see Table I), sphingosine breakdown may contribute to the formation of alkanes in sebaceous glands. Although the role of alkanes is not well understood, decreased alkane synthesis seems to be implicated in neurological disorders caused by demyelination (Bourre *et al.,* 1977, 1986).

With sphingenine as base, 2-*trans*-hexadecenal is formed. Some of this unsaturated fatty aldehyde can be reduced to 2-*trans*-hexadecenol, probably by the cytosolic NADH-dependent alcohol dehydrogenase, or oxidized to 2-*trans*-hexadecenoic acid, a reaction apparently occurring in the mitochondria. The major part of 2-*trans*-hexadecenal is, however, reduced to palmitaldehyde by the action of a specific enzyme, NADPH-dependent 2-*trans*-alkenal reductase, which has been isolated from rat liver cytosol

(Stoffel and Därr, 1974). This enzyme has also been found in *T. pyriformis* (Stoffel *et al.,* 1974). The formed palmitaldehyde then follows the same routes as described above for sphinganine.

Although 4*D*-hydroxysphinganine is mostly present in plants, yeasts, fungi, and protozoa, substantial amounts are found in the cerebrosides of kidney and intestine of vertebrates (see Karlsson, 1970a,b; Nishimura, 1987). About the origin of 4*D*-hydroxysphinganine present in the latter organs there still exists some discussion. According to Assmann and Stoffel (1972), it originates from dietary sources, but conversion of sphinganine to 4*D*-hydroxysphinganine has also been reported (Crossman and Hirschberg, 1984). After phosphorylation, this sphingoid base is cleaved to 2*D*-hydroxypalmitaldehyde and ethanolamine-phosphate. The latter undergoes the metabolic steps outlined above. Further metabolism of 2*D*-hydroxypalmitaldehyde is depicted in Fig 4. In yeasts, only one metabolite was recovered, identified as 2*D*-hydroxypalmitic acid (Karlsson *et al.,* 1967; Stoffel *et al.,* 1968a). In *in vivo* experiments with rats, 2*D*-hydroxypalmitic acid has never been detected. Instead, the major portion of the label is recovered in esterified pentadecanoic acid. Elongation of pentadecanoyl-CoA explains the label present in heptadecanoic acid, or even nonadecanoic acid, which are incorporated in neutral lipids and phospholipids. Since 2*D*-hydroxypalmitic acid is the major metabolite formed in rat liver homogenates on incubation with 4*D*-hydroxysphinganine, it is regarded as an intermediate with a rapid turnover *in vivo*. How the oxidative decarboxylation of 2-hydroxy fatty acids, a process which is apparently absent in yeasts, proceeds is not clear.

The breakdown of dietary or endogenously synthesized 4*D*-hydroxysphinganine could perhaps contribute to the presence of odd-numbered fatty acids, which are always found as traces in total fatty acid analyses.

An interesting aspect of sphingoid base breakdown is its link with ether lipid synthesis [for a review on ether lipids and their synthesis, see Snyder (1991)]. On intracerebral injection of [3-^{3}H]sphinganine in young rats, the label was recovered in sphingolipids and phosphatidylcholine and phosphatidylethanolamine. The tritium present in the latter phospholipids was linked almost exclusively via a vinylether bond (this class of phospholipids is normally referred to as plasmalogens, the vinylether bond of which is destroyed by strong acids, resulting in the release of a fatty aldehyde). This demonstrates that the labeled hexadecanol, incorporated in the plasmalogens, was derived from reduced palmitaldehyde generated by the lyase reaction. The initial steps of ether lipid synthesis, which occur in peroxisomes, are depicted in Fig. 3. Hexadecanol is exchanged with the acyl group of acyldihydroxyacetone-phosphate, which results in the formation of alkyldihydroxyacetone-phosphate, the first intermediate committed to ether lipid synthesis. Alkyldihydroxyacetone-phosphate is then reduced to 1-alkylglycerol-3-phosphate. This reaction takes place in peroxisomes as well as in endoplasmic reticulum. The further steps in the formation of ether lipids, including the desaturation to vinylether lipids, occur in the endoplasmic reticulum (Declercq *et al.,* 1984; see Snyder, 1991).

As mentioned above, it is not clear by which enzyme(s) palmitaldehyde is reduced in the cell. Since the catalytic site of the lyase is exposed to the cytosol (see Section IV), fatty alcohol dehydrogenase, which is a cytosolic enzyme in rat liver (Stoffel *et al.,* 1970b), is a possible candidate. In some way the formed hexadecanol has then to be transported into peroxisomes, since the formation of acyldihydroxyacetone-phosphate and the subsequent exchange reaction occur inside the peroxisomes (Declercq *et al.,* 1984; Hardeman and van den Bosch, 1988). Another possibility that we like to consider is that palmitaldehyde is handled by acyl-CoA reductase. Burdett *et al.* (1991) recently showed that this enzyme is present predominantly if not exclusively in peroxisomes. Moreover, its catalytic site is located at the cytosolic side of the membrane.

Incorporation of the produced fatty aldehyde into ether lipids is not unique to mammals. It has also been observed in *T. pyriformis* (Stoffel *et al.,* 1974), an organism that is rather rich in such lipids. It has even been suggested that degradation of sphingolipids, which constitute only a minor fraction of the phospholipids in this species, might fulfill a role in the supply of fatty alcohols for the formation of ether lipids (Stoffel *et al.,* 1974). This might also be true for certain mammalian tissues, like testis and Harderian gland. In these rat tissues, the highest peroxisomal dihydroxyacetone-phosphate acyltransferase activity (catalyzing the initial step in the synthesis of ether lipids) has been measured (Van Veldhoven and Mannaerts, 1985; P. P. Van Veldhoven, K. Gorgas, and G. P. Mannaerts, unpublished data), and, as shown in Table I, these tissues also display a high sphingosine-phosphate cleavage activity. If liver and intestinal mucosa, which are probably involved in the degradation of dietary sphingoid bases, are not considered, a close correlation is observed between the tissue activities of dihydroxyacetone-phosphate acyltransferase and sphingosine-phosphate lyase.

VII. Role in Sphingosine-Phosphate Attenuation

Several sphingolipids appear to possess bioactivity. Examples are sphingosine (see Merrill, 1991; Merrill and Stevens, 1989), lysosphingolipids (see Hannun and Bell, 1989), ceramide (see Kolesnick, 1991), glycosphin-

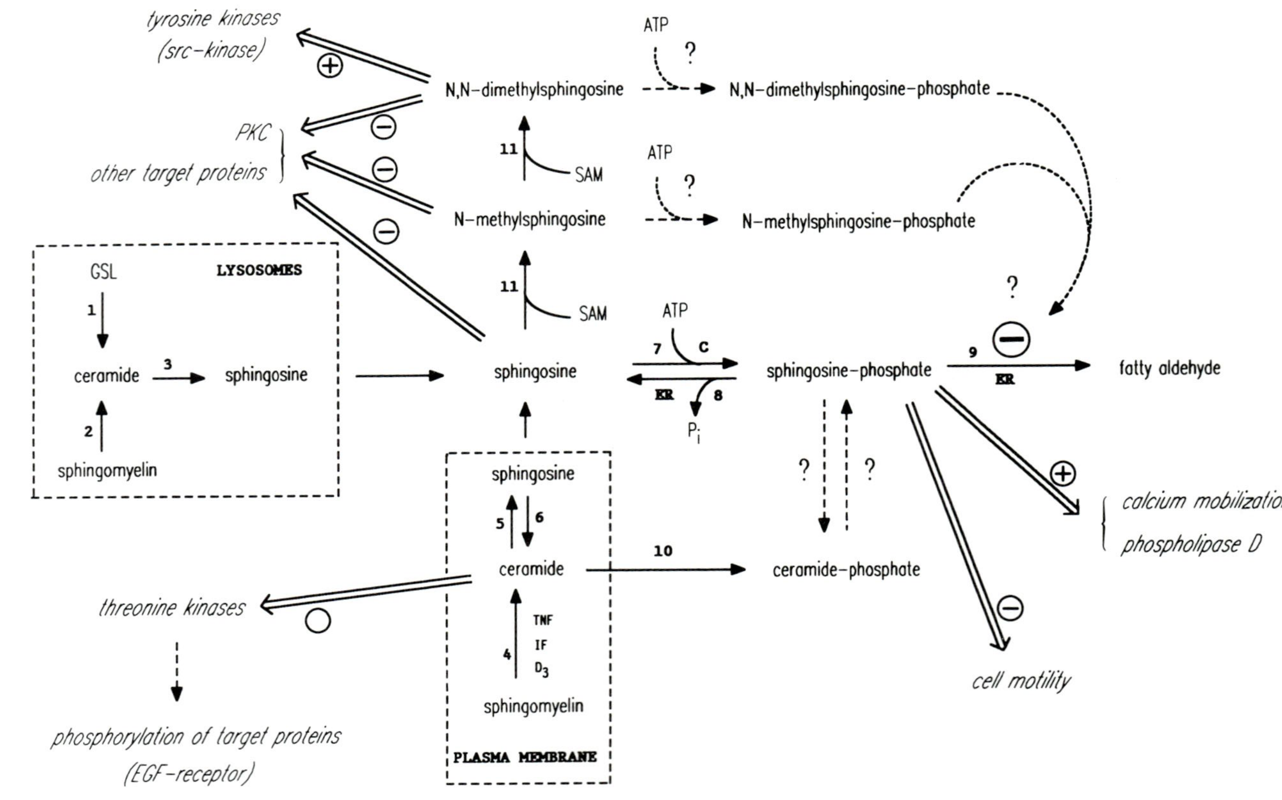
tyrosine kinases
(src-kinase)
N,N-dimethylsphingosine
ATP
?
N,N-dimethylsphingosine-phosphate
PKC
other target proteins
11
SAM
N-methylsphingosine
ATP
?
N-methylsphingosine-phosphate
GSL
LYSOSOMES
1
ceramide
3
sphingosine
2
sphingomyelin
11
SAM
sphingosine
ATP
7
C
ER
8
P_i
sphingosine-phosphate
?
9
ER
fatty aldehyde
?
?
calcium mobilization
phospholipase D
sphingosine
5
6
ceramide
10
ceramide-phosphate
threonine kinases
4
TNF
IF
D_3
sphingomyelin
PLASMA MEMBRANE
cell motility
phosphorylation of target proteins
(EGF-receptor)

golipids (see Cuello, 1990; Hakomori, 1990; Saito, Vol. 25 of this series), and *N*-methylated sphingosine (Igarashi *et al.*, 1989; Felding-Habermann *et al.*, 1990). More recently, sphingosine-phosphate has been added to this list (Ghosh *et al.*, 1990; Zhang *et al.*, 1991; Desai *et al.*, 1992; Sadahira *et al.*, 1992). In Fig. 5, the intimate metabolic interconnections between these bioactive sphingolipids have been outlined, as well as their proposed cellular targets.

Without doubt, most research has been focused on sphingosine. Work by different groups has proved the presence of free sphingoid bases in mouse and rat tissues (Kobayashi *et al.*, 1988; Merrill *et al.*, 1988) and in several cultured cells (Merrill *et al.*, 1986; Wilson *et al.*, 1988; Van Veldhoven *et al.*, 1989a). Moreover, different pools of sphingosine are likely to exist. Besides the newly synthesized sphingoid bases (Kobayashi *et al.*, 1989), probably associated with the endoplasmic reticulum (Kanfer, 1983a; Sweeley, 1991), and a catabolic pool inside the lysosomes, a third pool exists which is associated with the plasma membrane and is likely linked to signal transduction (Slife *et al.*, 1989). In the previous sections, it was tacitly assumed that the depicted degradation pathway: phosphorylation, followed by cleavage by sphingosine-phosphate lyase, applies to all sphingoid bases, including those derived from the lysosomal degradation of more complex sphingolipids. Although likely, this has not been proved. In only a few studies has the fate of sphingomyelin and glycosphingolipids been followed after administration to animals (Nilsson, 1968; Ogura and Handa, 1988; Ghidoni *et al.*, 1983, 1986; Stein *et al.*, 1991) or addition to

---

FIG. 5. Metabolic interconversions of sphingosine, sphingosine-phosphate, and other bioactive derivatives, and their possible cellular targets. Sphingosine generated in the lysosomes or at the plasma membrane is phosphorylated in the cytosol to sphingosine-phosphate, which activates intracellular Ca^{2+} release. This signal may be attenuated in several ways. Sphingosine-phosphate can be dephosphorylated or cleaved to a fatty aldehyde and ethanolamine-phosphate. Acylation to ceramide-phosphate is theoretically possible, but has not been demonstrated so far. Sphingosine itself can also be metabolized to ceramide, or it can be *N*-methylated. If phosphorylation of these *N*-methyl derivatives occurs, this might result in potent inhibitors of the lyase. The enzymes involved in the conversions of sphingolipids, ceramide, sphingosine, and sphingosine-phosphate are indicated by arabic numerals, while the subcellular sites, if known, are abbreviated as follows: C, cytosol; ER, endoplasmic reticulum. 1, Glycosphingolipid hydrolases; 2, acidic sphingomyelinase; 3, acidic ceramidase; 4, neutral sphingomyelinase; 5, neutral ceramidase; 6, *N*-sphingosine acyltransferase; 7, sphingosine kinase; 8, sphingosine-phosphatase; 9, sphingosine-phosphate lyase; 10, ceramide kinase; 11, *N*-methylsphingosine transferase. ATP, Adenosine triphosphate; D_3, vitamin D_3; EGF, epidermal growth factor; GSL, glycosphingolipids; IF, interferon; PKC, protein kinase C; SAM, *S*-adenosylmethionine; TNF, tumor necrosis factor.

cell cultures (Raghaven *et al.*, 1985; Trinchera *et al.*, 1988; Stein *et al.*, 1991). Sphingolipids, containing a [3-^{3}H]-sphingenine backbone, gave rise to the formation of tritiated water (Ghidoni *et al.*, 1986; Trinchera *et al.*, 1988) and also caused the labeling of ether-linked phospholipids (Ogura and Handa, 1988; Stein *et al.*, 1991). If the label was further down the sphingoid base moiety, part of the label was recovered in esterified palmitic acid (Nilsson, 1968; Raghaven *et al.*, 1985). These data suggest that sphingoid bases derived from lysosomal sphingolipid catabolism follow the same degradation pathway as intravenously injected sphingoid bases. In this connection, we should mention that it is not known how the intravenously injected sphingosine is removed from the circulation by the hepatocytes. If the sphingoid bases associate with lipoproteins, endocytosis is involved, whereas uptake of the free or the albumin-bound form may be mediated by passive diffusion [transbilayer movement of sphingosine has been observed in synthetic vesicles (Hope and Cullis, 1987)] or facilitated transport.

In most cell cultures fortified with radioactive sphingosine, only ceramide (and further metabolites) formation has been reported (Merrill *et al.*, 1986; Stevens *et al.*, 1990; Wilson *et al.*, 1988), but the recovery of label was not quantitative,[5] suggesting a breakdown of the sphingoid bases as well. The appearance of the phosphorylated bases in response to sphingenine was reported only recently in Swiss 3T3 cells (Zhang *et al.*, 1991) and B16 melanoma cells (Sadahira *et al.*, 1992). Importantly, it appears that in cultured cells several biological effects of sphingosine, which are known not to be mediated via protein kinase C, are caused by either ceramide (Goldkorn *et al.*, 1991) or sphingosine-phosphate (Zhang *et al.*, 1991; Desai *et al.*, 1992; Sadahira *et al.*, 1992). The latter metabolite appears to be responsible for the release of Ca^{2+} from intracellular pools, both in saponin-permeabilized smooth muscle cells (Ghosh *et al.*, 1990) and in intact Swiss 3T3 cells (Zhang *et al.*, 1991), for the formation of phosphatidate, probably through an activation of a phospholipase D in Swiss 3T3 cells (Desai *et al.*, 1992), and for the inhibition of cell motility and phagokinesis of B16 melanoma cells (Sadahira *et al.*, 1992). These findings suggest that sphingosine-phosphate may be an intracellular

[5] Discrepancies observed between cell-associated radioactivity and the label recovered in the base-hydrolyzed sphingolipid extracts can be explained by the use of [3-^{3}H]sphingenine (Merrill *et al.*, 1986; Stevens *et al.*, 1990), which will result in the formation of labeled water and, after alkaline hydrolysis, labeled water-soluble deacylated ether phospholipids, or, when [4,5-^{3}H]sphinganine is used (Wilson *et al.*, 1988), by the fact that the label, originally present in esterified fatty acids, will distribute in the aqueous phase after alkaline hydrolysis and phase separation.

messenger. Consequently, the discussed phosphorylation of sphingoid bases by sphingosine kinase, followed by their cleavage by sphingosine-phosphate lyase, would not merely be a catabolic reaction, but would play a role in the formation and attenuation of the sphingosine-phosphate signal. However, the Ca^{2+}-releasing effect of sphingenine-phosphate is not observed in all tissues. In skinned rabbit skeletal muscle fibers, sphingenine in the low μM range in fact blocks Ca^{2+} release (Sabbadini *et al.*, 1992), whereas in rat parotid acinar cells, unphysiologically high concentrations of sphingenine (50 to 200 μM) are needed to promote a rise in the intracellular Ca^{2+} concentration (Sugiya and Furuyama, 1991). In rabbit acinar pancreas cells, either intact or permeabilized, no effect of sphingenine or sphingenine-phosphate on the intracellular Ca^{2+} levels could be demonstrated, although, under the same experimental conditions, the findings of Zhang *et al.* (1991) were confirmed in intact Swiss 3T3 cells (van de Put *et al.*, 1993). Sphingenine-phosphate, taken up by Swiss 3T3 cells or generated in 3T3 cells (Zhang *et al.*, 1991) or B16 melanoma cells (Sadahira *et al.*, 1992) given sphingenine, appears to be quite stable. This is in contrast with the fast turnover of phosphorylated sphingoid bases in cultured fibroblasts (P. P. Van Veldhoven and G. P. Mannaerts, unpublished data) and other biological systems. The formation of these intermediary esters *in vivo* has seldom been demonstrated. One example is the presence of small amounts of 4*D*-hydroxysphinganine-phosphate in the intestinal mucosa of rats orally given 4*D*-hydroxysphinganine. When administered intravenously, a substantial amount of the phosphate ester was recovered in kidney (about 15 nmol/g of tissue) 24 hours after the injection (Assmann and Stoffel, 1972). The other examples of phosphate ester formation reflect unphysiological situations: in platelets which lack lyase activity (Stoffel *et al.*, 1970a) or on administration to rats of the (2*L*,3*L*) or (2*D*,3*D*) optical isomers of sphinganine which are phosphorylated but not cleaved (Stoffel and Bister, 1973).

Besides cleavage, dephosphorylation of sphingosine-phosphate might be another attenuation mechanism. Theoretically, *N*-acylation and *N*-methylation are also possible, but they have never been demonstrated. Especially in neuronal tissues, in which very high sphingosine-phosphatase activities are found (P. P. Van Veldhoven and G. P. Mannaerts, unpublished data; see Section II), dephosphorylation may be important. The importance of dephosphorylation is perhaps illustrated by the severe toxicity of sphinganine-1-phosphonate, which cannot be dephosphorylated but is still cleaved by the lyase. This compound arrests the growth of *Tetrahymena* and, when intravenously injected in rats, the animals die within 1 minute (Stoffel and Grol, 1974). Such effects are not seen with sphinganine-phosphate.

Whether the enzymes involved in the metabolism of sphingosine-phosphate are regulated remains unknown. Preliminary data indicate that the lyase in rat liver is not regulated by phosphorylation/dephosphorylation. We propose, however, that *N*-methylated sphingosine-phosphate could be a potent inhibitor of the enzyme (see Section III). The methylation of sphingenine has been observed in mouse brain homogenates (Igarashi and Hakomori, 1989), A431 cells (Igarashi *et al.*, 1990; Goldkorn *et al.*, 1991), and cultured murine lymphocytes (Felding-Habermann *et al.*, 1990). Some biological effects of sphingosine are claimed to be due to this metabolite (Igarashi *et al.*, 1989, 1990). Since most sphingosine-kinases do not discriminate between sphingoid base isomers, phosphorylation of mono- or di-*N*-methylated sphingosine is not unlikely. The synthetic compound, *N,N,N*-trimethylsphingenine, although efficiently taken up by B16 cells, appears, however, to be metabolically inert (Sadahira *et al.*, 1992).

Finally, phosphorylation of sphingosine is not the only pathway to generate bioactive sphingosine-phosphate. Hydrolysis of sphingenine-1-phosphocholine or deacylation of ceramide-phosphate are examples which can operate independently from the proposed sphingomyelin/ceramide/sphingosine cycle (see Hannun and Bell, 1989). Sphingenine-1-phosphocholine is a potent mitogen for different cell lines (Desai and Spiegel, 1991), and it mimics the effect of sphingenine-phosphate in permeabilized smooth muscle cells (Ghosh *et al.*, 1990). It is hydrolyzed by phospholipase D, at least by the enzyme of bacterial origin (Van Veldhoven *et al.*, 1989a). Ceramide kinase activity is associated with synaptic vesicles isolated from rat brain (Bajjalieh *et al.*, 1989) and is also present in HL-60 cells, where it is responsible for the conversion of sphingomyelin-derived ceramide to ceramide-phosphate (Dressler and Kolesnick, 1990). However, despite the multiple possible metabolic pathways, the formation of sphingosine-phosphate in response to a physiological stimulus still awaits to be proved.

Acknowledgments

This work was supported in part by grants from the Geconcerteerde Onderzoeksacties van de Vlaamse Gemeenschap, the Belgian Fonds voor Geneeskundig Wetenschappelijk Onderzoek, and the Belgian Fonds de la Recherche Fondamentale Collective. One of us (PPVV) was supported in part by a grant (Krediet voor Jonge Navorsers) from the Belgian Nationaal Fonds voor Wetenschappelijk Onderzoek. We greatly appreciate the willingness of our colleagues of the university to provide us with the different cell lines described in Table II: from the Faculty of Medicine, Dr. J.-J. Cassiman and Dr. G. David (Center for Human Genetics), Dr. J. Auwercx (Laboratory for Experimental Medicine and Endocrinology and Dr. H. Desmedt (Department of Physiology) and, from the School of Pharmacy, Dr. P. Declercq (Laboratory for Clinical Chemistry.

References

Antonenkov, V. D., Pirozhkov, S. V., and Panchenko, L. F. (1985). *Eur. J. Biochem.* **149,** 159–167.

Assmann, G., and Stoffel, W. (1972). *Hoppe-Seyler's Z. Physiol. Chem.* **353,** 971–979.

Bajjalieh, S. M., Martin, T. F. J., and Floor, E. (1989). *J. Biol. Chem.* **264,** 14354–14360.

Barenholz, Y., and Gatt, S. (1967). *Biochem. Biophys. Res. Commun.* **32,** 588–594.

Barenholz, Y., and Gatt, S. (1968). *Biochemistry* **7,** 2603–2609.

Beaufay, H., Amar-Costesec, A., Feytmans, E., Thinès-Sempoux, D., Wibo, M., Robbi, M., and Berthet, J. (1974). *J. Cell Biol.* **61,** 188–200.

Bourre, J. M., Cassagne, C., Larroquere-Regnier, S., and Darriet, D. (1977). *J. Neurochem.* **29,** 645–648.

Bourre, J.-M., Boiron, F., Cassagne, C., Dumont, O., Leterrier, F., Metzger, H., and Viret, J. (1986). *Neurochem. Pathol.* **4,** 29–42.

Burdett, K., Larkins, L. K., Das, A. K., and Hajra, A. K. (1991). *J. Biol. Chem.* **266,** 12201–12206.

Cheesbrough, T. M., and Kolattukudy, P. E. (1984). *Proc. Natl. Acad. Sci. U.S.A.* **81,** 6613–6617.

Cheesbrough, T. M., and Kolattukudy, P. E. (1988). *J. Biol. Chem.* **263,** 2738–2743.

Cook, H. W. (1991). *In* "Biochemistry of Lipids, Lipoproteins, and Membranes" New Comprehensive Biochemistry, Vol. 20 (D. E. Vance and J. E. Vance, eds.), pp. 141–169. Elsevier, Amsterdam.

Crossman, M. W., and Hirschberg, C. B. (1984). *Biochim. Biophys. Acta* **795,** 411–416.

Cuello, A. C. (1990). *Adv. Pharmacol.* **217,** 1–50.

Declercq, P. E., Haagsman, H. P., Van Veldhoven, P., Debeer, L. J., Van Golde, L. M. G., and Mannaerts, G. P. (1984). *J. Biol. Chem.* **259,** 9064–9075.

Desai, N. N., and Spiegel, S. (1991). *Biochem. Biophys. Res. Commun.* **181,** 361–366.

Desai, N. N., Zhang, H., Olivera, A., Mattie, M. E., and Spiegel, S. (1992). *J. Biol. Chem.* **267,** 23122–23128.

Dressler, K. A., and Kolesnick, R. N. (1990). *J. Biol. Chem.* **265,** 14917–14921.

Enzyme Nomenclature (1984). "Recommendations of the Nomenclature Committee of the International Union of Biochemistry on the Nomenclature and Classification of Enzymes." Academic Press, Orlando, FL.

Felding-Habermann, B., Igarishi, Y., Fenderson, B. A., Park, L. S., Radin, N. S., Inokuchi, J., Strassmann, G., Handa, K., and Hakomori, S.-I. (1990). *Biochemistry* **29,** 6314–6322.

Gatt, S. (1966). *J. Biol. Chem.* **241,** 3724–3730.

Gatt, S., and Barenholz, Y. (1968). *Biochem. Biophys. Res. Commun.* **32,** 588–594.

Ghidoni, R., Sonnino, S., Chigorno, V., Venerando, B., and Tettamanti, G. (1983). *Biochem. J.* **213,** 321–329.

Ghidoni, R., Trinchera, M., Venerando, B., Fiorilli, A., Sonnino, S., and Tettamanti, G. (1986). *Biochem. J.* **237,** 147–155.

Ghosh, T. K., Bian, J., and Gill, D. L. (1990). *Science* **248,** 1653–1656.

Godcheaux, W., III, and Leadbetter, E. R. (1980). *J. Bacteriol.* **144,** 592–602.

Goldkorn, T., Dressler, K., Muindi, J., Radin, N. S., Mendelsohn, J., Menaldino, D., Liotta, D., and Kolesnick, R. N. (1991). *J. Biol. Chem.* **266,** 16092–16097.

Hakomori, S. (1990). *J. Biol. Chem.* **265,** 18713–18716.

Hannun, Y. A., and Bell, R. M. (1989). *Science* **243,** 500–507.

Hardeman, D., and van den Bosch, H. (1988). *Biochim. Biophys. Acta* **963,** 1–9.

Hirschberg, C. B., Kisič, A., and Schroepfer, G. J., Jr. (1970). *J. Biol. Chem.* **245,** 3084–3090.
Hope, M. J., and Cullis, P. R. (1987). *J. Biol. Chem.* **262,** 4360–4366.
Igarashi, Y., and Hakomori, S. (1989). *Biochem. Biophys. Res. Commun.* **164,** 1411–1416.
Igarashi, Y., Hakomori, S., Toyokuni, T., Dean, B., Fujita, S., Sugimoto, M., Ogawa, T., El-Ghendy, K., and Racker, E. (1989). *Biochemistry* **28,** 6796–6800.
Igarashi, Y., Kitamura, K., Toyokuni, T., Dean, B., Fenderson, B., Ogawa, T., and Hakomori, S.-I. (1990). *J. Biol. Chem.* **265,** 5385–5389.
Kanfer, J. N. (1983a). *In* "Handbook of Lipid Research" (J. N. Kanfer and S.-I. Hakomori, eds.), Vol. 3, pp. 167–247. Plenum, New York.
Kanfer, J. N. (1983b). *In* "Handbook of Lipid Research" (J. N. Kanfer and S.-I. Hakomori, eds.), Vol. 3, pp. 249–325. Plenum, New York.
Karlsson, K.-A. (1970a). *Lipids* **5,** 878–891.
Karlsson, K.-A. (1970b). *Chem. Phys. Lipids* **5,** 6–43.
Karlsson, K.-A., Samuelson, B. E., and Steen, G. O. (1967). *Acta Chem. Scand.* **21,** 2566–2567.
Keenan, R. W. (1972). *Biochim. Biophys. Acta* **270,** 383–396.
Keenan, R. W., and Haegelin, B. (1969). *Biochem. Biophys. Res. Commun.* **37,** 888–894.
Keenan, R. W., and Maxam, A. (1969). *Biochim. Biophys. Acta* **176,** 348–356.
Keenan, R. W., and Okabe, K. (1968). *Biochemistry* **7,** 2696–2701.
Kobayashi, T., Mitsuo, K., and Goto, I. (1988). *Eur. J. Biochem.* **172,** 747–752.
Kobayashi, T., Shinnoh, N., and Goto, I. (1989). *Eur. J. Biochem.* **186,** 493–499.
Kolesnick, R. N. (1991). *Prog. Lipid Res.* **30,** 1–38.
Lee, T. (1979). *J. Biol. Chem.* **254,** 2892–2896.
Louie, D. D., Kisič, A., and Schroepfer, G. J., Jr. (1976). *J. Biol. Chem.* **251,** 4557–4564.
Lumeng, L., and Li, T.-K. (1979). *In* "Methods in Enzymology" (D. McCormick and L. Wright, eds.), Vol. 62, pp. 574–582. Academic Press, New York.
Mannaerts, G. P., and Debeer, L. J. (1979). *In* "Lipoprotein Metabolism and Endocrine Regulation" (L. W. Hessel and H. M. J. Krans, eds.), pp. 271–278. Elsevier/North-Holland Biomedical Press, Amsterdam.
Mannaerts, G. P., and Van Veldhoven, P. P. (1991). *In* "Inborn Errors of Metabolism" (J. Schaub, F. Van Hoof, and H. L. Vis, eds.), Nestlé Nutr. Workshop Ser., Vol. 24, pp. 1–18. Raven Press, New York.
Mannaerts, G. P., Van Veldhoven, P., Van Broekhoven, A., Vandebroeck, G., and Debeer, L. (1982). *Biochem. J.* **204,** 17–23.
Matsuo, Y. (1957). *J. Am. Chem. Soc.* **79,** 2011–2015.
Merrill, A. H., Jr. (1991). *J. Bioenerg. Biomembr.* **23,**83–104.
Merrill, A. H., Jr., and Stevens, V. L. (1989). *Biochim. Biophys. Acta* **1010,** 131–139.
Merrill, A. H., Jr., Sereni, A. M., Stevens, V. L., Hannun, Y. A., Bell, R. M., and Kinkade, J. M., Jr. (1986). *J. Biol. Chem.* **261,** 12610–12615.
Merrill, A. H., Jr., Wang, E., Mullins, R. E., Jamison, W. C. L., Nimkar, S., and Liotta, D. C. (1988). *Anal. Biochem.* **171,** 373–381.
Nakayasu, H., Mihara, K., and Sato, R. (1978). *Biochem. Biophys. Res. Commun.* **83,** 697–703.
Nilsson, Å. (1968). *Biochim. Biophys. Acta* **164,** 575–584.
Nilsson, Å. (1969). *Biochim. Biophys. Acta* **176,** 339–347.
Nilsson, Å. (1970). *Acta Chem. Scand.* **24,** 598–604.
Nishimura, K. (1987). *Comp. Biochem. Physiol.* **86B,** 149–154.
Nomenclature of Lipids (1977). *Eur. J. Biochem.* **79,** 11–21.
Ogura, K., and Handa, S. (1988). *J. Biochem. (Tokyo)* **104,** 87–92.

Panchenko, L. F., Pirozhkov, S. V., and Antonenkov, V. D. (1985). *Biochem. Pharmacol.* **34,** 471–479.

Raghaven, S., Krusell, A., Lyerla, T. A., Bremer, E. G., and Kolodny, E. H. (1985). *Biochim. Biophys. Acta* **834,** 238–248.

Riendeau, D., and Meighem, E. (1985). *Experientia* **41,** 707–713.

Rock, C. O. (1977). *In* "Lipid Metabolism in Mammals" (F. Snyder, ed.), Vol. 2, pp. 311–321. Plenum, New York.

Roitman, A., Barenholz, Y., and Gatt, S. (1967). *Isr. J. Chem.* **5,** 134.

Sabbadini, R. A., Betto, R., Teresi, A., Fachechi-Cassano, G., and Salviati, G. (1992). *J. Biol. Chem.* **267,** 15475–15484.

Sadahira, Y., Ruan, F., Hakomori, S., and Igarashi, Y. (1992). *Proc. Natl. Acad. Sci. U.S.A.* **89,** 9686–9690.

Shimojo, T., and Schroepfer, G. J., Jr. (1976). *Biochim. Biophys. Acta* **431,** 433–446.

Shimojo, T., Akino, T., Miura, Y., and Schroepfer, G. J., Jr. (1976). *J. Biol. Chem.* **251,** 4448–4457.

Slife, C. W., Wang, E., Hunter, R., Wang, S., Burgess, C., Liotta, D. C., and Merrill, A. H., Jr. (1989). *J. Biol. Chem.* **264,** 10731–10377.

Snyder, F. (1991). *In* "Biochemistry of Lipids, Lipoproteins, and Membranes" New Comprehensive Biochemistry, Vol. 20 (D. E. Vance and J. E. Vance, eds.), pp. 241–267. Elsevier, Amsterdam.

Stein, Y., Oette, K., Dabach, Y., Hollander, G., Ben-Naim, M., and Stein, O. (1991). *Biochim. Biophys. Acta* **1084,** 87–93.

Stevens, V. L., Nimkar, S., Jamison, W. C. L., Liotta, D. C., and Merrill, A. H., Jr. (1990). *Biochim. Biophys. Acta* **1051,** 37–45.

Stoffel, W. (1970). *Chem. Phys. Lipids* **5,** 139–158.

Stoffel, W. (1973a). *Mol. Cell. Biochem.* **1,** 147–155.

Stoffel, W. (1973b). *Chem. Phys. Lipids* **11,** 318–334.

Stoffel, W., and Assmann, G. (1970). *Hoppe-Seyler's Z. Physiol. Chem.* **351,** 1041–1049.

Stoffel, W., and Assmann, G. (1972). *Hoppe-Seyler's Z. Physiol. Chem.* **353,** 965–970.

Stoffel, W., and Bister, K. (1973). *Hoppe-Seyler's Z. Physiol. Chem.* **354,** 169–181.

Stoffel, W., and Därr, W. (1974). *Hoppe-Seyler's Z. Physiol. Chem.* **355,** 54–60.

Stoffel, W., and Grol, M. (1974). *Chem. Phys. Lipids* **13,** 372–388.

Stoffel, W., and Henning, R. (1968). *Hoppe-Seyler's Z. Physiol. Chem.* **349,** 1400–1404.

Stoffel, W., and Scheid, A. (1969). *Hoppe-Seyler's Z. Physiol. Chem.* **350,** 1593–1604.

Stoffel, W., and Sticht, G. (1967a). *Hoppe-Seyler's Z. Physiol. Chem.* **348,** 941–943.

Stoffel, W., and Sticht, G. (1967b). *Hoppe-Seyler's Z. Physiol. Chem.* **348,** 1345–1351.

Stoffel, W., and Sticht, G. (1967c). *Hoppe-Seyler's Z. Physiol. Chem.* **348,** 1561–1569.

Stoffel, W., Sticht, G., and LeKim, D. (1968a). *Hoppe-Seyler's Z. Physiol. Chem.* **349,** 1149–1156.

Stoffel, W., Sticht, G., and LeKim, D. (1968b). *Hoppe-Seyler's Z. Physiol. Chem.* **349,** 1745–1748.

Stoffel, W., Sticht, G., and LeKim, D. (1969a). *Hoppe-Seyler's Z. Physiol. Chem.* **350,** 63–68.

Stoffel, W., LeKim, D., and Sticht, G. (1969b). *Hoppe-Seyler's Z. Physiol. Chem.* **350,** 1233–1241.

Stoffel, W., Assmann, G., and Binczek, E. (1970a). *Hoppe-Seyler's Z. Physiol. Chem.* **351,** 635–642.

Stoffel, W., LeKim, D., and Heyn, G. (1970b). *Hoppe-Seyler's Z. Physiol. Chem.* **351,** 875–883.

Stoffel, W., Heimann, G., and Hellenbroich, B. (1973a). *Hoppe-Seyler's Z. Physiol. Chem.* **354,** 562–566.

Stoffel, W., Hellenbroich, B., and Heimann, G. (1973b). *Hoppe-Seyler's Z. Physiol. Chem.* **354,** 1311–1316.

Stoffel, W., Bauer, E., and Stahl, J. (1974). *Hoppe-Seyler's Z. Physiol. Chem.* **355,** 61–74.

Sugiya, H., and Furuyama, S. (1991). *FEBS Lett.* **286,** 113–116.

Sweeley, C. C. (1991). *In* "Biochemistry of Lipids, Lipoproteins, and Membranes" New Comprehensive Biochemistry, Vol. 20 (D. E. Vance and J. E. Vance, eds.), pp. 327–361. Elsevier, Amsterdam.

Trinchera, M., Wiesmann, U., Pitto, M., Acquotti, D., and Ghidoni, R. (1988). *Biochem. J.* **252,** 375–379.

Turner, A. J., and Whittle, S. R. (1981). *Biochem. Soc. Trans.* **9,** 279–281.

Vance, D. E. (1985). *In* "Biochemistry of Lipids, Lipoproteins, and Membranes" New Comprehensive Biochemistry, Vol. 20 (D. E. Vance and J. E. Vance, eds.), pp. 205–240. Elsevier, Amsterdam.

van de Put, F. H. M. M., Van Veldhoven, P. P., and Willems, P. H. G. M. (1993). Submitted for publication.

Van Veldhoven, P. P., and Mannaerts, G. P. (1985). *Biochem. J.* **227,** 737–741.

Van Veldhoven, P. P., and Mannaerts, G. P. (1991). *J. Biol. Chem.* **266,** 12502–12507.

Van Veldhoven, P. P., Bishop, W. R., and Bell, R. M. (1989a). *Anal. Biochem.* **183,** 177–189.

Van Veldhoven, P. P., Foglesong, R. J., and Bell, R. M. (1989b). *J. Lipid Res.* **30,** 611–616.

Weiss, B. (1957). *J. Am. Chem. Soc.* **79,** 5553–5557.

Wilson, E., Wang, E., Mullins, R. E., Uhlinger, D. J., Liotta, D. C., Lambeth, J. D., and Merrill, A. H., Jr. (1988). *J. Biol. Chem.* **263,** 9304–9309.

Zhang, H., Desai, N. N., Olivera, A., Seki, T., Brooker, G., and Spiegel, S. (1991). *J. Cell Biol.* **114,** 155–167.

Part II

TRAFFICKING AND TOPOLOGY OF SPHINGOLIPID METABOLISM

ADVANCES IN LIPID RESEARCH, VOL. 26

Intracellular Transport of Ceramide and Its Metabolites at the Golgi Complex: Insights from Short-Chain Analogs

ANNE G. ROSENWALD AND RICHARD E. PAGANO

Department of Embryology
Carnegie Institution of Washington
Baltimore, Maryland 21210

I. Introduction

Nearly 100 years ago, Camillo Golgi (1898) described his observations of a distinctive "endocellular apparatus" in nerve cells. Until the middle of this century, the very existence of what is now known as the Golgi apparatus or Golgi complex remained a point of contention among cell biologists. Since then, a variety of lines of experimentation have shown that it is indeed a discrete intracellular organelle (Farquhar and Palade, 1981), and over the past several years, it has become increasingly clear that this organelle plays a central role in the metabolism, sorting, and transport of both proteins and lipids in cells (Pfeffer and Rothman, 1987; van Meer, 1989; Pagano, 1990a; Rothman and Orci, 1990, 1992; Schwarzmann and Sandhoff, 1990).

Determining the mechanisms by which lipids are moved through the Golgi complex and other organelles has been a challenging problem. Classical biochemical, immunological, and cytochemical techniques, which have been valuable for the study of protein transport in cells, have been less useful in the study of lipids (Pagano, 1990a). In the past several years, new techniques for studying lipid transport and sorting have been developed. One such method involves the use of lipids bearing either a short-chain fluorescent or radioactive fatty acid in place of a long-chain fatty acid (see Fig. 1). These lipids can be readily integrated into intracellular membranes from exogenous sources, and, in many instances, are metabolized similarly to their endogenous long-chain counterparts. Furthermore, the use of the short-chain fluorescent lipids permits determination of the intracellular distribution of the lipid and its metabolites within a living cell by high-resolution fluorescence microscopy. These data can then be correlated with data obtained from classical biochemical analyses (Pagano and Sleight, 1985; Pagano, 1990a).

In this article, we review the current knowledge about metabolism and transport of sphingolipids, especially with regard to the Golgi complex, and emphasize the results obtained with short-chain fluorescent and radioactive analogs of three sphingolipids, ceramide (Cer), glucosylceramide (GlcCer), and sphingomyelin (SM). We discuss the biophysical properties of the sphingolipid analogs, the uptake and initial distribution of these analogs in cells, the metabolism of sphingolipids at the Golgi complex, and the movement of sphingolipids into and out of this organelle.

II. Biophysical Properties of Short-Chain Sphingolipid Analogs

The structures of the sphingolipid analogs discussed in this review are shown in Fig. 1. In each case, a long-chain base is *N*-acylated with a short-chain fluorescent or radioactive fatty acid. In general, the presence of these short-chain fatty acids renders the analogs more water-soluble than their endogenous counterparts. As a result, these molecules can undergo spontaneous transfer between donor and acceptor membranes and thus can be readily integrated into cellular membranes (Struck and Pagano, 1980).

In the case of the fluorescent analogs, half-times for spontaneous interbilayer transfer in artificial lipid vesicles can be quantified, and their ability to undergo spontaneous transbilayer movement (''flip-flop'') can be assessed using assays based on resonance energy transfer (Nichols and Pagano, 1982; Pagano and Sleight, 1985; Pagano and Martin, 1988; Pagano, 1989) or other changes in the spectral properties of the molecules (Table I). For lipids labeled with (*N*-[6-(7-nitrobenz-2-oxa-1,3-diazol-4-yl)amino]hexanoic acid (C_6-NBD), interbilayer transfer occurs by dissociation of lipid monomers from donor membranes, diffusion through the aqueous phase, and association with acceptor membranes (Nichols and Pagano, 1982). Increasing the chain length of the fluorescent fatty acid from 6 to 12 carbon

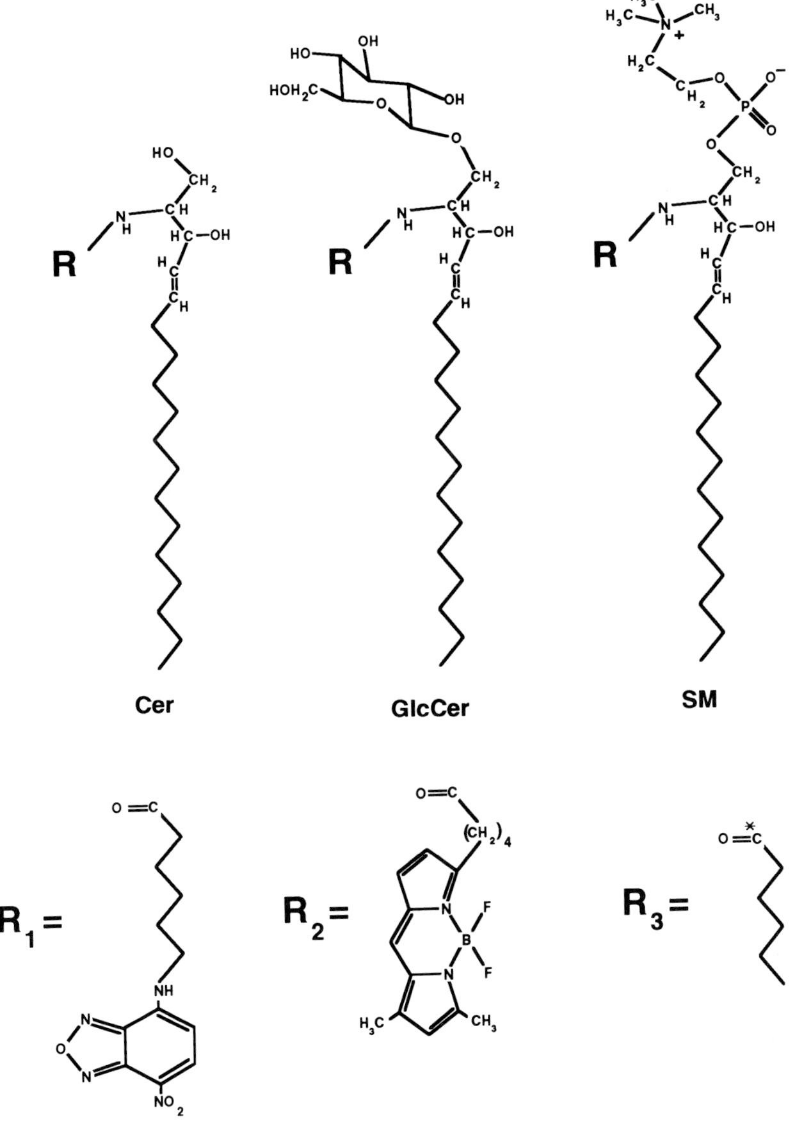

FIG. 1. Structures and abbreviations of some short-chain analogs of ceramide (Cer), glucosylceramide (GlcCer), and sphingomyelin (SM) discussed in this review. Fatty acyl groups are R_1, *N*-[6-(7-nitrobenz-2-oxa-1,3-diazol-4-yl)amino]hexanoyl, C_6-NBD-; R_2, *N*-[5-(5,7-dimethyldipyrrometheneboron difluoride)-1-pentanoyl], C_5-DMB-; and R_3, [1-^{14}C]hexanoyl, [^{14}C]hex.

Table I
CHARACTERISTICS OF FLUORESCENT SPHINGOLIPID ANALOGS IN LIPOSOMES

Analog	$t_{1/2}$ for interbilayer transfer at 20°C (min)[a]	Transbilayer movement in liposomes[b]
C_6-NBD-Cer	0.42	+
C_6-NBD-SM	0.04	–
C_6-NBD-GlcCer	0.11	–
C_5-DMB-Cer	3.20	+
C_5-DMB-SM	0.50	–[c]
C_5-DMB-GlcCer	n.d.[d]	–[c]
(Palmitoyl)-SM	$(1–1.4 \times 10^5)$[e]	n.d.[d]

[a] Half-times for equilibration of C_6-NBD-lipids between PC vesicles and the ability of these lipids to undergo transbilayer movement in lipid vesicles were measured as described (Nichols and Pagano, 1982; Pagano and Martin, 1988). Half-times for the equilibration and transbilayer movement of C_5-DMB-lipids were determined by analyzing time-dependent changes in the spectral properties of these lipids when vesicles containing high concentrations of the probe were mixed with nonfluorescent lipid vesicles (Bai and Pagano, 1993).

[b] +, Transbilayer movement occurs; –, no transbilayer movement detected.

[c] Half-times for transbilayer movement are >150 minutes at 22°C.

[d] n.d., Not determined.

[e] Taken from Frank *et al.* (1983).

atoms increases the half-time for spontaneous transfer of the lipid by approximately 200-fold (Nichols and Pagano, 1982, 1983; Pagano, 1983). Biophysical studies (Chattopadhyay and London, 1987; Wolf *et al.*, 1992) have demonstrated that the NBD-fluorophore may not be fully intercalated in the membrane bilayer, but rather can "loop back" to the bilayer/water interface, presumably because of the polar nature of this fluorophore.

We have also studied the behavior of lipids labeled with 5-(5,7-dimethyl-dipyrrometheneboron difluoride)-1-pentanoic acid (C_5-DMB). The half-times for spontaneous transfer of C_5-DMB-lipids in liposomes are approximately an order of magnitude longer than the corresponding C_6-NBD analogs (Table I), suggesting that the C_5-DMB-labeled analogs are more firmly embedded in the bilayer than the NBD-labeled compounds. This difference in spontaneous transfer is also observed in cellular membranes where C_6-NBD-lipids can be readily removed from labeled membranes during incubation with an excess of acceptor [defatted bovine serum albumin (BSA) or nonfluorescent liposomes; Struck and Pagano, 1980], while

C_5-DMB-sphingolipids are not readily removed by this procedure (Pagano *et al.,* 1991). While this resistance to "back-exchange" may be explained in part by differences in the rates of spontaneous transfer of the lipids as assayed in liposomes, additional factors such as the association of the C_5-DMB-lipids with other components in cellular membranes or aggregation of the probe in membranes must be evaluated.

In addition, sphingolipid analogs containing the C_5-DMB fluorophore are useful because the spectral properties of these lipids are altered depending on their molar density in membranes. At low concentrations, the fluorescence emission maximum is green (~515 nm), while at high concentrations, the fluorescence emission maximum is red (~620 nm). This property is advantageous for studying the distribution and transport of these lipids and their metabolites in living cells (Pagano *et al.,* 1991), as this spectral shift can be used to differentiate membranes which are labeled to different extents by a C_5-DMB-lipid. A particularly striking example is seen when cells are labeled with C_5-DMB-Cer. This lipid accumulates at the Golgi apparatus and emits red fluorescence, while other intracellular membranes, containing lower concentrations of the probe, emit green fluorescence (Fig. 2). When combined with fluorescence ratio imaging microscopy to determine the relative amounts of red and green fluorescence in a given region of a cell, the concentration of the fluorescent lipids in an intracellular compartment can be estimated (Pagano *et al.,* 1991).

No quantitative measurements of the transfer rates or transbilayer movement of the short-chain radioactive sphingolipid analogs have been performed. However, all studies to date suggest that these lipids behave in a manner analogous to the corresponding fluorescent sphingolipids with respect to their metabolism in cells (Rosenwald *et al.,* 1992) and subcellular membrane fractions, and their ability to undergo spontaneous transfer between membranes (Futerman *et al.,* 1990; Futerman and Pagano, 1991).

III. Sphingolipid Targeting and Synthesis

A. Fluorescent Ceramide Analogs Vitally Stain the Golgi Apparatus

When cells are incubated with C_6-NBD-Cer, a striking concentration of fluorescence is seen in a perinuclear structure that was determined to be the Golgi apparatus (Lipsky and Pagano, 1983, 1985b). Subsequent studies have demonstrated that this distribution of intracellular fluorescence is the result of a series of steps as shown in Fig. 3A: (1) monomer transfer of the fluorescent lipid from an appropriate donor (liposomes containing fluorescent Cer or complexes of BSA and fluorescent Cer) to the plasma membranes of the recipient cells, (2) transbilayer movement from the outer to the inner leaflet of the plasma membrane, (3) spontaneous transfer of the fluorescent lipid to intracellular membranes, and finally (4) concentration of the lipid at the Golgi complex.

Several parameters have been shown to be important in targeting fluorescent Cers to the Golgi complex. Work from our laboratory has shown that structural features of the sphingosine backbone and the nature of the short acyl chain influence targeting of these molecules to the Golgi apparatus (Pagano and Martin, 1988). Since accumulation of C_6-NBD-Cer at the Golgi complex occurs at low temperature (4°C), endocytosis of the lipid is not involved in internalization of the fluorescent Cer nor in its targeting to the Golgi complex. Remarkably, targeting of C_6-NBD-Cer to the Golgi apparatus also occurs in fixed cells, implying a physical interaction between Cer molecules and the membranes of this organelle. Fixation protocols which extract endogenous lipids from cells abrogate the accumulation, suggesting that hydrophobic interactions are important for this phenomenon (Pagano *et al.,* 1989). Moreover, C_6-NBD-Cer appears to target to the trans aspects of the Golgi complex (Pagano *et al.,* 1989). The potential significance of this observation is discussed more fully below (see Section IV,B).

C_5-DMB-Cer also exhibits localization to the Golgi apparatus (Pagano *et al.,* 1991). As mentioned in Section II and as seen in Fig. 2, concentration of C_5-DMB-Cer at the Golgi apparatus of living cells results in a striking change in fluorescence emission, from green to red. Although C_5-DMB-Cer concentrates at the Golgi complex in living cells (Pagano *et al.,* 1991), C_5-DMB-Cer, in contrast to C_6-NBD-Cer, cannot stain the Golgi complex of fixed cells, suggesting that an active process is required to target C_5-DMB-Cer to the Golgi complex. This may be related to the fact that this lipid has a slower rate of spontaneous transfer than the corresponding NBD analog and/or that this lipid has stronger interactions with cellular membrane components, as discussed previously (Section II).

It is not yet known whether [^{14}C]hex-Cer accumulates at the Golgi complex as the fluorescent Cers do. However, indirect evidence suggests that at least a portion of the radioactive Cer is found at the Golgi complex

---

FIG. 2. Fluorescence micrograph of a human skin fibroblast after incubation with C_5-DMB-Cer. Cells were incubated with a BSA complex containing 5 μM C_5-DMB-Cer for 30 minutes at 2°C, washed, then further incubated for 30 minutes at 37°C, and photographed. Samples were excited with blue light and photographed in either the (A) green + red (>520nm) or (B) red (>590 nm) regions of the spectrum.

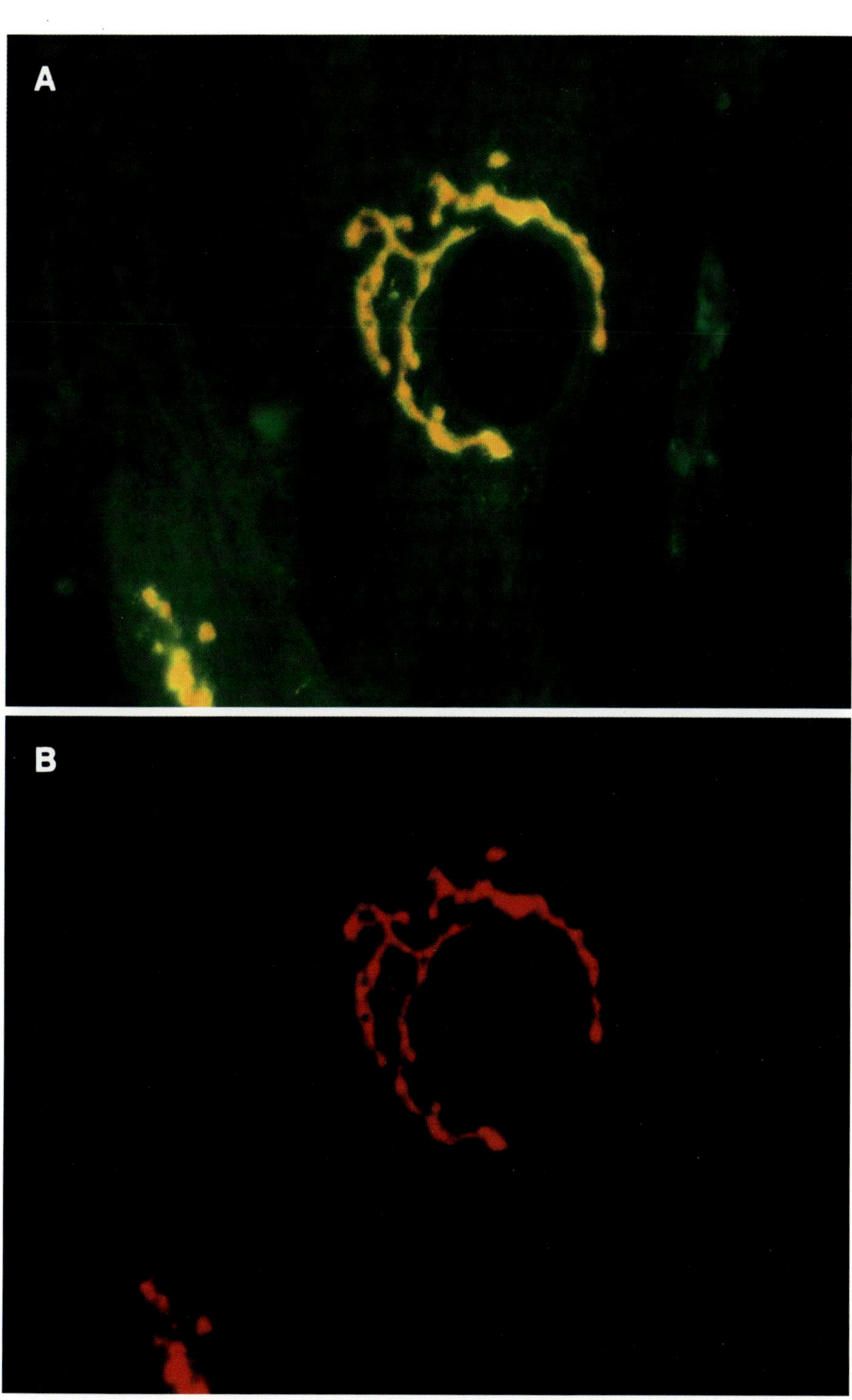
A
B

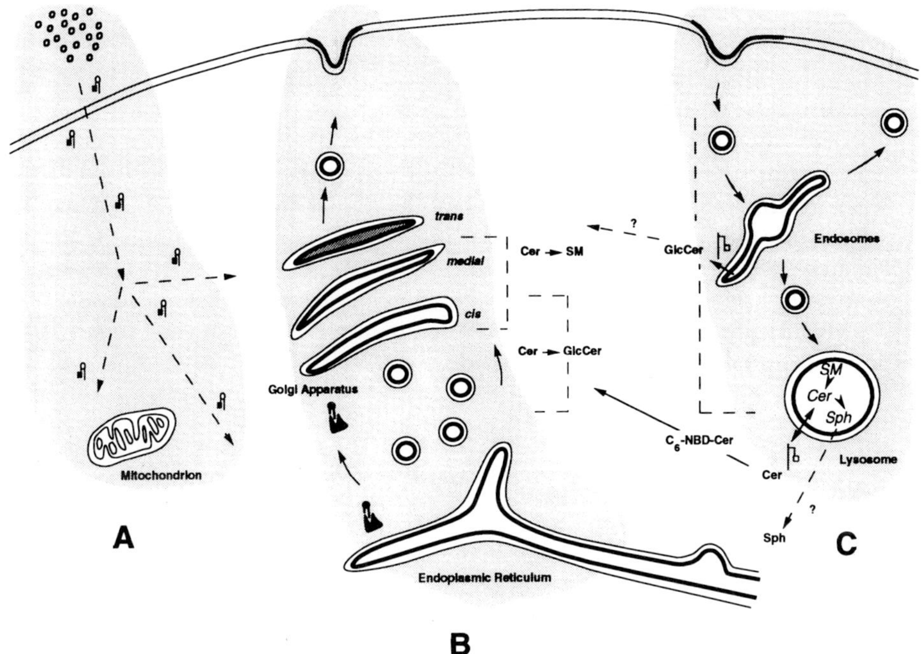

FIG. 3. Model for the uptake, transport, and metabolism of short-chain sphingolipid analogs in animal cells. (A) Delivery of short-chain Cer analogs to cells from an exogenous source. Cer moves by monomeric transfer from an appropriate donor and intercalates into the plasma membrane bilayer. At the plasma membrane it undergoes spontaneous transbilayer movement to the cytoplasmic surface of the plasma membrane, followed by monomeric transfer to label other intracellular membranes. Cer also accumulates in the late, or trans-Golgi compartment. (B) Metabolism of Cer at the Golgi complex and movement of the metabolites along the secretory pathway. Metabolism of Cer to SM occurs in the early, or cis,medial-Golgi compartments, while metabolism to GlcCer is more widely distributed, occurring in both the cis,medial-Golgi and perhaps in a compartment intermediate between the ER and the Golgi apparatus (see text). Newly synthesized SM and GlcCer move from the Golgi apparatus to the plasma membrane by a vesicle-mediated process. Transport of Cer from the ER to the Golgi apparatus may occur by vesicle movement and/or by protein carriers (see text). (C) Endocytosis of lipids from the plasma membrane. Short-chain SM and/or short-chain GlcCer delivered to the outer leaflet of the plasma membrane either from exogenous sources as in (A) or from *de novo* synthesis as in (B) are internalized to the endocytic compartment. Most of this lipid returns to the plasma membrane (''recycles''), while for SM, a fraction of the lipid travels to the lysosome, where it is degraded to Cer. Short-chain Cer then travels to the Golgi apparatus where it can be metabolized to SM and/ or GlcCer. Endogenous (long-chain) Cer may be further degraded to sphingosine prior to efflux from the lysosome. Short-chain GlcCer analogs may be sorted away from the lysosomal degradation pathway in some cells (see text). Cer, Ceramide; GlcCer, glucosylceramide; SM, sphingomyelin; Sph, sphingosine.

since it is a substrate for synthesis of [^{14}C]hex-SM and [^{14}C]hex-GlcCer (see Section III,B).

The C_6-NBD-Cer analog has been used extensively to study the behavior of the Golgi complex in living cells. For example, C_6-NBD-Cer has been used to document movements of Golgi elements in live cells by video microscopy (Cooper *et al.*, 1990). This analog has also been used to visualize fragmentation of the Golgi complex upon microtubule disruption by nocodozole (Lipsky and Pagano, 1985a) and reclustering of the Golgi elements upon removal of the drug (Ho *et al.*, 1989). C_6-NBD-Cer has also been used to study the role of the Golgi apparatus in bile production by hepatocytes (Crawford *et al.*, 1991). Finally, it has been used to study lipid sorting in polarized cells (see Section IV,F).

B. Sphingolipid Synthesis at the Golgi Complex: Enzyme Activities, Subcellular Location, and Topology

Cer is the immediate precursor in the synthesis of GlcCer, galactosylceramide (GalCer), and SM. GlcCer and GalCer in turn are the substrates for the synthesis of higher-order glycolipids. (The higher-order glycosylation reactions are not discussed here since they are the topics of other articles in this volume.) Like endogenous Cer, the short-chain fluorescent and radioactive Cers are metabolized to GlcCer and SM *in vivo* and *in vitro*. These reactions and where they occur are summarized in Fig. 3B.

C_6-NBD-Cer is metabolized to both C_6-NBD-GlcCer and C_6-NBD-SM in cultured cells (Lipsky and Pagano, 1983; 1985a; Koval and Pagano, 1989, 1990) and in rat liver Golgi membrane fractions (Futerman *et al.*, 1990; Futerman and Pagano, 1991). In contrast, C_5-DMB-Cer is converted primarily to C_5-DMB-SM in several different cell types (Pagano *et al.*, 1991), and in intact and detergent-solubilized rat liver Golgi fractions (Y. Kamisaka and R. E. Pagano, unpublished), suggesting that this analog is a poor substrate for GlcCer synthase. It has also been shown that [^{14}C]hex-Cer is converted to [^{14}C]hex-GlcCer and [^{14}C]hex-SM in living cells (Rosenwald *et al.*, 1992) and in rat liver homogenates and Golgi apparatus subfractions (Futerman *et al.*, 1990; Futerman and Pagano, 1991).

Two results using intact cells suggest that C_6-NBD-Cer is metabolized to products at the Golgi complex (Lipsky and Pagano, 1985a). First, the conversion of fluorescent Cer to fluorescent SM and GlcCer is unaffected by incubation of cells containing C_6-NBD-Cer with monensin, an ionophore which disrupts vesicular traffic from the Golgi apparatus to the plasma membrane (reviewed in Mollenhauer *et al.*, 1990). Second, when cells containing C_6-NBD-Cer are incubated at $<20°C$, synthesis of C_6-

NBD-SM and C_6-NBD-GlcCer continues, but transport of these products to the plasma membrane does not occur.

Direct proof that SM and GlcCer are synthesized at the Golgi apparatus has been obtained by analyzing the activities in purified subcellular fractions. Using highly purified subcellular fractions from rat liver and [^{14}C]hex-Cer as a substrate, it was shown that SM synthase activity [Cer + phosphatidylcholine (PC) → SM + diacylglycerol] is localized primarily at the Golgi apparatus (>87% of recovered activity), although a portion is found at the plasma membrane (<13% of the recovered activity) that cannot be accounted for by contamination with Golgi fractions (Futerman *et al.,* 1990). The activity in Golgi fractions is enriched approximately 85-fold relative to the homogenate. A similar conclusion was reached by Jeckel *et al.* (1990) using a truncated Cer analog (see the article by Jeckel and Wieland in this volume). Subfractionation of the Golgi complex showed that SM synthase activity is enriched in compartments corresponding to cis and medial aspects of the Golgi apparatus as defined by cofractionation with mannosidase II antigen (Futerman *et al.,* 1990).

Rat liver GlcCer synthase activity (Cer + UDP-Glc → GlcCer + UDP) using [^{14}C]hex-Cer as a substrate is also found primarily in Golgi fractions (Futerman and Pagano, 1991). Similar results have been obtained by Coste *et al.* (1985) using subcellular fractions of porcine submaxillary glands as an enzyme source and either endogenous Cer or exogenously added long-chain Cer as substrate. Interestingly, the GlcCer synthase activity of rat liver does not show the same subcellular distribution as SM synthase, but is found in cis/medial elements of the Golgi apparatus and in a light, smooth vesicle fraction that may correspond to the intermediate compartment of *cis*-Golgi network (Futerman and Pagano, 1991). These results are at variance with those of Jeckel *et al.* (1992), who found that GlcCer synthase activity, using a truncated Cer analog, is present in distal (trans) Golgi elements. The reason for this discrepancy is unknown.

The GlcCer synthase and SM synthase of rat liver have been found to differ with respect to the topologies of their active sites. Incubation of Golgi vesicles with trypsin has no effect on SM synthase activity unless the vesicles are first permeabilized by swelling in water (Futerman *et al.,* 1990), implying that the active site of this enzyme faces the lumen of these vesicles. In contrast, the active site of GlcCer synthase appears to face the cytosol, since enzyme activity is destroyed on incubation of vesicles with trypsin or chymotrypsin. Another explanation of these results, namely, that protease treatment destroys a potential UDP-Glc transporter in the Golgi membrane could not be ruled out directly. However, the rate of sugar nucleotide uptake by Golgi vesicles did not account for the rate of GlcCer synthesis activity. Taken together, these data suggest that the active site of GlcCer synthase is at the cytosolic surface (Futerman and

Pagano, 1991). A similar conclusion was reached by Coste *et al.* (1986) using Golgi fractions from porcine submaxillary glands as an enzyme source, and by Jeckel *et al.* (1992) using permeabilized Chinese hamster ovary cells. Interestingly, Trinchera *et al.* (1991) have shown that the active site of the transferase which adds galactose to GlcCer to form lactosylceramide (LacCer) also faces the cytosolic surface of the Golgi apparatus. These results predict the existence of a GlcCer and/or LacCer translocase or "flippase," since the higher order glycosylation reactions are thought to occur on the lumenal side of the Golgi apparatus (see article by Sandhoff and van Echten in this volume), which is topologically equivalent to the outer leaflet of the plasma membrane where sphingolipids are localized (Op den Kamp, 1979). In support of this model, recent data from Kobayashi *et al.* (1992) have shown that C_6-NBD-GlcCer and C_6-NBD-SM synthesized *de novo* from C_6-NBD-Cer appear to travel from the trans-Golgi network to the apical surface of MDCK cells in vesicles, and that the fluorescent lipids are localized to the lumen of these vesicles.

To summarize, it appears that GlcCer synthase and SM synthase reside in partially distinct subcompartments of the Golgi apparatus and that their active sites are on different sides of the bilayer (see Fig. 3B). Since short-chain Cer analogs can undergo transbilayer movement across the plasma membrane bilayer to enter cells and across the Golgi membrane as well, these analogs are readily accessible to both enzymes. Assuming endogenous Cers have the same ability to undergo transbilayer movement in intracellular membranes *in vivo,* this arrangement of enzymes may influence the relative levels of SM and glycosphingolipids.

To date, little is known about the molecular nature of these enzymes. Both GlcCer synthase and SM synthase from rat liver appear to be integral membrane proteins, as they cannot be easily removed from membranes with high salt or high pH. Further, the enzymes can be solubilized fully only by incubation with detergents (Y. Kamisaka and R. E. Pagano, unpublished). In other experiments, solubilization of GlcCer synthase was reported by Durieux *et al.* (1990) from porcine submaxillary gland Golgi membranes. Also, solubilization of SM synthase from CHO cells and reconstitution of the activity by detergent dialysis has been reported by Hanada *et al.* (1991).

IV. Intracellular Traffic of Sphingolipids

A. Overview of Lipid Traffic

Theoretically, lipids may move between intracellular membranes by one of a variety of mechanisms: diffusion of lipid monomers, either freely in solution or mediated by transport proteins; vesicular movement between compartments; or lateral diffusion between membranes through intermembrane bridges. For reviews on the general topic of lipid transport, see Dawidowicz (1987), Sleight (1987), Bishop and Bell (1988), van Meer (1989), Pagano (1990a), and Voelker (1991). For reviews on the metabolism and intracellular movement of sphingolipids, see Merrill and Jones (1990), Pagano (1990b), Schwarzmann and Sandhoff (1990), and Koval and Pagano (1991).

B. Transport of Precursors from the Endoplasmic Reticulum to the Golgi Complex

Cell fractionation studies have established that the majority of PC biosynthesis occurs at the ER (Bell *et al.*, 1981; Dawidowicz, 1987). In addition, a recent study by Mandon *et al.* (1992) confirms earlier studies localizing the enzymes involved in the formation of 3-oxosphinganine, sphinganine, and dihydroceramide to the ER. Since the synthesis of SM and GlcCer occurs at the Golgi complex, transport of the lipid precursors PC and Cer from the ER to the Golgi complex is required for synthesis of these lipids.

In principle, newly synthesized lipids can move from the ER to the Golgi apparatus by vesicle transport or by protein carriers (Fig. 3B). Transport vesicles, which may reflect the lipid composition of the ER, could account for the movement of PC from the ER to the Golgi apparatus, where it is utilized in SM biosynthesis, since vesicle movement is required for transporting newly synthesized glycoproteins from the ER to the Golgi apparatus and PC is a major lipid constituent of both organelles. However, such a mechanism is more difficult to envision for Cer since, like other intermediates in lipid biosynthesis, its concentration in cells is thought to be very low, while the concentration of Cer metabolites at the Golgi apparatus is relatively high ($\sim$10–15 mol%). This raises the possibility that Cer may be selectively transported from the ER to the Golgi apparatus. This could be accomplished if transport vesicles enriched in Cer formed from specialized domains of the ER. Alternatively, selective transport of Cer from the ER to the Golgi apparatus could occur via a protein-mediated carrier or lipid transfer protein (Crain and Zilversmit, 1980; Rueckert and Schmidt, 1990). The use of fluorescent sphingolipid analogs to activate 5-[^{125}-I]iodonaphthyl-1-azide to derivatize proteins may be useful for identification of such proteins (Rosenwald *et al.*, 1991).

Finally, studies at the EM level using C_6-NBD-Cer and C_5-DMB-Cer have shown that these lipids preferentially label a subset of Golgi compartments, most likely corresponding to the trans-Golgi stacks (Pagano *et* 1989, 1991). This raises the intriguing possibility that Cer might be delivered directly from the ER to a late compartment of the Golgi apparatus. Since metabolism of Cer to SM and GlcCer occurs in early Golgi compartments, retrograde movement of Cer within the Golgi apparatus would then be required for its subsequent metabolism. Future studies using the short-chain fluorescent or radioactive lipid analogs to monitor the movement of Cer and its metabolites between subcompartments of the Golgi apparatus may help resolve this interesting paradox.

C. Sphingolipid Transport from the Golgi Complex to the Plasma Membrane

Several recent experiments suggest that sphingolipids are transported from the Golgi complex to the plasma membrane by a vesicular route much like that taken by proteins in cells (Fig. 3B). It has been found that, under conditions where protein transport is inhibited, sphingolipid transport is also inhibited. For example, temperatures (<20°C) that block vesicular transport of proteins also block sphingolipid transport (Lipsky and Pagano, 1985a; van Echten and Sandhoff, 1989). Also, incubation of cells with the ionophore monensin, which blocks protein transport in the Golgi complex (Mollenhauer *et al.*, 1990, and references therein), blocks transport of newly synthesized sphingolipids (Lipsky and Pagano, 1985a; van Echten and Sandhoff, 1989). In addition, mitotic cells do not transport proteins (Warren *et al.*, 1983; Hesketh *et al.*, 1984; Featherstone *et al.*, 1985; Preston *et al.*, 1985) or sphingolipids (Kobayashi and Pagano, 1988; Kreiner and Moore, 1990) to the cell surface.

Efforts are ongoing to determine the molecular mechanisms required to transport sphingolipids. A genetic approach taken by Puoti *et al.* (1991) showed that certain yeast mutants which display temperature-sensitive defects in the secretion of proteins (*sec* mutants) are unable to transport sphingolipids at restrictive temperatures. Using a biochemical approach, in cultured mammalian cells it has been found that secretion of a truncated SM derivative (synthesized at the Golgi apparatus from a truncated Cer; see the article by Jeckel and Wieland in this volume) and a glycosylated peptide follow similar pathways *in vivo* (Wieland *et al.*, 1987; Karrenbauer *et al.*, 1990). In permeabilized mammalian cells, secretion of both the truncated SM and the glycosylated peptide requires ATP and cytosolic factors and is inhibited by GTPγS (Helms *et al.*, 1990). Finally, a fluorescent analog of SM (C_6-NBD-SM, synthesized *de novo* from C_6-NBD-Cer) is localized in vesicles released from permeabilized MDCK cells (Bennet *et al.*, 1988).

Recently, we demonstrated that inhibition of protein and sphingolipid transport results from incubation of cells with a sphingolipid synthesis inhibitor (1-phenyl-2-decanoylamino-3-morpholino-1-propanol or PDMP) (Rosenwald *et al.*, 1992; for a general discussion of PDMP, see the article by Radin *et al.* in this volume). Upon incubation of cells with low concentrations of PDMP (2.5–10 μM), conversion of [^{14}C]hex-Cer to [^{14}C]hex-GlcCer is inhibited. At higher concentrations ($\geq 25\ \mu M$), [^{14}C]hex-SM synthesis is also decreased. As a result of inhibition of these two enzymes, [^{14}C]hex-Cer accumulates. Incubation of cells at these higher concentrations of PDMP results in inhibition of transport of vesicular stomatitis virus glycoprotein (VSV-G protein) through medial and trans compartments of the Golgi apparatus and from the Golgi complex to the plasma membrane. In addition, transport of fluorescent sphingolipids (synthesized *de novo* at the Golgi complex) to the plasma membrane is inhibited by PDMP. Inhibition of transport of both sphingolipids and proteins exhibits a concentration dependence with respect to PDMP. Recent studies have demonstrated that high concentrations of exogenous Cer also inhibit glycoprotein traffic through the secretory pathway (Rosenwald and Pagano, 1993).

D. Intra-Golgi Transport of Sphingolipids

It is likely that agents which affect transport of sphingolipids from the Golgi apparatus to the plasma membrane may also have effects on intra-Golgi transport. Currently, it is difficult, for instance, to dissect the effects of an inhibitor like monensin on a single step in the transport of sphingolipids through the Golgi apparatus. However, the following pieces of evidence suggest that transport of sphingolipids through the Golgi apparatus also occurs by a vesicular mechanism. As discussed elsewhere in this volume, it appears that glycosphingolipid synthesis occurs in the Golgi apparatus. Using the membrane traffic inhibitor Brefeldin A, at least two distinct compartments in the synthesis of sphingolipids are defined (Brüning *et al.*, 1992; van Echten *et al.*, 1990; Young *et al.*, 1990). Since Brefeldin A is thought to disrupt vesicular traffic (reviewed in Klausner *et al.*, 1992), these results argue that vesicular traffic is required to transport sphingolipid precursors from proximal compartments to more distal ones. Perhaps the most compelling evidence, however, for intra-Golgi vesicular transport comes from the experiments of Wattenberg (1990), which show that a single transport step in the synthesis of sphingolipids [LacCer to sialyllactosylceramide (GM_3)] is dependent on the same parameters that are known to be important for the transport of proteins through the Golgi complex (Rothman and Orci, 1990, 1992); these parameters include ATP, cytosolic factors, intact membranes, and elevated temperatures. This reaction, like steps in protein transport, is inhibited by GTPγS and antibodies directed against NEM-sensitive factor (NSF).

E. Transport of Sphingolipids from the Plasma Membrane to Internal Compartments

Transport of sphingolipids from the plasma membrane to intracellular compartments appears to follow the endocytic pathway at least in part (Fig. 3C). Using a short-chain fluorescent analog of SM (C_6-NBD-SM) to label the plasma membrane, previous work from this laboratory demonstrated that the majority of the internalized fluorescent lipid returns to the plasma membrane, in compartments identical to those defined by the "recycling" transferrin receptor. However, a portion of the internalized probe appears to be degraded to C_6-NBD-Cer, most likely in the lysosomal compartment. This was deduced from the fact that Niemann-Pick A cells, which are defective in lysosomal sphingomyelinase, exhibit much less degradation of C_6-NBD-SM to C_6-NBD-Cer and accumulate NBD fluorescence in lysosomes (Koval and Pagano, 1990). C_6-NBD-Cer is then transported to the Golgi apparatus, where it can serve as a substrate for SM and GlcCer synthesis (Koval and Pagano, 1989; 1990). The transport of C_6-NBD-Cer, and possibly other short-chain Cers, from the lysosomes to the Golgi complex may be unique to short-chain Cers because of their ability to undergo rapid spontaneous monomeric transfer. Endogenous Cers do not appear to have the ability to leave the lysosome. Indeed, efflux of material derived from endogenous Cer appears to be dependent on hydrolysis by a lysosomal ceramidase (Kolesnick, 1991).

Experiments similar to those done by Koval and Pagano (1989; 1990) were performed by Hoekstra and co-workers, using C_6-NBD-GlcCer (Kok *et al.*, 1989). This lipid labels the plasma membrane of cells in a similar manner to that seen with C_6-NBD-SM. In undifferentiated HT29 cells, C_6-NBD-GlcCer, in contrast to what has been found for C_6-NBD-SM, is apparently sorted to the Golgi complex (Kok *et al.*, 1991). The intracellular site and mechanisms responsible for this sorting reaction remain to be determined.

F. Sorting of Sphingolipids in Polarized Cells

Various studies have shown that glycosphingolipids are enriched on the apical surface of polarized epithelia (reviewed in Simons and Fuller, 1985; Simons and van Meer, 1988). In keeping with this finding, van Meer and colleagues have shown that C_6-NBD-GlcCer (synthesized *de novo* from C_6-NBD-Cer) is enriched on the apical surfaces of two polarized epithelial cell lines in culture, MDCK cells (van Meer *et al.*, 1987) and CaCO2 cells (van't Hoff and van Meer, 1990). This sorting of sphingolipid analogs appears to be relatively insensitive to changes in the fatty acyl moiety on the analogs and to changes in the sphingosine backbone (van't Hof *et al.,* 1992). The mechanism for sphingolipid sorting remains to be determined. However, it has been hypothesized that intermolecular hydrogen bonding between glycosphingolipids, forming microdomains in the plane of the membrane bilayer, may be important in directing newly synthesized glycosphingolipids (Simons and van Meer, 1988) and glycosylphosphatidylinositol-linked proteins (Brown and Rose, 1992) to the apical plasma membrane domain.

For some cells, such as MDCK cells, proteins destined for different plasma membrane domains are transported to those domains directly from the trans-Golgi network ("direct pathway") (Simons and Fuller, 1985; Hubbard *et al.,* 1989). In hepatocytes, however, it appears that proteins destined for the apical (bile canalicular) surface first appear on the basolateral (sinusoidal) surface, and are then retrieved by endocytosis and delivered to the apical surface ("indirect pathway") (Hubbard *et al.,* 1989). An Important question is whether glycosphingolipids follow a similar indirect pathway in hepatocytes. As a first step, Crawford *et al.* (1991) have studied the metabolism and intracellular distribution of C_6-NBD-Cer in hepatocyte couplets, but the route of transport of fluorescent metabolites to the plasma membrane was not addressed. A polarized, hepatocyte-like cell line has been described (Cassio *et al.,* 1991), and may be a promising experimental system for studying "direct" versus "indirect" pathways of lipid transport.

V. Summary and Perspectives

The use of fluorescent and radioactive short-chain sphingolipid analogs has helped to define potential pathways for sphingolipid metabolism and transport. However, interpretation of the results obtained with these analogs requires some caution, since their high rates of spontaneous transfer may obscure other processes which might be occurring for longer chain endogenous sphingolipids.

Current evidence suggests that most sphingolipid traffic occurs by vesicular mechanisms, both along the secretory pathway and along the endocytic pathway. However, this model may be too simplistic, and there are many areas where further experimentation is needed before we have a complete understanding of the mechanisms by which sphingolipids are metabolized and transported in cells. There are several curious observations that need to be accounted for in a complete model of sphingolipid metabolism and transport. These include determination of the *in vivo* role of cytosolic glycolipid transfer proteins (Sasaki, 1990), because most glycosphingolipids are thought to be present on the lumen of intracellular membranes and at the outer leaflet of the plasma membrane, and therefore should be inaccessible to these transfer proteins. In addition, Matyas and Morré (1987) have found evidence for the appearance of pulse-labeled gangliosides at the ER and nuclear envelope. It is not yet clear how lipids are moved to these sites or what their functional roles are. For those aspects of protein traffic and sphingolipid traffic that appear to coincide, it will be important to show that the molecular components which play a role in traffic of proteins also have a role in traffic of sphingolipids. However, it is important to consider the possibility that these pathways may not be entirely overlapping, and that there may be components important for sphingolipid traffic alone. One such component is the potential for transbilayer movement of Cer, GlcCer, and LacCer, which may be an important aspect of glycosphingolipid metabolism and traffic within cells. Finally, future goals must also include determination of the mechanisms by which lipid traffic is regulated in cells.

Acknowledgments

We thank the members of the Pagano laboratory past and present for their assistance with this work. AGR is supported by Damon Runyon-Walter Winchell Cancer Fund Fellowship DRG-1067. The work in REP's laboratory is supported by USPHS grant R37 GM-22942.

References

Bai, J., and Pagano, R. E. (1993). In preparation.

Bell, R. M., Ballas, L. M., and Coleman, R. A. (1981). *J. Lipid Res.* **22,** 391–403.

Bennet, M. K., Wandinger-Ness, A., and Simons, K. (1988). *EMBO J.* **7,** 4075–4085.

Bishop, W. R., and Bell, R. M. (1988). *Annu. (Cambridge, Mass.) Rev. Cell Biol.* **4,** 579–610.

Brown, D. A., and Rose, J. K. (1992). *Cell* **68,** 533–544.

Brüning, A., Karrenbauer, A., Schnabel, E., and Wieland, F. T. (1992). *J. Biol. Chem.* **267,** 5052–5055.

Cassio, D., Hamon-Benais, C., Guérin, M., and Lecoq, O. (1991). *J. Cell Biol.* **115,** 1397–1408.

Chattopadhyay, A., and London, E. (1987). *Biochemistry* **26,** 39–45.

Cooper, M. S., Cornell-Bell, A. H., Chernjavsky, A., Dani, J. W., and Smith, S. J. (1990). *Cell (Cambridge, Mass.)* **61,** 135–145.

Coste, H., Martel, M. B., Azzar, G., and Got, R. (1985). *Biochim. Biophys. Acta* **814,** 1–7.

Coste, H., Martel, M. B., and Got, R. (1986). *Biochim. Biophys. Acta* **858,** 6–12.

Crain, R. C., and Zilversmit, D. B. (1980). *Biochim. Biophys. Acta* **620,** 37–48.

Crawford, J. M., Vintner, D. W., and Gollan, J. L. (1991). *Am. J. Physiol.* **260,** G119–G132.

Dawidowicz, E. A. (1987). *Annu. Rev. Biochem.* **56,** 43–61.

Durieux, I., Martel, M. B., and Got, R. (1990). *Biochim. Biophys. Acta* **1024,** 263–266.

Farquhar, M. G., and Palade, G. E. (1981). *J. Cell Biol.* **91,** 77s–103s.
Featherstone, C., Griffiths, G., and Warren, G. (1985). *J. Cell Biol.* **101,** 2036–2046.
Frank, A., Barenholz, Y., Lichtenberg, D., and Thompson, T. E. (1983). *Biochemistry* **22,** 5647–5651.
Futerman, A. H., and Pagano, R. E. (1991). *Biochem. J.* **280,** 295–302.
Futerman, A. H., Stieger, B., Hubbard, A. L., and Pagano, R. E. (1990). *J. Biol. Chem.* **265,** 8650–8657.
Golgi, C. (1898). *Arch. Ital. Biol.* **30,** 278–286; appears in translation in *J. Microsc.* (*Oxford*) **155,** 9–14 (1989).
Hanada, K., Horii, M., and Akamatsu, Y. (1991). *Biochim. Biophys. Acta* **1086,** 151–156.
Helms, J. B., Karrenbauer, A., Wirtz, K. W. A., Rothman, J. E., and Wieland, F. T. (1990). *J. Biol. Chem.* **265,** 20027–20032.
Hesketh, T. R., Beaven, M. A., Rogers, J., Jurke, B., and Warren, G. B. (1984). *J. Cell Biol.* **98,** 2250–2254.
Ho, W. C., Allan, V. J., van Meer, G., Berger, E. G., and Kreis, T. E. (1989). *Eur. J. Cell Biol.* **48,** 250–263.
Hubbard, A. L., Steiger, B., and Bartles, J. R. (1989). *Annu. Rev. Physiol.* **51,** 755–770.
Jeckel, D., Karrenbauer, A., Birk, R., Schmidt, R. R., and Wieland, F. T. (1990). *FEBS Lett.* **261,** 155–157.
Jeckel, D., Karrenbauer, A., Burger, K. N. J., van Meer, G., and Wieland, F. T. (1992). *J. Cell Biol.* **117,** 259–267.
Karrenbauer, A., Jeckel, D., Just, W., Birk, R., Schmidt, R. R., Rothman, J. E., and Wieland, F. T. (1990). *Cell (Cambridge, Mass.)* **63,** 259–267.
Klausner, R. D., Donaldson, J. G., and Lippincott-Schwartz, J. (1992). *J. Cell Biol.* **116,** 1071–1080.
Kobayashi, T., and Pagano, R. E. (1988). *J. Biol. Chem.* **264,** 5966–5973.
Kobayashi, T., Pimplikar, S. W., Parton, R. G., Bhakdi, S., and Simons, K. (1992). *FEBS Lett.* **300,** 227–231.
Kok, J. W., Eskelinen, S., Hoekstra, K., and Hoekstra, D. (1989). *Proc. Natl. Acad. Sci. U.S.A.* **86,** 9896–9900.
Kok, J. W., Babia, T., and Hoekstra, D. (1991). *J. Cell Biol.* **114,** 231–239.
Kolesnick, R. N. (1991). *Prog. Lipid Res.* **30,** 1–38.
Koval, M., and Pagano, R. E. (1989). *J. Cell Biol.* **108,** 2169–2181.
Koval, M., and Pagano, R. E. (1990). *J. Cell Biol.* **111,** 429–442.
Koval, M., and Pagano, R. E. (1991). *Biochim. Biophys. Acta* **1082,** 113–125.
Kreiner, T., and Moore, H.-P. H. (1990). *Cell Regul.* **1,** 415–424.
Lipsky, N. G., and Pagano, R. E. (1983). *Proc. Natl. Acad. Sci. U.S.A.* **80,** 2608–2612.
Lipsky, N. G., and Pagano, R. E. (1985a). *J. Cell Biol.* **100,** 27–34.
Lipsky, N. G., and Pagano, R. E. (1985b). *Science* **228,** 745–747.
Mandon, E. C., Ehses, I., Rother, J., van Echten, G., and Sandhoff, K. (1992). *J. Biol. Chem.* **267,** 11144–11148.
Matyas, G. R., and Morré, D. J. (1987). *Biochim. Biophys. Acta* **921,** 599–614.
Merrill, A. H., Jr., and Jones, D. D. (1990). *Biochim. Biophys. Acta* **1044,** 1–12.
Mollenhauer, H. H., Morré, D. J., and Rowe, L. D. (1990). *Biochim. Biophys. Acta* **1031,** 225–246.
Nichols, J. W., and Pagano, R. E. (1982). *Biochemistry* **21,** 1720–1726.
Nichols, J. W., and Pagano, R. E. (1983). *J. Biol. Chem.* **258,** 5368–5371.
Op den Kamp, J. A. F. (1979). *Annu. Rev. Biochem.* **48,** 47–71.
Pagano, R. E. (1983). *In* "Liposome Letters" (A. D. Bangham, ed.), pp. 83–96. Academic Press, New York.

Pagano, R. E. (1989). *Methods Cell Biol.* **29,** 75–85.
Pagano, R. E. (1990a). *Curr. Opin. Cell Biol.* **2,** 652–663.
Pagano, R. E. (1990b) *Biochem. Soc. Trans.* **18,** 361–366.
Pagano, R. E., and Martin, O. C. (1988). *Biochemistry* **27,** 4439–4445.
Pagano, R. E., and Sleight, R. G. (1985). *Science* **229,** 1051–1057.
Pagano, R. E., Sepanski, M. A., and Martin, O. C. (1989). *J. Cell Biol.* **109,** 2067–2079.
Pagano, R. E., Martin, O. C., Kang, H. C., and Haugland, R. P. (1991). *J. Cell Biol.* **113,** 1267–1279.
Pfeffer, S. R., and Rothman, J. E. (1987). *Annu. Rev. Biochem.* **56,** 829–852.
Preston, S. G., Regula, C. S., Sager, P. R., Pearson, C. B., Daniels, L. S., Brown, P. A., and Berlin, R. E. (1985). *J. Cell Biol.* **101,** 1086–1093.
Puoti, A., Desponds, C., and Conzelmann, A. (1991). *J. Cell Biol.* **113,** 515–525.
Rosenwald, A. G., and Pagano, R. E. (1993). *J. Biol. Chem.* **268,** 4577–4579.
Rosenwald, A. G., Pagano, R. E., and Raviv, Y. (1991). *J. Biol. Chem.* **266,** 9814–9821.
Rosenwald, A. G., Machamer, C. E., and Pagano, R. E. (1992). *Biochemistry* **31,** 3581–3590.
Rothman, J. E., and Orci, L. (1990). *FASEB J.* **4,** 1460–1468.
Rothman, J. E., and Orci, L. (1992). *Nature (London)* **355,** 409–415.
Rueckert, D. G., and Schmidt, K. (1990). *Chem. Phys. Lipids* **56,** 1–20.
Sasaki, T. (1990). *Experientia* **46,** 611–616.
Schwarzmann, G., and Sandhoff, K. (1990). *Biochemistry* **29,** 10865–10871.
Simons, K., and Fuller, S. D. (1985). *Annu. Rev. Cell Biol.* **1,** 243–288.
Simons, K., and van Meer, G. (1988). *Biochemistry* **27,** 6197–6202.
Sleight, R. G. (1987). *Annu. Rev. Physiol.* **49,** 193–208.
Struck, D. K., and Pagano, R. E. (1980). *J. Biol. Chem.* **255,** 5404–5410.
Trinchera, M., Fabbri, M., and Ghidoni, R. (1991). *J. Biol. Chem.* **266,** 20907–20912.
van Echten, G., and Sandhoff, K. (1989). *J. Neurochem.* **52,** 207–214.
van Echten, G., Iber, H., Stotz, H., Takatsuki, A., and Sandhoff, K. (1990). *Eur. J. Cell Biol.* **51,** 135–139.
van Meer, G. (1989). *Annu. Rev. Cell Biol.* **5,** 247–275.
van Meer, G., Stelzer, E. H. K., Wijnaendts-van-Resandt, R. W., and Simons, K. (1987). *J. Cell Biol.* **105,** 1623–1635.
van't Hof, W., and van Meer, G. (1990). *J. Cell Biol.* **111,** 977–986.
van't Hof, W., Silvius, J., Wieland, F. T., and van Meer, G. (1992). *Biochem. J.* **283,** 913–917.
Voelker, D. R. (1991). *Microbiol. Rev.* **55,** 543–560.
Warren, G., Featherstone, C., Griffiths, G., and Burke, B. (1983). *J. Cell Biol.* **97,** 1623–1628.
Wattenberg, B. W. (1990). *J. Cell Biol.* **111,** 421–428.
Wieland, F. T., Gleason, M. L., Serafini, T. A., and Rothman, J. E. (1987). *Cell (Cambridge, Mass.)* **50,** 289–300.
Wolf, D. E., Winiski, A. P., Ting, A. E., Bocian, K. W., and Pagano, R. E. (1992). *Biochemistry* **31,** 2865–2873.
Young, W. W., Jr., Lutz, M. S., Mills, S. E., and Lechler-Osborn, S. (1990). *Proc. Natl. Acad. Sci. U.S.A.* **87,** 6838–6842.

ADVANCES IN LIPID RESEARCH, VOL. 26

Ganglioside Metabolism—Topology and Regulation

KONRAD SANDHOFF AND GERHILD VAN ECHTEN

Institut für Organische Chemie und Biochemie
Universität Bonn
D-5300 Bonn 1, Germany

I. Introduction

Glycosphingolipids (GSL) are amphiphilic plasma membrane components, characteristic of cells of vertebrate tissues (Wiegandt, 1982, 1985; Svennerholm, 1984; Ledeen and Yu, 1982; van Echten and Sandhoff, 1989).

The hydrophobic ceramide moiety anchors the GSL in the outer leaflet of the membrane bilayer (Gahmberg and Hakomori, 1973; Thompson and Tillack, 1985), so that the hydrophilic oligosaccharide residue faces the extracellular space.

GSL form cell- and species-specific patterns on cellular surfaces, and, together with membrane glycoproteins and proteoglycans, constitute the carbohydrate-rich glycocalyx of cells, which is thought to play a role in cell differentiation, morphogenesis, and oncogenic transformation (Hakomori, 1984a; Fenderson *et al.*, 1985).

On the one hand, GSL were found to be specific binding sites for viruses (Markwell *et al.*, 1981), toxins (Walton *et al.*, 1988; Bock *et al.*, 1985), and bacteria (Svennerholm, 1984; Karlsson, 1989); on the other hand, a few GSL and their derivatives were implicated in the modulation of membrane-bound receptors and enzymes (Hanai *et al.*, 1988; Bremer *et al.*, 1986). Their precise physiological role in cell adhesion, cell differentiation, and morphogenesis has not yet been proven; recently, however, sialyl Lewis X (SLe^x) and other determinants on GSL were found to be ligands for selectins, e.g., the endothelial leukocyte adhesion molecule 1 (ELAM-1) (Phillips *et al.*, 1990; Walz *et al.*, 1990).

Gangliosides are a group of GSL which contain one or more sialic acid units in their oligosaccharide chain (Fig. 1). They are highly enriched in nervous tissue, in which their more complex derivatives, namely, di-, tri-, and tetrasialogangliosides, are particularly prevalent (Hansson *et al.*, 1977; Ledeen, 1985; van Echten and Sandhoff, 1989).

During ontogenesis, the ganglioside content of nervous tissue increases dramatically and the GSL pattern of neuronal cell types changes specifically (Seyfried *et al.*, 1984). Although most of the gangliosides are concentrated on the plasma membrane, GSL biosynthesis and degradation are localized intracellularly. Whereas ganglioside biosynthesis starts on the membranes of the endoplasmic reticulum and continues on the Golgi membranes, catabolism occurs after endocytosis in the lysosomal compartment.

The great stability of specific GSL patterns suggests the maintenance of a balanced glycosphingolipid profile and demands a stringent control of GSL metabolism as well as of intracellular transport of GSL.

II. Pathways of Ganglioside Biosynthesis, Subcellular Localization, Topology, and Exocytotic Membrane Flow

A. From Serine to Lactosylceramide

Our current knowledge of glycosphingolipid biosynthesis and intracellular traffic of GSL has been derived from metabolic studies. The mechanism by which gangliosides are transported from sites of synthesis to other membranes has not been defined so far. It was generally assumed (Fishman and Brady, 1976; Mooré *et al.*, 1979) that GSL formation is coupled, like the formation of glycoproteins, to a vesicular membrane flow, from the ER through the cisternae of the Golgi complex to the plasma membrane (Fig. 2). However, the involvement of glycolipid binding and/or transfer proteins in the transport of GSL cannot be excluded as long as the function of several such proteins has not been identified (Yamada and Sasaki, 1982; Watanabe *et al.*, 1980; Abe *et al.*, 1984; Wong *et al.*, 1984; Thompson *et al.*, 1986; Tiemeyer *et al.*, 1989).

Glycosphingolipid biosynthesis starts with the condensation of serine and palmitoyl-CoA (Fig. 3), which is catalyzed by the pyridoxal phos-

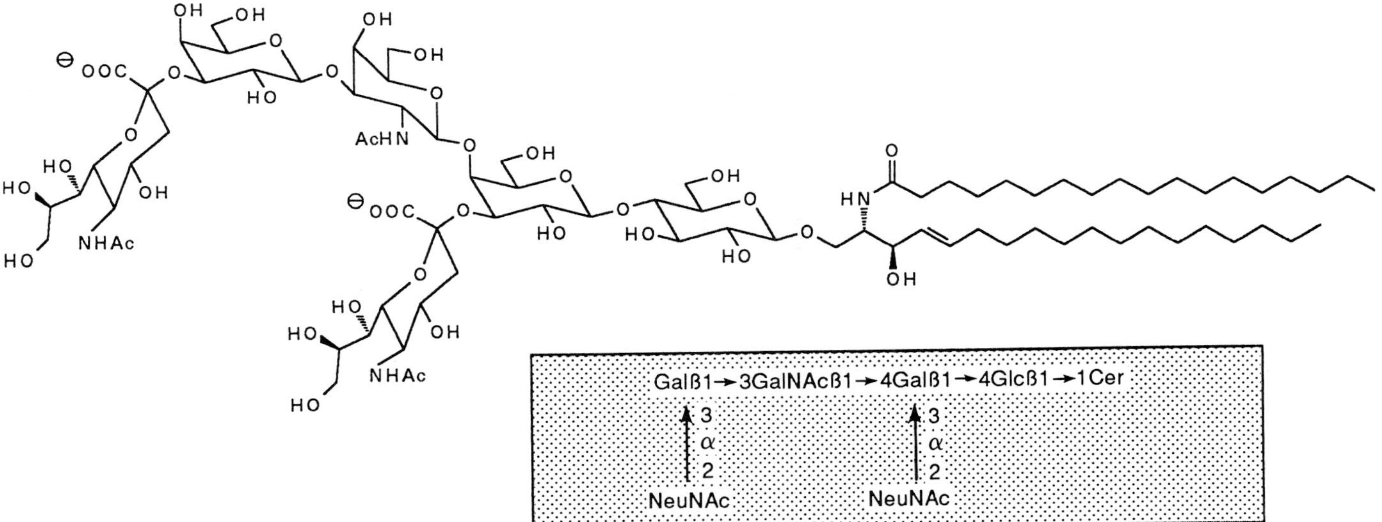

FIG. 1. Structure of ganglioside GD_{1a}.

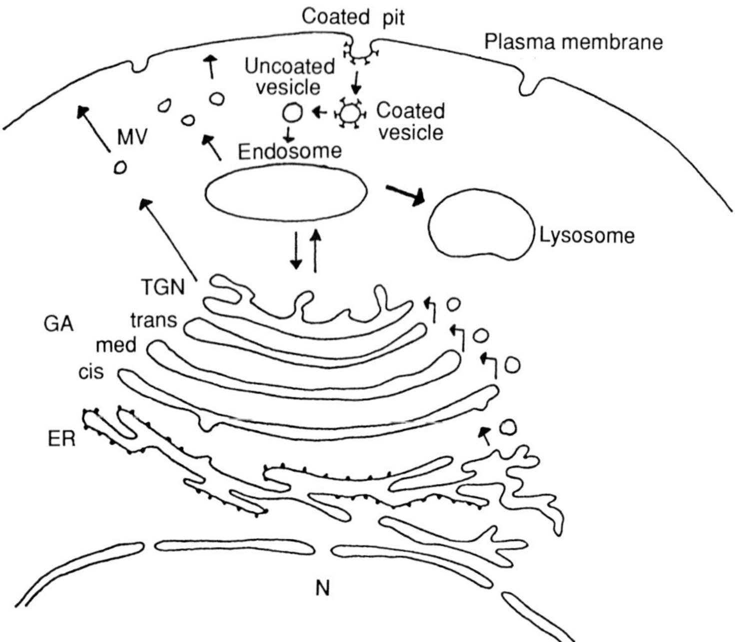

FIG. 2. Major compartments of glycosphingolipid metabolism (Schwarzmann and Sandhoff, 1990). Biosynthesis of glycosphingolipids is catalyzed by membrane-bound transferases of the endoplasmic reticulum (ER) and cisternae of the Golgi apparatus (GA) and trans-Golgi network (TGN). Glycolipids of the plasma membrane can reach the lysosomal compartment by a vesicular endocytotic pathway; their catabolism is facilitated by exohydrolases and glycolipid-binding proteins (activators) in the lysosomes. MV, Membrane vesicles; N, nucleus.

phate-dependent serine palmitoyltransferase (SPT), yielding 3-dehydrosphinganine (Mandon *et al.*, 1991; Stoffel *et al.*, 1968; Braun and Snell, 1968). This ketone is afterward rapidly reduced to sphinganine by a NADPH-dependent reductase (Stoffel *et al.*, 1968). The introduction of the 4-*trans* double bond occurs after addition of amide-linked fatty acid (Fig. 3). This sequence was proposed previously (Ong and Brady, 1973;

FIG. 3. Scheme for sphingolipid biosynthesis from serine to lactosylceramide. All enzymatic steps (except desaturation, which is not yet clear, and formation of LacCer, which is contradictory) take place on the cytosolic leaflet of ER or Golgi membranes. The serine palmitoyltransferase (SPT) contains pyridoxal phosphate (PLP) as a cofactor. 3-Dehydrosphinganine reductase contains reduced nicotinamide adenine dinucleotide phosphate (NADPH) as a cofactor.

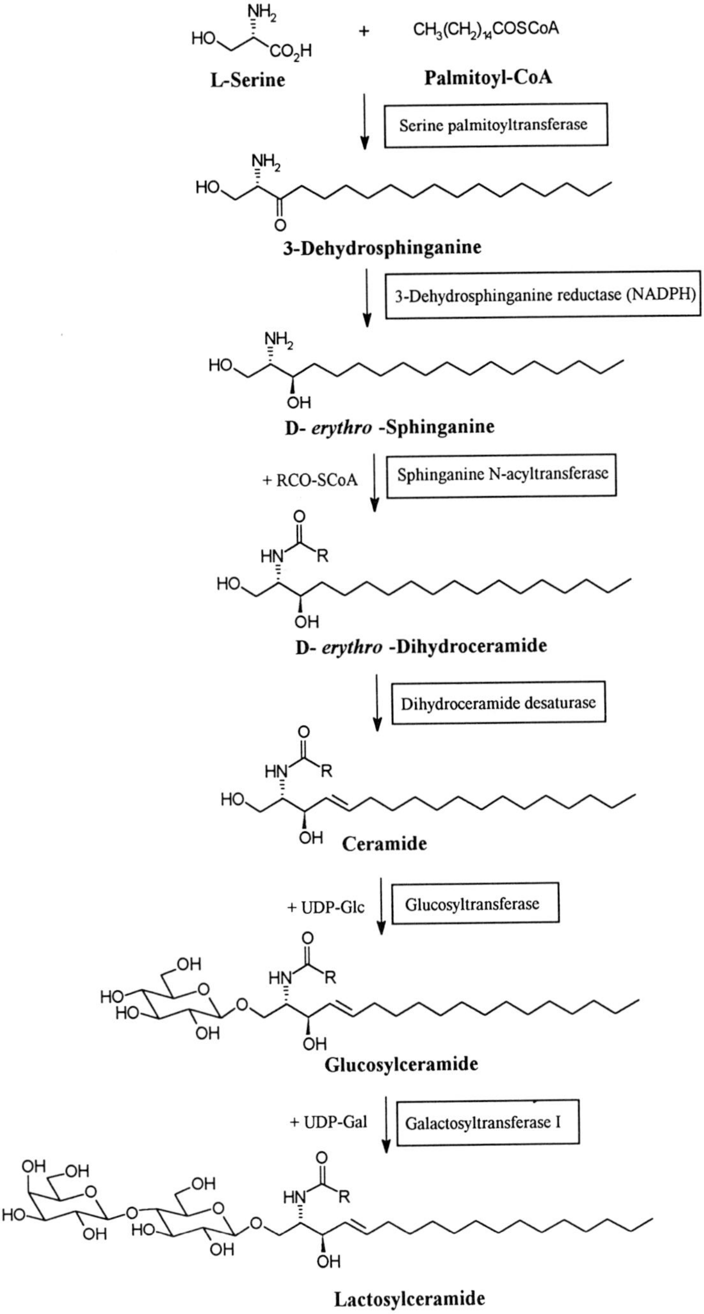

L-Serine + Palmitoyl-CoA ($CH_3(CH_2)_{14}COSCoA$)

Serine palmitoyltransferase

3-Dehydrosphinganine

3-Dehydrosphinganine reductase (NADPH)

D- *erythro* -Sphinganine

+ RCO-SCoA

Sphinganine N-acyltransferase

D- *erythro* -Dihydroceramide

Dihydroceramide desaturase

Ceramide

+ UDP-Glc

Glucosyltransferase

Glucosylceramide

+ UDP-Gal

Galactosyltransferase I

Lactosylceramide

Stoffel and Bister, 1974; Merrill and Wang, 1986) and demonstrated recently by Rother *et al.* (1992); thus sphingosine is not an intermediate of the *de novo* biosynthetic pathway.

All enzymatic steps involved in ceramide biosynthesis appear to be located on the cytosolic face of the endoplasmic reticulum (Mandon *et al.*, 1992).

The next step in the biosynthesis of gangliosides, the glucosylation of ceramide, is localized in the Golgi apparatus. Coste *et al.* (1985, 1986), and, recently, Trinchera *et al.* (1991), Futerman and Pagano, (1991), and Jeckel *et al.* (1992) showed that ceramide glucosyltransferase is accessible from the cytoplasmic side of Golgi vesicles. Thus, the question of ceramide transport from the ER to the Golgi arises. One possibility is ceramide transport via vesicular membrane flow (van Meer, 1989). There is, however, evidence suggesting that nonvesicular transport is involved which derives from studies using NBD-ceramide (Lipsky and Pagano, 1985a,b; Pagano and Sleight, 1985; Pagano, 1988) or from studying sphingolipid transport in mitotic HeLa cells (Collins and Warren, 1992).

The following step, formation of lactosylceramide (LacCer), the common precursor of all glycosphingolipid families was also reported to occur on the cytosolic side of the Golgi apparatus (Trinchera *et al.*, 1991). However, the finding that mutant CHO cells, with an intact enzyme (galactosyltransferase I) but lacking the translocator for UDP-Gal into the Golgi lumen, have extremely reduced levels of LacCer (Deutscher and Hirschberg, 1986), strongly suggests a luminal topology for LacCer-synthase.

B. From Lactosylceramide to Complex Gangliosides[1]

Transfer of subsequent sugars to LacCer to yield various trihexosylceramides (e.g., ganglioside G_{M3}), thus defining the core structure and diversity of the respective GSL series (ganglio-, globo-, lacto-, and mucostructures), presumably occurs on the luminal side of the Golgi vesicles.

[1] GlcCer, glucosylceramide is Glcβ1→1Cer, and LacCer, lactosylceramide is Galβ1→4Glcβ1→1Cer. The gangliosides mentioned in this section are as follows: G_{M3}, NeuAcα2→3Galβ1→4Glcβ1→1Cer; G_{D3}, NeuAcα2→8NeuAcα2→3Galβ1→4Glcβ1→1Cer; G_{T3}, NeuAcα2→8NeuAcα2→8NeuAcα2→3Galβ1→4Glcβ1→1Cer; G_{A2}, GalNAcβ1→4Galβ1→4Glcβ1→1Cer; G_{M2}, GalNAcβ1→4(NeuAcα2→3)Galβ1→4Glcβ1→1Cer; G_{D2}, GalNAc β1→4(NeuAcα2→8NeuAcα2→3)Galβ1→4Glcβ1→1Cer; G_{T2}, GalNAcβ1→4(NeuAcα2→8NeuAcα2→8NeuAcα2→3)Galβ1→4Glcβ1→1Cer; G_{A1} Galβ1→3GalNAcβ1→ 4Galβ1→ 4Glc β1→1Cer; G_{M1}, Galβ1→3GalNAcβ1→4(NeuAcα2→3)Galβ1→4Glcβ1→1Cer; G_{D1b}, Galβ1→3GAlNAcβ1→4(NeuAcα2→8NeuAcα2→3)Galβ1→4Glcβ1→1Cer; G_{T1c}, Galβ1→3GalNAcβ1→4(NeuAcα2→8NeuAcα2→8NeuAcα2→3)Galβ1→4Glcβ1→1Cer; G_{M1b}, NeuAcα2→3Galβ1→3GalNAc β1→4Galβ1→4Galβ14Glcβ1→1Cer; G_{D1a}, NeuAcα2→3Galβ1→3GalNAcβ1→4(NeuAcα2→3)

A subsequent transfer of LacCer from the cytosolic to the luminal side of the Golgi membranes is therefore required, but has not been experimentally proven so far.

The sequential addition of monosaccharide or sialic acid residues to the growing oligosaccharide chain, yielding G_{M3} and more complex gangliosides, is catalyzed by membrane-bound glycosyltransferases, which have been shown to be restricted to the luminal surface of the Golgi apparatus (Carey and Hirschberg, 1981; Creek and Morré, 1981; Fleischer, 1981; Yusuf *et al.*, 1983a,b; Trinchera *et al.*, 1991).

Recent studies demonstrated that transfer of the same sugar residue to analogous glycolopid acceptors, differing only in the number of neuraminic acid residues bound to the inner galactose of the oligosaccharide chain, is in several instances catalyzed by one and the same glycosyltransferase in rat liver Golgi (Fig. 4).

It has been shown that only one 2-acetamido-2-deoxy-D-galactopyranoside (GalNAc)-transferase catalyzes the reaction from LacCer, G_{M3}, G_{D3}, and G_{T3} to G_{A2}, G_{M2}, G_{D2}, and G_{T2}, respectively (Pohlentz *et al.*, 1988; Iber *et al.*, 1992b) and only one single galactosyltransferase is responsible for the formation of G_{A1}, G_{M1a}, and G_{D1b} from G_{A2}, G_{M2}, and G_{D2}, respectively (Iber *et al.*, 1989).

Likewise, one single sialyltransferase IV converts G_{A1}, $G_{M1a,}$ and G_{D1b} to $G_{M1b,}$ G_{D1a}, and G_{T1b}, respectively (Pohlentz *et al.*, 1988) and, again, one and the same sialyltransferase V is responsible for the reaction G_{M1b}, G_{D1a}, G_{T1b}, and G_{Q1c} to G_{D1c}, G_{T1a}, G_{Q1b}, and G_{P1c}, respectively (Iber and Sandhoff, 1989; Iber *et al.*, 1992b).

Moreover, kinetic studies showed that the glycosyltransferases, which catalyze the first steps of the ganglioside biosynthesis, are more substrate specific than enzymes forming the more complex gangliosides. Sialyltransferases I and II are quite specific for their respective substrates, whereas sialyltransferase IV is able to convert G_{M1a} to G_{D1a} as well as LacCer to G_{M3} and possibly also GalCer to G_{M4} (Iber *et al.*, 1991; Iber, 1991), and

---

Galβ1→4Glcβ1→1Cer; G_{T1b}, NeuAcα2→3Galβ1→3GalNAcβ1→4(NeuAcα2→8NeuAcα2→3) Galβ1→4Glcβ1→1Cer; G_{Q1c}, NeuAcα2→3Galβ1→3GalNAcβ1→4(NeuAcα2→8NeuAcα2→8-NeuAcα2→3)Galβ1→4Glcβ1→1Cer; G_{D1c}, NeuAcα2→8NeuAcα2→3Galβ1→3GalNAcβ1→4-Galβ1→4Glcβ1→1Cer; G_{T1a}, NeuAcα2→8NeuAcα2→3Galβ1→3GalNAcβ1→4(NeuAcα2→3) Galβ1→4Glcβ1→1Cer; G_{Q1b}, NeuAcα2→8NeuAcα2→3Galβ1→3GalNAcβ1→4(NeuAcα2→8NeuAcα2→3)Galβ1→4Glcβ1→1Cer; G_{P1c}, NeuAcα2→8NeuAcα2→3Galβ1→3GalNAcβ1→4(NeuAcα2→8NeuAcα2→8NeuAcα2→3)Galβ1→4Glcβ1→1Cer. Gal, D-Galactose; Glc, glucose; GalNAc, 2-acetamido-2-deoxy-D-galactopyranoside; Cer, ceramide; NeuAc, *N*-acetylneuraminic acid.NBD-C_6-Cer, *N*-[*N*-(7-Nitro-2,1,3-benzoxadiazol-4-yl)-*E*-aminohexanoyl]-D-*erythro*-sphingosine.

sialytransferase V is able to convert G_{D1a} to G_{T1a} as well as G_{D3} to G_{T3} and possibly G_{T3} to G_{Q3} (Iber *et al.*, 1992b; Iber, 1991).

The observation that sialyltransferase V is able to convert G_{D3} to G_{T3} *in vitro* could enable mammalian cells to synthesize c-series gangliosides (see Fig. 4) using their normal enzyme equipment. Quite recently, ganglioside G_{P1c} was indeed detected in low quantitites, additionally to the main a- and b-series gangliosides, in human brain tissue (Miller-Prodraza *et al.*, 1991).

It is however, not yet clear where exactly in the Golgi stack the individual glycosylation reactions take place. If the respective steps of ganglioside biosynthesis are localized in different Golgi compartments and GSL biosynthesis is coupled with the vesicle-bound membrane flow of the growing molecules through different Golgi compartments, inhibitors of exocytotic membrane flow should attenuate formation of complex gangliosides.

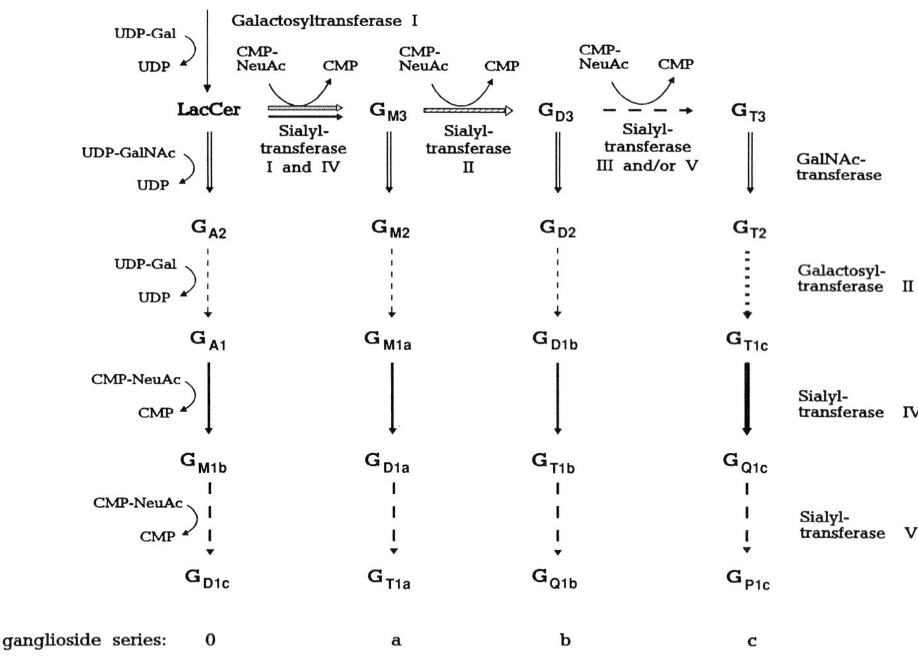

FIG. 4. General scheme for ganglioside biosynthesis (Iber *et al.*, 1992b). All the steps are catalyzed by glycosyltransferases of Golgi membranes. The terminology used for gangliosides is that recommended by Svennerholm (1963). Gangliosides belonging to different series differ in the number of sialic acid residues at the inner galactose: O-series, no sialic acid residue at the inner Gal; a-series, one sialic acid residue at the inner Gal; b-series, two sialic acid residues at the inner Gal, c-series, three sialic acid residues at the inner Gal.

Whereas biosynthetic labeling of intermediates before the respective block should be increased, labeling of complex GSL beyond the drug-induced transport block should be decreased.

One of the first drugs used was the cationic ionophore monensin, which was tested on neurotumor cells (Miller-Prodraza and Fishman, 1984), fibroblasts (Saito *et al.*, 1984), and primary cultured neurons (Hogan *et al.*, 1988; van Echten and Sandhoff, 1989). Monensin primarily impedes vesicular membrane flow between the proximal and the distal Golgi cisternae (for review, see Tartakoff, 1983). In the presence of this drug, incorporation of [^{14}C]galactose into glucosylceramide (GlcCer), LacCer, G_{M3}, G_{D3}, and G_{M2} was significantly increased, while labeling of more complex gangliosides, like G_{M1a}, G_{D1a}, G_{D1b}, G_{T1b}, and G_{Q1b}, decreased remarkably (van Echten and Sandhoff, 1989).

An impressive example of uncoupling ganglioside and sphingomyelin biosynthesis was observed in the presence of the antibiotic brefeldin A (BFA) (Fig. 5), which causes vesiculation of the Golgi and allows the fusion of cis, medial, and, to some extent, also trans elements of Golgi with endoplasmic reticulum (ER) (Fujiwara *et al.*, 1988; Doms *et al.*, 1989; Lippincott-Schwartz *et al.*, 1989, 1990).

Under the influence of this drug, the biosynthetic labeling of primary cultured neurons with different precursors of GSL, such as [^{14}C]galactose, [^{14}C]serine, [^{3}H]sphingosine, and [^{3}H]palmitic acid, resulted in a dramatic reduction of label in gangliosides G_{M1a}, G_{D1a}, G_{D1b}, G_{T1b}, G_{Q1b}, and, to a smaller extent of sphingomyelin (SM), whereas labeling of GlcCer, LacCer, G_{M3}, and G_{D3} increased drastically (van Echten *et al.*, 1990a). Similar results were obtained in CHO cells (Young *et al.*, 1990).

Opposite results, namely, an increase of SM biosynthesis after BFA treatment of CHO cells, was recently reported (Brüning *et al.*, 1992). To clarify this contradiction, we have pretreated in parallel experiments primary cultured cerebellar neurons and fibroblasts with BFA (1 μg/ml) and then fed the cells with NBD-C_6-Cer. For cultured neurons, we reproduced exactly our previously published results (van Echten *et al.*, 1990a), while for fibroblasts we obtained an almost 2-fold increase of NBD-C_6-SM synthesis after BFA treatment (Stotz, 1990). Interestingly, this increase of SM biosynthesis in fibroblasts in the presence of BFA was not observed (on the contrary, there was a 40% reduction of SM formation) when tritiated sphingosine instead of NBD-C_6-Cer was administered as a precursor to the cells (Manheller, 1990). Thus, caution is indicated in the interpretation of data obtained in different cell systems and with different sphingolipid precursors.

Taken together, our studies with BFA suggest that G_{M3} and G_{D3} are synthesized in the early Golgi compartment, whereas complex ganglio-

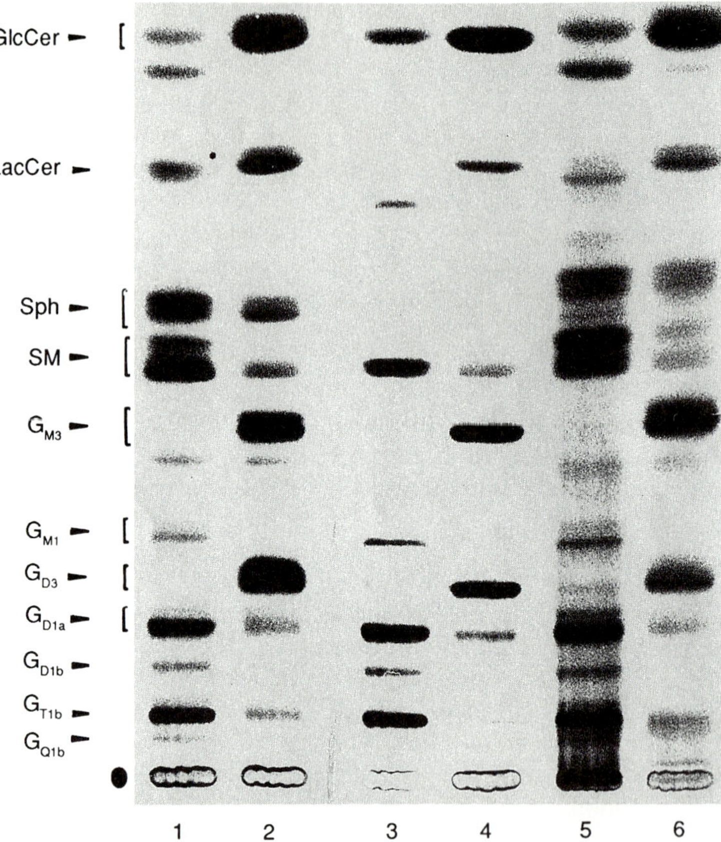

FIG. 5. The effect of brefeldin A (BFA) on the incorporation of radioactive precursors into the sphingolipids of cerebellar cells (van Echten *et al.*, 1990a). Cells were incubated for 3 hours in medium containing no addition (lanes 1, 3, 5), or 1 μg/ml BFA (lanes 2, 4, 6). The radioactive precursors were then added to the culture medium: [^{3}H]palmitic acid (lanes 1, 2), [^{3}H]sphingosine (lanes 3, 4); [^{14}C]serine (lanes 5, 6); after 21 hours, cells were harvested. Sphingolipids were extracted, purified, separated by TLC, and detected by fluorography. The mobilities of standard sphingolipids applied to the chromatogram are shown.

sides such as G_{M1a}, G_{D1a}, G_{D1b}, G_{T1b}, and G_{Q1b} are formed in a late Golgi compartment beyond the BFA-induced block. Attempts to subfractionate Golgi from rat liver (Trinchera and Ghidoni, 1989) and from primary cultured neurons (Iber *et al.*, 1992a) support this suggestion.

Iber *et al.* (1992a) suggested moreover that the different glycosyltransferase activities are not exclusively localized either in the cis- or in the trans-Golgi/TGN compartment; it seems more likely that the activity of sialyltransferase I continuously decreases, whereas that of sialyltransferases IV and V continuously increases from the early to the late Golgi compartments.

It is noteworthy that Matyas and Morré (1987) reported low levels of ganglioside biosynthetic enzymes in highly purified ER fractions of rat liver. This finding is not unexpected, since the ER is the organelle of enzyme biogenesis. Moreover, these authors propose that gangliosides synthesized in the Golgi apparatus may be transported not only to the plasma membrane, but also to the ER and to other internal endomembranes.

III. Endocytosis of Gangliosides

A. Direct Glycosylation of Plasma Membrane-Derived Gangliosides

Studies with electron spin resonance spectroscopy (Schwarzmann *et al.,* 1983, 1986) showed that exogenously labeled gangliosides can insert into the plasma membrane of cultured cells and mix with the pool of endogenous glycosphingolipids.

There is some experimental evidence that gangliosides are slowly internalized via endocytosis. Most of them reach the lysosomal compartment, where they are degraded to their components (monosaccharides, fatty acids, and sphingoids), which can afterward leave the lysosomes and be reused for lipid synthesis (for review, see Schwarzmann and Sandhoff, 1990).

Metabolic studies with radiolabeled gangliosides like G_{M2} and G_{M1a} and their respective amides, as well as electron microscopy studies employing biotinylated G_{M1a} (Schwarzmann *et al.,* 1987), showed that at least a few ganglioside molecules escape transport into lysosomes and are diverted to the Golgi complex, where they are directly glycosylated to more complex GSL. However, direct glycosylation without prior degradation was never clearly demonstrated for the plasma membrane-derived ganglioside precursors like GlcCer, LacCer, and G_{M3} (for review, see Schwarzmann and Sandhoff, 1990). Possibly, these GSL reach the Golgi too, but at a side (trans-Golgi/TGN) distal to the relevant glycosyltransferases, which are located in cis or medial Golgi (see above) and therefore cannot interact with these substrates.

B. Topology and Mechanism of Lysosomal Glycolipid Degradation; Role of Sphingolipid-Binding Proteins

Components and fragments of the plasma membrane (PM) reach the lysosomal compartment mainly by an endocytic membrane flow through the early and late endocytic reticulum (Griffiths *et al.*, 1988). During this vesicular membrane flow, molecules are subjected to a sorting process which directs some of the molecules to the lysosomal compartment and others to the Golgi or even back to the PM (Koval and Pagano, 1989, 1990; Wessling-Resnick and Braell, 1990; Kok *et al.*, 1991). It remains, however, an open question whether components of the PM will be included as components of the lysosomal membrane after successive steps of vesicle budding and fusion along the endocytotic pathway. We think that it is quite unlikely that the components of the lysosomal membrane originating from the PM should be more or less selectively degraded by the lysosomal enzymes.

Alternatively, the observation of multivesicular bodies at the level of the early and late endosomal reticulum (Kok *et al.*, 1991; Hopkins *et al.*, 1990; McKanna *et al.*, 1979) suggests that parts of the endosomal membranes—possibly those enriched in components derived from the PM—bud off into the endosomal lumen and thus form intraendosomal vesicles. These vesicles enriched in PM components could be delivered by successive processes of membrane fission and fusion along the endocytic pathways directly into the lysosol for final degradation of their components (Fig. 6). Thus, glycoconjugates originating from the outer leaflet of the PM would enter the lysosol on the outer leaflet of endocytic vesicles, facing the digestive juice of the lysosomes. This hypothesis is supported by the accumulation of multivesicular storage bodies in Kupffer cells of patients with a complete deficiency of the *sap* (sphingolipid activator protein) precursor protein with a combined activator protein deficiency (Harzer *et al.*, 1989; Schnabel *et al.*, 1992) and by the observation that the epidermal growth factor receptor derived from the plasma membrane and internalized into lysosomes of hepatocytes is not integrated into the lysosomal membrane (Renfrew and Hubbard, 1991).

Degradation of GSL occurs by stepwise action of specific acid hydrolases. Several of these enzymes need the assistance of small glycoprotein cofactors, the so-called sphingolipid activator proteins (SAPs) (Fürst and Sandhoff, 1992) to attack their lipid substrates.

Since the discovery of sulfatide activator protein (Mehl and Jatzkewitz, 1964), several other factors have been described, but their identity, specificity, and function often remained unclear. When sequence data became available, it turned out that only two genes code for the five SAPs known

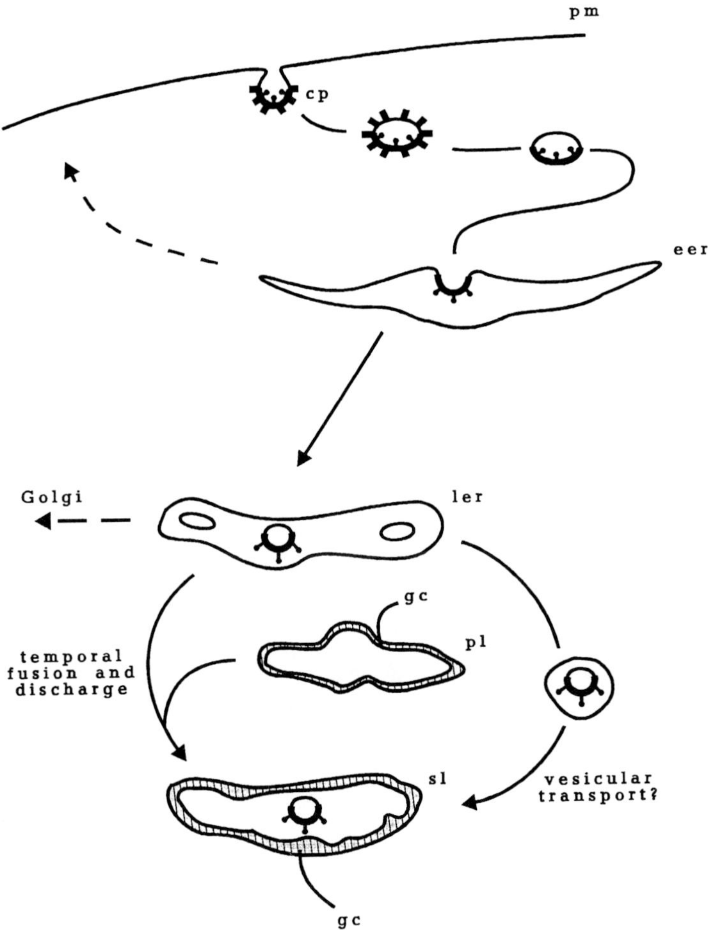

FIG. 6. A new model for the topology of endocytosis and lysosomal digestion of GSL derived from the plasma membrane (Fürst and Sandhoff, 1992). During endocytosis, glycolipids of the plasma membrane (pm) are supposed to end up in intraendosomal vesicles (multivesicular bodies), from where they are discharged into the lysosomal compartment. cp, Coated pit; eer, early endosomal reticulum; gc, glycocalyx; ler, late endosomal reticulum; pl, primary lysosome; sl, secondary lysosome; ⫯, glycolipid; —>, proposed pathway of endocytosis of GSL derived from the pm into the lysosomal compartment; --->, other intracellular routes for GSL derived from the pm.

so far (Fürst and Sandhoff, 1992). One gene carries the genetic information for the G_{M2}-activator and the second for the *sap*-precursor which is processed to four homologous proteins, including sulfatide activator protein (*sap*-B) and glucosylceramidase activator protein (*sap*-C).

Several experimental data (Meier *et al.*, 1991) suggest the mechanism of action of G_{M2}-activator. Hexosaminidase A is a water-soluble enzyme which acts on substrates of the membrane surface only if they extend far enough into the aqueous phase (Fig. 7). Like a razor blade or a lawnmower, the enzyme recognizes and cleaves all substrates (e.g., G_{D1a}-GalNAc) which stick out far enough into the aqueous space. However, those GSL substrates with oligosaccharide headgroups too short to be reached by the water-soluble enzyme cannot be degraded. Their degrada-

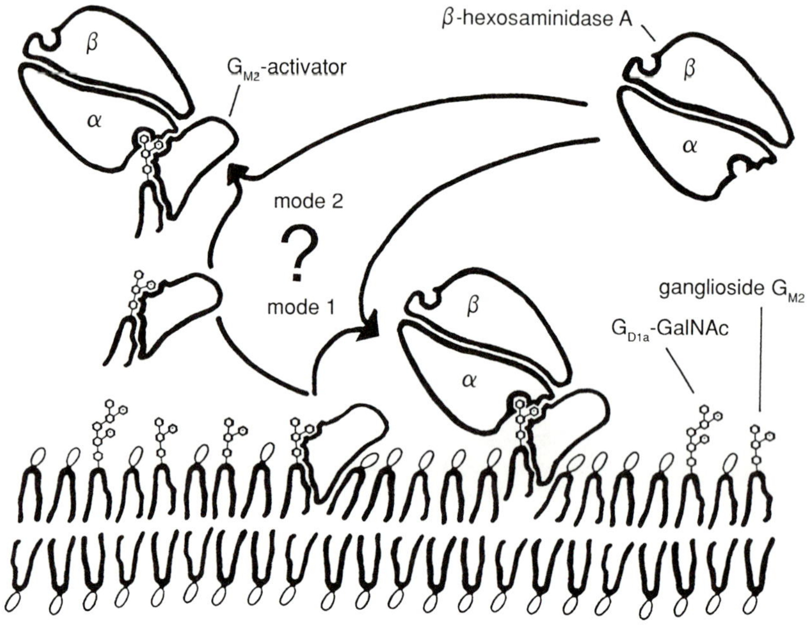

FIG. 7. Model for the G_{M2}-activator-stimulated degradation of ganglioside G_{M2} by human hexosaminidase A (Fürst and Sandhoff, 1992). Water-soluble hexosaminidase A does not degrade membrane-bound G_{M2} in the absence of G_{M2}-activator or appropriate detergents, but it degrades analogs of G_{M2} which contain a short acyl residue or no acyl residue (lysoganglioside G_{M2}). They are less firmly bound to the lipid bilayer and more water-soluble than G_{M2}. Ganglioside G_{M2} bound to a lipid bilayer, e.g., of an intralysosomal vesicle (see Fig. 6), is hydrolyzed in the presence of the G_{M2}-activator. The G_{M2}-activator binds one G_{M2} molecule and lifts it a few angstroms out of the membrane. This activator/lipid complex can be reached and recognized by water-soluble hexosaminidase A, which cleaves the substrate. However, it is also possible that the activator/lipid complex leaves the membrane and the enzymatic reaction takes place in free solution. The terminal GalNAc residue of membrane-incorporated G_{D1a}-GalNAc protrudes from the membrane far enough to be accessible to hexosaminidase A without an activator.

tion requires a second component, the G_{M2}-activator, a specialized GSL-binding protein which complexes the substrate (e.g., G_{M2}), lifts it from the membrane, and presents it to the hexosaminidase A for degradation. While G_{M2}-activator and hexosaminidase A represent a selective and precisely tuned machinery for the degradation of only a few structurally similar sphingolipids, *sap*-B stimulates the degradation of many lipids by several enzymes of human, plant, and even bacterial origin (Li *et al.*, 1988). Thus, *sap*-B seems to act as a kind of physiological detergent with broad specificity, and it solubilizes glycolipid substrates (Vogel *et al.*, 1991).

Unlike G_{M2}-activator and *sap*-B, *sap*-C is reported to form complexes with membrane-associated enzymes and apparently activates them (Berent and Radin, 1981; Ho and Light, 1973; Ho, 1975; Ho and Rigby, 1975).

GSL with short hydrophilic headgroups are degraded by a two-component system: an enzyme and an activator protein. This seems to be much safer for the stability of short-chain GSL outside the lysosomal compartment, e.g., on the cell surface. There, GSL are protected against premature degradation by lysosomal exohydrolases escaping from the cells into the extracellular fluid by low concentrations of both proteins involved in each degradation step and by an unfavorably high pH value.

C. Pathobiochemistry of Inherited Enzyme and Activator Protein Deficiencies

As already mentioned, final degradation of sphingolipids occurs in the lysosome. Here, they are degraded in a stepwise manner, starting at the hydrophilic end of the molecules (Fig. 8). As indicated in Fig. 8, almost every one of the degrading hyrolases can be deficient in a human lipid storage disease. The inherited deficiency of one of these ubiquitously occurring enzymes causes the lysosomal storage of its substrates. The diseases resulting from these defects are rather heterogeneous from the biochemical as well as from the clinical point of view (for review, see Moser *et al.*, 1989; Barranger and Ginns, 1989; O'Brien, 1989; Sandhoff *et al.*, 1989).

The analysis of sphingolipid storage diseases without detectable hydrolase deficiency resulted in the identification of several point mutations in the G_{M2}-activator gene (reviewed by Fürst and Sandhoff, 1992) and in the *sap*-precursor gene (Fig. 9). Interestingly enough, mutations affecting the *sap*-C domain (Gaucher factor) resulted in a variant form of Gaucher's disease, mutations affecting the *sap*-B domain (sulfatide activator) resulted in variant forms of metachromatic leukodystrophy, whereas a mutation in the start codon ATG of the *sap*-precursor resulted in a defect of several

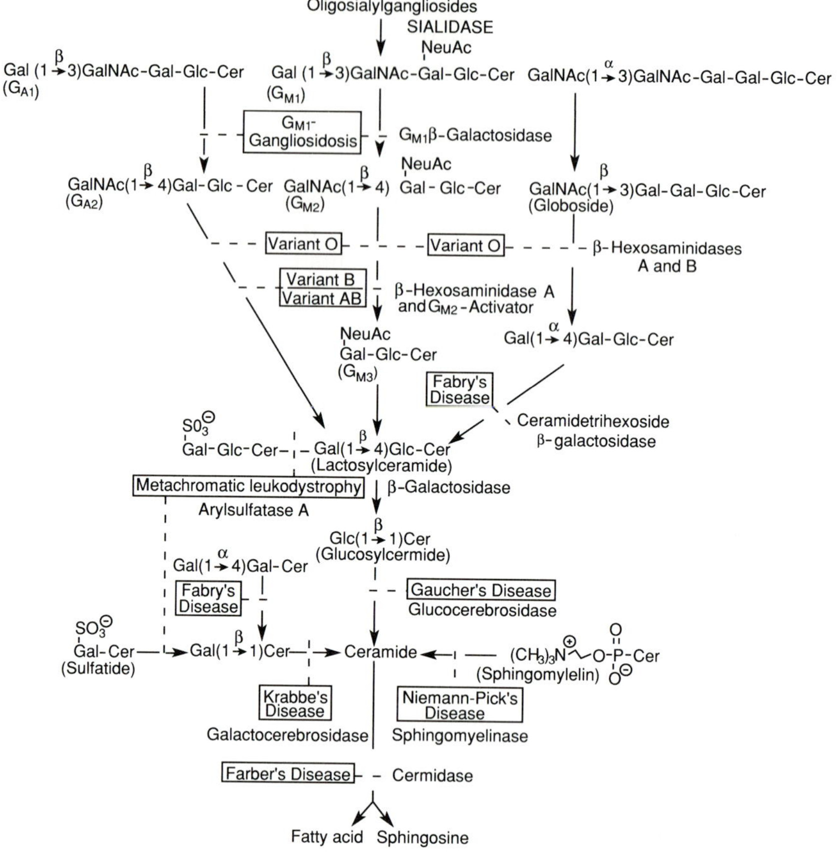

FIG. 8. Degradation scheme of sphingolipids denoting metabolic blocks of known diseases (Sandhoff and Christomanou, 1979). Variant B of infantile G_{M2}-gangliosidosis, Tay-Sachs disease; variant O of G_{M2}-gangliosidosis, hexosaminidase β-subunit deficiency, Sandhoff disease; variant AB of infantile G_{M2}-gangliosidosis.

saps and in a simultaneous storage of ceramide, glucosylceramide, lactosylceramide, and G_{M3} in the patient's tissue.

IV. Regulation of Ganglioside Biosynthesis

The formation of cell- and species-specific GSL patterns on cell surface is most likely controlled primarily at the transcriptional level of the respective glycosyltransferases (Hashimoto *et al.*, 1983; Nakakuma *et al.*, 1984; Nagai *et al.*, 1986).

It was shown that terminal glycosylation of glycoproteins and glycosphingolipids differs during development, differentiation, and oncogenic transformation (Hakomori, 1984b; Feizi, 1985; Rademacher *et al.*, 1988). In transformed cells it could be demonstrated that changes in glycosylation of glycoconjugates correspond to the respective changes in the expression of the relevant glycosyltransferases (Coleman *et al.*, 1975; Nakaishi *et al.*, 1988a,b; Matsuura *et al.*, 1989; Ruan and Lloyd, 1992).

Several studies have reported a successful purification of some glycolipid glycosyltransferases (Honke *et al.*, 1988; Sweeley *et al.*, 1988; Basu *et al.*, 1990; Gu *et al.*, 1990; Yu *et al.*, 1990; Melkerson-Watson and Sweeley, 1991), but further studies are necessary to obtain structural data of these glycosyltransferases and their respective genes to provide better understanding of the regulation of ganglioside biosynthesis at the transcriptional level. A pioneering step in this direction is the expression and

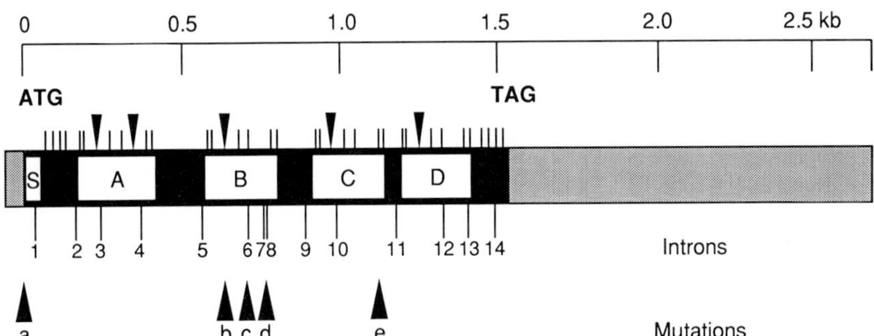

FIG. 9. Structure of the *sap*-precursor cDNA (Fürst and Sandhoff, 1992). The cDNA of *sap*-precursor codes for a sequence of 524 amino acids (or of 527 amino acids; see Holtschmidt *et al.*, 1991), including a signal peptide of 16 amino acids (termed s, for the entry into the ER) (Nakano *et al.*, 1989; Fürst and Sandhoff, 1992). The four domains on the precursor, termed saposins A–D by O'Brien *et al.* (1988), correspond to the mature proteins found in human tissues: A, *sap*-A or saposin A; B, *sap*-B or saposin B or SAP-1 or sulfatide activator; C, *sap*-C or SAP-2 or saposin C or glucosylceramidase activator protein, and D, *sap*-D or saposin D or component C. The positions of cysteine residues are marked by vertical bars and those of the *N*-glycosylation sites by arrowheads. The positions of the 14 introns and of the known mutations leading to diseases are also given: (a) A-1 → T (Met-1 → Leu) (Schnabel *et al.*, 1992), (b) C-650 → T (Thr-217 → Ile) (Rafi *et al.*, 1990; Kretz *et al.*, 1990), (c) 33-bp insertion after G-777 (11 additional amino acids after Met-259) (Zhang *et al.*, 1990, 1991), (d) G-722 → C (Cys-241 → Ser) (Holtschmidt *et al.*, 1991), (e) G-1154 → T (Cys-385 → Phe) (Schnabel *et al.*, 1991).

cloning of cDNAs of GalNAc-transferases, recently published by Nagata *et al.* (1992).

In addition to the genetic level, some evidence for an epigenetic regulation of GSL biosynthesis have been given. Recent studies provide strong evidence that sphingosine, the long-chain base backbone of all GSL, seems to regulate the first step of sphingolipid biosynthesis, formation of 3-dehydrosphinganine catalyzed by serine palmitoyltransferase (SPT) (Mandon *et al.*, 1991; van Echten *et al.*, 1990b). The decrease of SPT activity observed in cultured neurons after feeding sphingosine or azidosphingosine is not caused by a direct feedback inhibition; rather, it is a downregulation caused by an unknown pathway involving a sphingosine-binding protein, either triggering a modification of SPT or repressing the transcription of its mRNA or inhibiting its translation. The key regulatory role of SPT was also suggested in studies on the regulation of epidermal sphingolipid synthesis by permeability barrier (Holleran *et al.*, 1991). Holleran *et al.* demonstrated that, 5–7 hours after epidermal barrier disruption with acetone, incorporation of ^3H_2O into sphingolipids as well as SPT activity increased specifically.

Feedback control of GSL biosynthesis was suggested only by *in vitro* studies. Thus, using rat liver Golgi it was shown that G_{M2}-synthase was most strongly inhibited by G_{D1a}, while G_{D3}-synthase was inhibited most by G_{Q1b}, indicating that the regulating steps in the synthesis of a- and b-series gangliosides are preferentially inhibited by their respective products (Yusuf *et al.*, 1987; Nores and Caputto, 1984). Similar effects were reported by Shukla *et al.* (1991), who observed a feedback control of GlcCer-synthase by complex gangliosides in brain microsomes. Another enzyme in the GSL series, G_{M3}-synthase of rat liver Golgi, was also found to be inhibited by gangliosides, though at very high concentrations (Richardson *et al.*, 1977). An autoregulation of G_{D3}-synthase by G_{D3} was recently reported (Iber *et al.*, 1992b). Feedback control of GSL biosynthesis, however, has not yet been observed *in vivo* or in cell culture.

Our hypothesis is that the sequence LacCer $\rightarrow$ GM_3 $\rightarrow$ GD_3 $\rightarrow$ GT_3 (Fig. 4) might be essential for the regulation of ganglioside biosynthesis. According to this model, regulation of biosynthesis of the main ganglioside series in mammals, a and b, is linked with the control of GalNAc-transferase and sialyltransferase II, converting GM_3 either into GM_2 (a-series) or into GD_3 (b-series), respectively.

It has been shown that a reversible shift from a- to b-series ganglioside biosynthesis resulted after lowering the pH from 7.4 to 6.2 in the culture medium of murine cerebellar cells (Iber *et al.*, 1990). This effect can be explained by the different pH profiles of the two key regulatory glycosyltransferases involved. At pH 6.2, sialyltransferase II, a key enzyme in the biosynthesis of b-series, is more active than GalNAc-transferase, the first enzyme in the synthesis of a-series gangliosides, whereas at pH 7.4 the reverse is true. The physiological relevance of this phenomenon is unclear. In this context it is, however, interesting that undifferentiated cells (Seyfried *et al.,* 1984) and transformed cells are rich in b-series gangliosides GD_3 and GD_2 (Cheresh *et al.*, 1986).

Moreover, GD_3-synthase (sialyltransferase II) activity is maintained by different hormones and epidermal growth factor or activated by steroid sex hormones in isolated or cultured rat hepatocytes, respectively, whereas that of GM_3-synthase (sialyltransferase I) is not (Mesarič and Decker, 1990a,b).

V. Concluding Remarks

Ever since the gangliosides were discovered more than 50 years ago as tissue constituents (Klenk, 1942), much research has been done on these substances, especially during the past two decades. We now understand, as described in this article, many aspects of their metabolism, synthesis in the Golgi complex, transport from there to the plasma membrane, and their final migration to lysosomes for being degraded. However, the details of these processes have still to be spelt out clearly. We still do not know for certain the exact topology of the transferase enzymes in Golgi cisternae. The exact mechanism of precursor flow from ER to Golgi, the flow of the growing gangliosides through to the successive cisternae of the Golgi complex, and the transport of the finished products to plasma membrane are still far from clear. Whatever may be the topology in normal cells, it now seems possible that the topology is changed in transformed cells (C. Zacharias, H. K. M. Yusuf, G. van Echten, and K. Sandhoff, unpublished observation).

While the regulation of biosynthesis of most of the cellular constituents is now unraveled, the exact mechanism that regulates ganglioside biosynthesis in different cells and in cells at different periods of development remains a fertile field for further research. Finally, much circumstantial evidence has been produced from different laboratories to indicate various cellular functions of the gangliosides. The exact physiological roles which these mysterious molecules really play can only be pinpointed by future research.

Acknowledgments

We thank Prof. Harun Yusuf (Dhaka, Bangladesh) for reading the manuscript. The work of the authors quoted in this paper was supported by funds from the Deutsche Forschungsgemeinschaft (SFB 284), Fonds der Chemischen Industrie, and by Fidia Pharmaforschung GmbH (München, Germany).

References

Abe, A., Yamada, K., Sakagami, T., and Sasaki, T. (1984). *Biochim. Biophys. Acta* **778,** 239–244.

Barranger, Y. A., and Ginns, E. J. (1989). *In* "The Metabolic Basis of Inherited Disease" (C. R. Scriver, A. L. Beaudet, W. S. Sly, and D. Valle, eds.), 6th ed., Vol. 2, pp. 1677–1698. McGraw-Hill, New York.

Basu, S., Ghosh, S., Basu, M., Weng, S., Das, K. K., Hawes, J. W., Li, Z., and Zhang, B. (1990). *FASEB J.* **4**(7), A2140.

Berent, S. L., and Radin, N. S. (1981). *Biochim. Biophys. Acta* **664,** 572–582.

Bock, K., Breimer, M. E., Brignole, A., Hansson, C. G., Karlsson, K. A., Larson, G., Leffler, H., Samuelsson, B. E., Strömberg, N., Svanborg Edén, C., and Thurin, J. (1985). *J. Biol. Chem.* **260,** 8545–8551.

Braun, P. E., and Snell, E. E. (1968). *J. Biol. Chem.* **243,** 3775–3783.

Bremer, E. G., Schlessinger, J., and Hakomori, S.-I. (1986). *J. Biol. Chem.* **261,** 2434–2440.

Brüning, A., Karrenbauer, A., Schnabel, E., and Wieland, F. T. (1992). *J. Biol. Chem.* **267,** 5052–5055.

Carey, D. J., and Hirschberg, C. W. (1981). *J. Biol. Chem.* **256,** 989–993.

Cheresh, D. A., Pierschbacher, M. D., Herzig, M. A., and Mujoo, K. (1986). *J. Cell Biol.* **102,** 688–696.

Coleman, P. L., Fishman, P. H., Brady, R. O., and Todaro, G. J. (1975). *J. Biol. Chem.* **250,** 55–60.

Collins, R. N., and Warren, G. (1992). *J. Biol. Chem.* **267,** 24906–24911.

Coste, H., Martel, M.-B., Azzar, G., and Got, R. (1985). *Biochim. Biophys. Acta* **814,** 1–7.

Coste, H., Martel, M.-B., and Got, R. (1986). *Biochim. Biophys. Acta* **858,** 6–12.

Creek, K. E., and Morrré, D. J. (1981). *Biochim. Biophys. Acta* **643,** 292–305.

Deutscher, S. L., and Hirschberg, C. B. (1986). *J. Biol. Chem.* **261,** 96–100.

Doms, R. W., Russ, G., and Yewdell ,J. W. (1989). *J. Cell Biol.* **109,** 61–72.

Feizi, T. (1985). *Nature* (*London*) **314,** 53–57.

Fenderson, B. A., Zehavi, U., and Hakomori, S.-I. (1985). *J. Exp. Med.* **160,** 1591–1596.

Fishman, P. H., and Brady, R. O. (1976). *Science* **194,** 906–915.

Fleischer, B. (1981). *J. Cell Biol.* **89,** 246–255.

Fujiwara, T., Oda, K., Yokota, S., Takatsuki, A., and Ikehara, Y. (1988). *J. Biol. Chem.* **263,** 18545–18552.

Fürst, W., and Sandhoff, K. (1992). *Biochim. Biophys. Acta* **1126,** 1–16.

Futerman, A. H., and Pagano, R. E. (1991). *Biochem. J.* **280,** 295–302.

Gahmberg, C. G., and Hakomori, S.-I. (1973). *J. Biol. Chem.* **248,** 4311–4317.

Griffiths, G., Hoflack, B., Simons, K., Mellman, I., and Kornfeld, S. (1988). *Cell* (*Cambridge, Mass.*) **52,** 329–341.

Gu, X.-B., Gu, T.-J., and Yu, R. K. (1990). *Biochem. Biophys. Res. Commun.* **166,** 387–393.

Hakomori, S.-I. (1984a). *Trends Biochem. Sci.* **9,** 453–458.

Hakomori, S.-I. (1984b). *Annu. Rev. Immunol.* **2,** 103–126.

Hanai, N., Nores, G. A., MacLeod, C., Torres-Mendez, C.-R., and Hakomori, S.-I. (1988). *J. Biol. Chem.* **263,** 10915–10921.

Hansson, H. A., Holmgren, J., and Svennerholm, L. (1977). *Proc. Natl. Acad. Sci. U.S.A.* **74,** 3782–3786.

Harzer, K., Paton, B. C., Poulos, A., Kustermann-Kuhn, B., Roggendorf, W., Grisar, T., and Popp, M. (1989). *Eur. J. Pediatr.* **149,** 31–39.

Hashimoto, Y., Suzuki, A., Yamakawa, T., Miyashita, N., and Moriwaki, K. (1983). *J. Biochem.* (*Tokyo*) **94,** 2043–2048.

Ho, M. W. (1975). *FEBS Lett.* **53,** 243–247.
Ho, M. W., and Light, N. D. (1973). *Biochem. J.* **136,** 821–823.
Ho, M. W., and Rigby, M. (1975). *Biochim. Biophys. Acta* **397,** 267–273.
Hogan, M. V., Saito, M., and Rosenberg, A. (1988). *J. Neurosci. Res.* **20,** 390–394.
Holleran, W. M., Feingold, K. R., Mao Qiang, M., Gao, W. N., Lee, J. M., and Elias, P. M. (1991). *J. Lipid Res.* **32,** 1151–1158.
Holtschmidt, H., Sandhoff, K., Kwon, H. Y., Harzer, K., Nakano, T., and Suzuki, K. (1991). *J. Biol. Chem.* **266,** 7556–7560.
Honke, K., Gasa, S., and Makita, A. (1988). *Abstr., Rinsho-Ken Int. Conf., 3rd,*Tokyo, p. 126.
Hopkins, C. R., Gibson, A., Shipman, M., and Miller, K. (1990). *Nature* (*London*) **346,** 335–339.
Iber, H. (1991). Ph.D. Thesis, Institut für Organische Chemie und Biochemie, Universität Bonn, Germany.
Iber, H., and Sandhoff, K. (1989). *FEBS Lett.* **254,** 124–128.
Iber, H., Kaufmann, R., Pohlentz, G., Schwarzmann, G., and Sandhoff, K. (1989). *FEBS Lett.* **248,** 18–22.
Iber, H., van Echten, G., Klein, R. A., and Sandhoff, K. (1990). *Eur. J. Cell Biol.* **52,** 236–240.
Iber, H., van Echten, G., and Sandhoff, K. (1991). *Eur. J. Biochem.* **195,** 115–120.
Iber, H., van Echten, G., and Sandhoff, K. (1992a). *J. Neurochem.* **58,** 1533–1537.
Iber, H., Zacharias, C., and Sandhoff, K. (1992b). *Glycobiology* **2,** 137–142.
Jeckel, D., Karrenbauer, A., Burger, K. N. J., van Meer, G., and Wieland, F. T. (1992). *J. Cell Biol.* **117,** 259–267.
Karlsson, K.-A. (1989). *Annu. Rev. Biochem.* **58,** 309–350.
Klenk, E. (1942). *Hoppe-Seyler's Z. Physiol. Chem.* **273,** 76–86.
Kok, J. W., Babia, T., and Hoekstra, D. (1991). *J. Cell Biol.* **114,** 231–239.
Koval, M., and Pagano, R. E. (1989). *J. Cell Biol.* **108,** 2169–2181.
Koval, M., and Pagano, R. E. (1990). *J. Cell Biol.* **111,** 429–442.
Kretz, K. A., Carson, G. S., Morimoto, S., Kishimoto, Y., Fluharty, A. L., and O'Brien, J. S. (1990). *Proc. Natl. Acad. Sci. U.S.A.* **87,** 2541–2544.
Ledeen, R. W. (1985). *Trends Neurosci.* **8,** 169–174.
Ledeen, R. W., and Yu, R. K. (1982). *In* "Methods in Enzymology" (V. Ginsberg, ed.), Vol. 83, pp. 139–189. Academic Press, New York.
Li, S.-C., Sonnino, S., Tettamanti, G., and Li, Y.-T. (1988). *J. Biol. Chem.* **263,** 6588–6591.
Lippincott-Schwartz, J., Yuan, L. C., Bonifacino, J. S., and Klausner, R. D. (1989). *Cell* (*Cambridge, Mass.*) **56,** 801–813.
Lippincott-Schwartz, J., Donaldson, J. G., Schweizer, A., Berger, E. G., Hauri, H. P., Yuan, L. C., and Klausner, R. D. (1990). *Cell* (*Cambridge, Mass.*) **60,** 821–836.
Lipsky, N. G., and Pagano, R. E. (1985a). *J. Cell Biol.* **100,** 27–34.
Lipsky, N. G., and Pagano, R. E. (1985b). *Science* **228,** 745–747.
Mandon, E. C., van Echten, G., Birk, R., Schmidt, R. R., and Sandhoff, K. (1991). *Eur. J. Biochem.* **198,** 667–674.
Mandon, E. C., Ehses, I., Rother, J., van Echten, G., and Sandhoff, K. (1992). *J. Biol. Chem.* **267,** 11144–11148.
Manheller, W. (1990). Ph.D. Thesis, Institut für Organische Chemie und Biochemie, Universität Bonn, Germany.
Markwell, M. A. K., Svennerholm, L., and Paulson, J. C. (1981). *Proc. Natl. Acad. Sci. U.S.A.* **78,** 5406–5410.
Matsuura, H., Greene, T., and Hakomori, S.-I. (1989). *J. Biol. Chem.* **264,** 10472–10476.

Matyas, G. R., and Morré, D. J. (1987). *Biochim. Biophys. Acta* **921,** 599–614.
McKanna, J. A., Haigler, H. T., and Cohen, S. (1979). *Proc. Natl. Acad. Sci. U.S.A.* **76,** 5689–5693.
Mehl, E., and Jatzkewitz, H. (1964). *Hoppe-Seyler's Z. Physiol. Chem.* **339,** 260–276.
Meier, E. M., Schwarzmann, G., Fürst, W., and Sandhoff, K. (1991). *J. Biol. Chem.* **266,** 1879–1887.
Melkerson-Watson, L. J., and Sweeley, C. C. (1991). *J. Biol. Chem.* **266,** 4448–4457.
Merrill, A. H., and Wang, E. (1986). *J. Biol. Chem.* **261,** 3764–3769.
Mesarič, M., and Decker, K. (1990a). *Biochem. Biophys. Res. Commun.* **171,** 132–137.
Mesarič, M., and Decker, K. (1990b). *Biochem. Biophys. Res. Commun.* **171,** 1188–1191.
Miller-Prodraza, H., and Fishman, P. H. (1984). *Biochim. Biophys. Acta* **804,** 44–51.
Miller-Prodraza, H., Mansson, J. E., and Svennerholm, L. (1991). *FEBS Lett.* **288,** 212–214.
Morré, D. J., Kartenbeck, J., and Franke, W. W. (1979). *Biochim. Biophys. Acta* **559,** 71–152.
Moser, W., Moser, A. B., Chen, W. W., and Schram, A. W. (1989). *In* "The Metabolic Basis of Inherited Disease" (C. R. Scriver, A. L. Beaudet, W. S. Sly, and D. Valle, eds), 6th ed., Vol. 2, pp. 1645–1654. McGraw-Hill, New York.
Nagai, Y., Nakaishi, H., and Sanai, Y. (1986). *Chem. Phys. Lipids* **42,** 91–103.
Nagata, Y., Yamashino, S., Yodoi, J., Lloyd, K. O., Shiku, M., and Furukawa, K. (1992). *J. Biol. Chem.* **267,** 12082–12089.
Nakaishi, H., Sanai, Y., Shiroki, K., and Nagai, Y. (1988a). *Biochem. Biophys. Res. Commun.* **150,** 760–765.
Nakaishi, H., Sanai, Y., Shibuya, M., and Nagai, Y. (1988b). *Biochem. Biophys. Res. Commun.* **150,** 766–774.
Nakakuma, H., Sanai, Y., Shiroki, K., and Nagai, Y. (1984). *J. Biochem.* (*Tokyo*) **96,** 1471–1480.
Nakano, T., Sandhoff, K., Stümper, J., Christomanou, H., and Suzuki, K. (1989). *J. Biochem.* (*Tokyo*) **105,** 152–154.
Nores, G. A., and Caputto, R. (1984). *J. Neurochem.* **42,** 1205–1211.
O'Brien, J. S. (1989). *In* "The Metabolic Bases of Inherited Disease" (C. R. Scriver, A. L. Beaudet, W. S. Sly, and D. Valle, eds.), 6th ed., Vol. 2, pp. 1797–1806. McGraw-Hill, New York.
O'Brien, J. S., Kretz, K. A., Dewji, N., Wenger, D. A., Esch, F., and Fluharty, A. L. (1988). *Science* **241,** 1098–1101.
Ong, D. E. and Brady, R. N. (1973). *J. Biol. Chem.* **248,** 3884–3888.
Pagano, R. E. (1988). *Trends Biochem. Sci.* **13,** 202–205.
Pagano, R. E., and Sleight, R. G. (1985). *Science* **229,** 1051–1057.
Phillips, M. L., Nudelman, E., Gaeta, F. C. A., Pevez, M., Singhal, A. K., Hakomori, S.-I., and Paulson, J. C. (1990). *Science* **250,** 1130–1132.
Pohlentz, G., Klein, D., Schwarzmann, G., Schmitz, D., and Sandhoff, K. (1988). *Proc. Natl. Acad. Sci. U.S.A.* **85,** 7044–7048.
Rademacher, T. W., Parekh, R. B., and Dwek, R. A. (1988). *Annu. Rev. Biochem.* **57,** 785–838.
Rafi, M. A., Zhang, X.-L., De Gala, G., and Wenger, D. A. (1990). *Biochem. Biophys. Res. Commun.* **166,** 1017–1023.
Renfrew, C. A., and Hubbard, A. L. (1991). *J. Biol. Chem.* **266,** 21265–21273.
Richardson, C. L., Keenaan, T. W., and Morré, D. J. (1977). *Biochim. Biophys. Acta* **488,** 88–96.
Rother, J., van Echten, G., Schwarzmann, G., and Sandhoff, K. (1992). *Biochem. Biophys. Res. Commun.* **189,** 14–20.

Ruan, A., and Lloyd, K. O. (1992). *Cancer Res.* **52,** 5725–5731.

Saito, M., Saito, M., and Rosenberg, A. (1984). *Biochemistry* **23,** 1043–1046.

Sandhoff, K., and Christomanou, H. (1979). *Hum. Genet.* **50,** 107–143.

Sandhoff, K., Conzelmann, E., Neufeld, E. T., Kaback, M. M., and Suzuki, K. (1989). *In* "The Metabolic Basis of Inherited Disease" (C. R. Scriver, A. L. Beaudet, W. S. Sly, and D. Valle, eds.), 6th ed., Vol. 2, pp. 1807–1839. McGraw-Hill, New York.

Schnabel, D., Schröder, M., and Sandhoff, K. (1991). *FEBS Lett.* **284,** 57–59.

Schnabel, D., Schröder, M., Fürst, W., Klein, A., Hurwitz, R., Zenk, T., Weber, G., Harzer, K., Paton, B., Poulos, A., Suzuki, K., and Sandhoff, K. (1992). *J. Biol. Chem.* **267,** 3312–3315.

Schwarzmann, G., and Sandhoff, K. (1990). *Biochemistry* **29,** 10865–10871.

Schwarzmann, G., Hoffmann-Bleihauer, P., Schubert, J., Sandhoff, K., and Marsh, D. (1983). *Biochemistry* **22,** 5041–5048.

Schwarzmann, G., Hinrichs, U., Sonderfeld, S., Marsh, D., and Sandhoff, K. (1986). *NATO ASI Ser., Ser. A* **116,** 553–562.

Schwarzmann, G., Marsh, D., Herzog, V., and Sandhoff, K. (1987). *NATO ASI Ser., Ser. H* **7,** 217–229.

Seyfried, T. N., Bernard, D. J., and Yu, R. K. (1984). *J. Neurochem.* **43,** 1152–1162.

Shukla, G. S., Shukla, A., and Radin, N. S. (1991). *J. Neurochem.* **56,** 2125–2132.

Stoffel, W., LeKim, D., and Sticht, G. (1968). *Hoppe-Seyler's Z. Physiol. Chem.* **349,** 664–670.

Stoffel, W., and Bister, K. (1974). *Hoppe-Seyler's Z. Physiol. Chem.* **355,** 911–923.

Stotz, H. (1990). Ph.D. Thesis, Institut für Organische Chemie und Biochemie, Universität Bonn, Germany.

Svennerholm, L. (1963). *J. Neurochem.* **10,** 613–623.

Svennerholm, L. (1984). *In* "Cellular and Pathological Aspects of Glycoconjugate Metabolism" (H. Dreyfus, R. Massarelli, L. Freysz, and G. Rebel, eds.), Vol. 126, pp. 21–44. INSERM, France.

Sweeley, C. C., Usuki, S., Hollingsworth, R., and Swamy, M. (1988). *Abstr., Rinsho-Ken Int. Conf., 3rd,* Tokyo, p. 8.

Tartakoff, A. M. (1983). *Cell (Cambridge, Mass.)* **32,** 1026–1028.

Thompson, L. K., Horowitz, P. M., Bently, K. L., Thomas, D. D., Alderete, J. F., and Klebe, R. J. (1986). *J. Biol. Chem.* **261,** 5209–5214.

Thompson, T. E., and Tillack, T. W. (1985). *Annu. Rev. Biophys. Biophys. Chem.* **14,** 361–386.

Tiemeyer, M., Yasuda, Y., and Schnaar, R. L. (1989). *J. Biol. Chem.* **264,** 1671–1681.

Trinchera, M., and Ghidoni, R. (1989). *J. Biol. Chem.* **264,** 15766–15769.

Trinchera, M., Fabbri, M., and Ghidoni, R. (1991). *J. Biol. Chem.* **266,** 20907–20912.

van Echten, G., and Sandhoff, K. (1989). *J. Neurochem.* **52,** 207–214.

van Echten, G., Iber, H., Stotz, H., Takatsuki, A., and Sandhoff, K. (1990a). *Eur. J. Cell Biol.* **51,** 135–139.

van Echten, G., Birk, R., Brenner-Weiss, G., Schmidt, R. R., and Sandhoff, K. (1990b). *J. Biol. Chem.* **265,** 9333–9339.

van Meer, G. (1989). *Annu. Rev. Cell Biol.* **5,** 247–275.

Vogel, A., Schwarzmann, G., and Sandhoff, K. (1991). *Eur. J. Biochem.* **200,** 591–597.

Walton, K. M., Sandberg, K., Rogers, T. B., and Schnaar, R. L. (1988). *J. Biol. Chem.* **263,** 2055–2063.

Walz, G., Aruffo, A., Kolanus, W., Bevilacqua, M., and Seed, B. (1990). *Science* **250,** 1132–1135.

Watanabe, K., Hakomori, S.-I., Powell, M. E., and Jokota, M. (1980). *Biochem. Biophys. Res. Commun.* **92,** 638–646.

Wessling-Resnick, M., and Braell, W. A. (1990). *J. Biol. Chem.* **265,** 690–699.

Wiegandt, H. (1982). *Adv. Neurochem.* **4,** 149–223.

Wiegandt, H. (1985). *New Compr. Biochem.* **10,** 199–260.

Wong, M., Brown, R. E., Barenholz, Y., and Thompson, T. E. (1984). *Biochemistry* **23,** 6498–6505.

Yamada, K., and Sasaki, T. (1982). *J. Biochem. (Tokyo)* **92,** 457–464.

Young, W. W., Jr., Lutz, M. S., Mills, S. E., and Lechler-Osborn, S. (1990). *Proc. Natl. Acad. Sci. U.S.A.* **87,** 6838–6842.

Yu, R. K., Gu, T.-J., and Gu, X.-B. (1990). *FASEB J.* **4**(7), A2140.

Yusuf, H. K. M., Pohlentz, G., Schwarzmann, G., and Sandhoff, K. (1983a). *Eur. J. Biochem.* **134,** 47–54.

Yusuf, H. K. M., Pohlentz, G., and Sandhoff, K. (1983b). *Proc. Natl. Acad. Sci. U.S.A.* **80,** 7075–7079.

Yusuf, H. K. M., Schwarzmann, G., Pohlentz, G., and Sandhoff, K. (1987). *Hoppe-Seyler's Z. Physiol. Chem.* **368,** 455–462.

Zhang, X.-L., Rafi, M. A., De Gala, G., and Wenger, D. A. (1990). *Proc. Natl. Acad. Sci. U.S.A.* **87,** 1426–1430.

Zhang, X.-L., Rafi, M. A., De Gala, G., and Wenger, D. A. (1991). *Hum. Genet.* **87,** 211–215.

ADVANCES IN LIPID RESEARCH, VOL. 26

Truncated Ceramide Analogs as Probes for Sphingolipid Biosynthesis and Transport

DIETER JECKEL AND FELIX WIELAND

Institut für Biochemie I
Universität Heidelberg
D-6900 Heidelberg, Germany

I. Introduction

In mammalian cells the bulk of sphingolipids is found in the plasma membrane. According to a widely accepted view, both classes of ceramide-based lipids, sphingomyelin and glycosphingolipids, seem to be restricted to the outer leaflet of the plasma membrane so that their hydrophilic moieties are oriented to the extracellular space (for review, see van Meer, 1989). This specialized topology, as well as the fact that the various species of sphingolipids are synthesized at different sites within the cell, and the still cryptic functions of ceramide-based membrane lipids have excited the interest of cell biologists in recent years. According to our present knowledge, sphingosine and ceramide biosynthesis seems to occur in (or at) the endoplasmic reticulum (ER) (Walter *et al.*, 1983; Mandon *et al.*, 1992). Subsequent formation of sphingomyelin and glycosphingolipids takes place in the Golgi apparatus. It is highly likely that the resulting polar membrane constituents are then transported within vesicles that fuse with the plasma membrane. Consequently, a luminally oriented hydrophilic head group would become expressed at the cell surface.

To follow sphingolipid biosynthesis and transport, a variety of useful probes have been developed and successfully employed, among them the fluorescently labeled analogs of ceramide that can be loaded to membranes and eventually are converted into fluorescently labeled sphingomyelin and glucosylceramide. Thus, this type of probe for lipid biosynthesis yields membrane lipid analogs that remain intercalated into one of the leaflets of a bilayer but can be extracted from the outer (accessible) leaflet under mild conditions (without the use of detergent) by "backloading" to serum albumin or similar carriers (for review, see Pagano, 1990). Derivatives of this type of probe have the advantage in principle to behave like endogenous sphingolipids.

An approach tailored for the study of the topology of sphingolipid biosynthesis and vesicular transport of luminally oriented Golgi contents takes advantage of ceramide analogs that do not mimic endogenous ceramide in its hydrophobic intercalation into membranes but rather are water soluble and amphiphilic enough to penetrate through membranes easily. This characteristic is achieved by shortening both hydrophobic chains, the sphingosine and the fatty acyl moiety, to a length as short as eight carbon atoms each [ceramide C_8C_8, truncated ceramide (t-Cer)] and allows for quick access of high concentrations of these analogs to ceramide-converting enzymes even in intact cells. Like the fluorescent ceramide analogs, the truncated ceramides (C_8C_8 and some variations of this type) are mainly converted into glucosylceramide (GlcCer) and sphingomyelin derivatives, although higher glycolipids are also formed under certain conditions.

With this type of probe, our main focus is to investigate the activities and topologies of sphingolipid-synthesizing enzymes in isolated, intact organelles (bypassing the need to destroy membranes to allow access of the substrate to luminally oriented enzymes), to study vesicular transport processes in semi-intact cells (which allow the addition of non-membrane-permeable biochemicals), and to determine the rate of luminal efflux (bulk flow) from the Golgi to the plasma membrane.

II. Studies with Isolated Intact Organelles

A. A Simple Quantitative Assay for Sphingomyelin, Glucosylceramide, Lactosylceramide, and GM_3 Synthesis

With the t-Cer C_8C_8, radioactively labeled with ^3H in its octanoic acid residue, we have established a convenient and strikingly rapid quantitative assay for the biosynthesis of various sphingolipid derivatives (Jeckel *et al.*,

1992). In this assay, the activity of the membrane enzymes is determined *in vitro* in intact organelles without the use of detergents. In the following we describe a standard assay: isolated Golgi membranes (Tabas and Kornfeld, 1979) (10–20 μg membrane protein) are incubated in a plastic microtube in a total volume of 20 μl with 100 μM [^{3}H]ceramide C_8C_8 (specific radioactivity 150 μCi/μmol), 50 mM NaCl, 10 mM EDTA, 4 mM UDP-Glc, 10 mM Tris/maleate pH 7.4, under shaking at 37°C for 40 minutes. The reaction is stopped by the addition of 1 vol of isopropanol, the sample is centrifuged (10,000 *g*, 2 minutes), and an aliquot of the supernatant analyzed by thin-layer chromatography (TLC) on Whatman silica gel plates LK6 in butanone/acetone/water/formic acid (ratio of 30/3/5/0.04, v/v). The chromatograms are evaluated either by an automatic TLC-2D analyzer (digital autoradiograph; Berthold, Wildbad, Germany) or by fluorography according to Randerath (1970) and subsequent scraping of the spots and liquid scintillation counting. A corresponding evaluation is shown in Fig. 1. In addition to ceramide C_8C_8, the fastest moving band, two spots can be seen that migrate more slowly. (Incubation in the absence of UDP-Glc leads to only one additional spot, the one close to the origin.) To investigate the structures of these compounds, the sample was incubated with either sphingomyelinase or with α- or β-glucosidase or β-galactosidase. The results are shown in Fig. 1, lanes 3 to 6, and indicate that the bottom and the middle spot represent truncated sphingomyelin (t-Sph) and t-GlcCer, respectively. Digestion of this material by β-glucosidase showed that glucose is transferred from UDP to the t-Cer to result in t-β-GlcCer. With UDP-[^{14}C]Glc and unlabeled ceramide C_8C_8, an identical spot was obtained. The hexose liberated from this compound proved to be Glc. With UDP-Gal instead of UDP-Glc in the incubation only the spot at the bottom is observed, which represents sphingomyelin C_8C_8. Thus, rat liver Golgi membranes are unable to synthesize GalCer C_8C_8. In order to finally prove these structures, we isolated the substances for fast atom bombardment mass spectrometry. The spectrum of the substance with slow migration showed a prominent molecular peak $M^+ = 451$, consistent with a truncated sphingomyelin. The substance in the middle resulted in a strong molecular peak $M + H^+ = 448$, consistent with ceramide C_8C_8 linked to a neutral hexose (Karrenbauer *et al.*, 1990).

Synthesis of t-Sph was independent of externally added CDP-choline. This is in agreement with the results of Voelker and Kennedy (1982), who showed that the phosphorylcholine group of sphingomyelin derives from phosphatidylcholine.

In the *in vitro* system described, the synthesis of higher glycosphingolipids can be studied as well. To this end, EDTA must be omitted from the incubation. In the presence of 10 mM Mg^{2+}, simple addition of 5 mM UDP-Gal gives rise to an additional spot on the TLC, presumably the t-

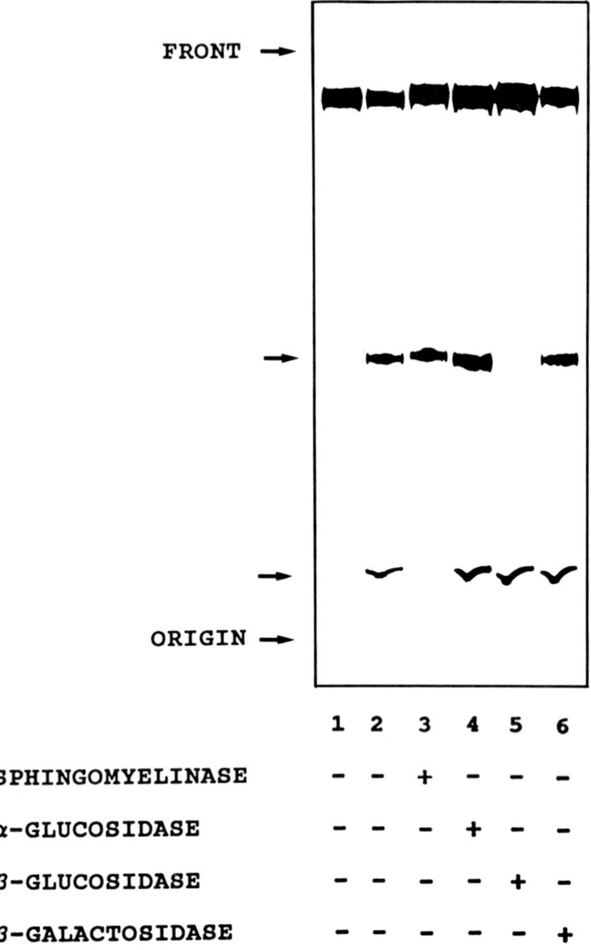

FIG. 1. Analysis of the products formed by Golgi membranes after addition of t-[^{3}H]Cer. After incubation of intact isolated Golgi membranes in the presence of t-[^{3}H]Cer and UDP-Glc, the reaction was stopped by addition of an equal volume of isopropanol, the sample was centrifuged, and microliter-aliquots of the supernatant analyzed by TLC. A fluorograph of the thin layer plate is shown. Lane 1, t-[^{3}H]Cer; lane 2, extract as described; lanes 3–6, sample as in lane 2 but treated with hydrolases as indicated. [Reproduced from *Cell* **63,** 259–267 (1990) by permission of Cell Press.]

LacCer derivative. Further addition of 4 m*M* CMP-neuraminic acid results in the synthesis of G_{M3} or G_{D3} (Scheme 1). Digestion with neuraminidase eliminates acid-containing spot and a new spot appears which comigrates with t-LacCer. Thus, the conditions to assay higher glycosphingolipid

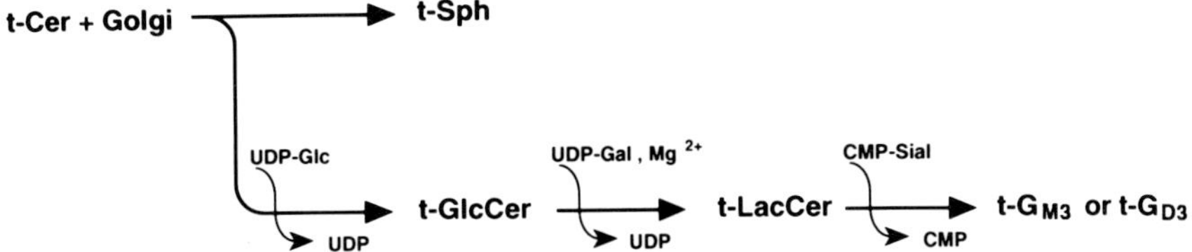

SCHEME 1. Formation of truncated sphingolipids in the *in vitro* system described.

formation are identical with those described for synthesis of t-Sph and t-GlcCer, except that additional precursors are added and the incubation time was prolonged to 120 minutes. For TLC separation of t-GlcCer, t-LacCer, and t-G_{M3}, we used chloroform/methanol/0.22% $CaCl_2$ in water (ratio of 65/35/8, v/v).

What is the minimal overall hydrophobicity of a ceramide analog that still allows conversion into sphingolipid derivatives? In a sphingosine C_8, the fatty acyl moiety must exceed a length of four carbon atoms: *N*-acetyl- and *N*-butyrylsphingosine C_8 are neither converted to the corresponding sphingomyelin nor to the GlcCer. Hexanoic acid in sphingosine C_8 leads to both products, with a slight preference for the formation of GlcCer C_8C_6 (A. Karrenbauer and F. T. Wieland, unpublished).

B. Topology of Sphingolipid-Generating Enzymes

1. *Sphingomyelin Is Synthesized in the cis-Golgi*

We have probed sphingolipid synthesis after partial separation of Golgi subcompartments by sucrose density centrifugation according to Trinchera and Ghidoni (1989). With this method, the authors were able to separate cis- and trans-Golgi markers partially, and observed separation of two different sialyltransferases involved in glycosphingolipid biosynthesis in the fractions characterized by a cis- and a trans-Golgi marker, respectively. Following this procedure, we have analyzed the resulting fractions for cis- and trans-Golgi marker activities as well as t-Sph synthesis, as depicted in Fig. 2 (where in panel C the protein profile and the sucrose gradient are shown). The two marker enzyme activities could be partially separated from each other (Fig. 2B), and t-Sph synthesis strictly followed the activity of the cis-Golgi marker UDP-*N*-acetylglucosaminylphosphotransferase (Fig. 2A). Esterase activity assayed as marker for the ER was clearly separated from the peak activity of t-Sph formation (Jeckel *et al.*, 1990). This preferential localization of sphingomyelin synthase to the early Golgi is confirmed by results of Futerman *et al.*, (1990), who observed that

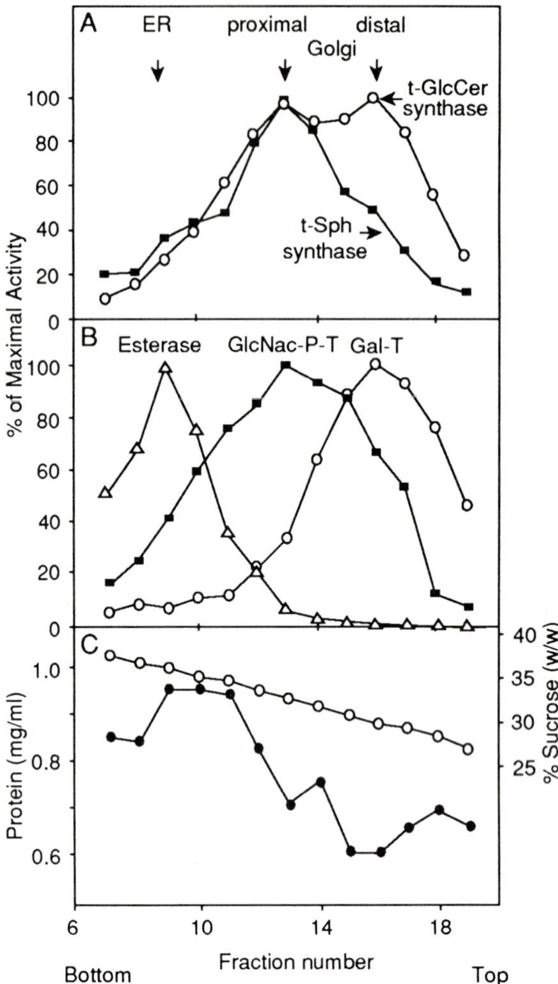

FIG. 2. Partial separation of t-GlcCer and t-Sph synthase activities in rat liver Golgi. Rat liver Golgi membranes were submitted to sucrose density gradient centrifugation as described (Trinchera and Ghidoni, 1989), and t-[^{3}H]Sph (■) and t-[^{3}H]GlcCer synthase (○) activities were assayed (A). The gradient was characterized by measuring the activities of marker enzymes for the ER (esterase, △), for the proximal Golgi (*N*-acetylglucosaminylphosphotransferase, (■), and for the distal Golgi (galactosyltransferase, ○) (B). In (C) the protein content (●) and the sucrose concentration (○) of the fractions are shown. [Reproduced from *J. Cell Biol.* **117,** 259–267 (1992) by permission of Rockefeller University Press.]

sphingomyelin synthase cofractionates with the proximal/medial Golgi marker enzyme mannosidase II.

2. *GlcCer Is Synthesized at Various Golgi Subfractions*

The sucrose density gradient described above was also used to localize the activity of t-GlcCer synthase within the rat liver Golgi apparatus (Fig. 2A). Like sphingomyelin synthase, GlcCer synthase activity follows the profile of the proximal Golgi marker enzyme UDP-*N*-acetylglucosaminyl-phosphotransferase. However, in addition, a second peak of t-GlcCer synthase is obtained that coincides with that of galactosyltransferase, a distal Golgi marker (Jeckel *et al.*, 1992). Futerman and Pagano (1991) mentioned that, in contrast to sphingosine synthesis, GlcCer synthesis was more widely distributed along the Golgi apparatus.

Taken together, these results indicate that the ER has no preferential activity in sphingomyelin biosynthesis. Rather, this activity is concentrated in the proximal Golgi compartment. Strikingly, GlcCer synthase activity is not restricted to only one Golgi subcompartment but is found in subfractions active in marker enzymes for the early as well as the late Golgi.

The assumption that sphingolipids are transported in vesicles has led to the widely accepted view that the active sites of the enzymes involved in their synthesis are oriented toward the luminal side of the organelle. Experimental evidence to confirm this view is available at least for those enzymes that are involved in higher glycosphingolipid biosynthesis. However, a luminal orientation of GlcCer synthase has been questioned by Coste *et al.*, (1986), who suggested that this key enzyme of glycosphingolipid biosynthesis resides at the cytosolic surface of the Golgi. This suggestion has not been widely accepted, probably due to problems intrinsic to the experiments carried out in this study. By use of the ceramide C_8C_8, including protease treatments and modification by a membrane-impermeable protein modification reagent, we could confirm these results. As is described below, we and others have established additional independent evidence that GlcCer is indeed formed at the cytosolic surface of various Golgi subcompartments.

III. Reconstitution of Steps along the Secretory Pathway

A. Criteria for Vesicular Transport of Truncated Sphingomyelin

Proteins destined for the biosynthesis of membranes as well as for secretion are translocated into the ER and subsequently transported in vesicles via the Golgi compartments to the plasma membrane (Rothman and Orci, 1992). This vesicular transport depends on temperature, on the presence of various cytosolic protein factors, and on ATP and GTP. Vesicular transport is inhibited by decreasing the temperature to 15°C, by incubation of cytosol with *N*-ethylmaleimide, by ATP depletion, and by replacement of GTP by GTP-γS, a nonhydrolyzable analog of GTP. The involvement of the Golgi apparatus in sphingolipid transport was demonstrated by Pagano *et al.*, who employed fluorescent analog of ceramide that are converted to sphingomyelin and glucosylceramide derivatives (Lipsky and Pagano, 1983, 1985). As mentioned above, sphingolipids exhibit a strong transmembrane asymmetry in the plasma membrane, with a strong preference for its exoplasmic surface (Allan and Walkin, 1988). This may be seen as a consequence of a luminal orientation of their biosynthetic enzymes in the Golgi, and this luminal orientation strongly suggests that these compounds are transported by a vesicular mechanism.

In an *in vitro* system, Wattenberg (1990) has reconstituted glycolipid transport between compartments of the Golgi apparatus and compared its rate with the rate of glycoprotein transport. He found nearly identical rates, and glycolipid transport in this system, like that of proteins, was dependent on cytosol and the presence of ATP. This was taken as an indication that glycosphingolipids and proteins may be transported within the same carrier vesicles.

We were interested to determine the rate of transport of bulk luminal contents of the Golgi *in vivo* as well as in a system that allows addition of chemicals to probe for biochemical characteristics of vesicular transport. The latter has been achieved in semi-intact cells (Beckers *et al.*, 1987; Simons and Virta, 1987). In order to generate a luminal membrane-impermeable marker for vesicular transport, we have exploited the amphiphilic character of the t-Cer. This compound partitions between water and the lipid phase of membranes and therefore gains quick access to the cytosol and lumen of organelles in intact or semi-intact cells. The t-Cer is converted to t-Sph in the lumen of the cis- or cis/medial-Golgi, where it becomes trapped due to its hydrophilic head group. Unlike endogenous sphingolipids which stay in the plasma membrane after fusion of the transport vesicle with the plasma membrane, the truncated derivatives with their short hydrophobic chains are not able to stay anchored in the outer leaflet, but rather are released into the medium. Accordingly, truncated sphingolipid found in the medium should have been transported in vesicular carriers (Karrenbauer *et al.*, 1990).

In the following, a brief review is given of experimental data that favor the idea that the truncated sphingolipid analog is indeed transported by a vesicular mechanism similar to that of secretory protein transport, and some additional and independent evidence is shown that the active site of GlcCer synthase faces the cytosol.

When intact CHO cells are incubated with t-[^{3}H]Cer, synthesis of t-[^{3}H]Sph and of t-[^{3}H]GlcCer is observed at 37°C as well as at 15°C. However, unlike at 37°C, at 15°C release of both truncated sphingolipid derivatives into the medium is markedly inhibited (Jeckel *et al.,* 1992). This observation would be in agreement with a vesicular transport mechanism of both sphingolipids (Matlin and Simons, 1983).

Additional criteria for a vesicular transport mechanism are (1) dependence on cytosolic factors (Balch *et al.,* 1984; Beckers *et al.,* 1987), (2) dependence on ATP (Balch *et al.,* 1984, 1986; Beckers *et al.,* 1987), and (3) inhibition of transport by the nonhydrolyzable GTP analog GTP-γS (Goud *et al.,* 1988; Melancon *et al.,* 1987). Obviously, these parameters cannot be determined in intact cells due to the impermeability of plasma membranes for such biochemicals. To circumvent this problem, we have used CHO-15B cells rendered permeable by introducing holes into their plasma membranes according to Beckers *et al.,* (1987). This is achieved by mechanically shearing off parts of the plasma membranes of intact adhered cells, yielding "semi-intact" cells in which single steps of vesicular secretory protein transport have been reconstituted (Beckers *et al.,* 1987; Helms *et al.,* 1990). When studied in such semi-intact cells, transport of t-Sph and t-GlcCer is strikingly different: depletion of ATP leads to an inhibition of t-Sph transport, whereas t-GlcCer remains unaffected. Likewise, the addition of GTP-γS leads to an inhibition of t-Sph transport, whereas t-GlcCer transport remains unaffected. Transport of t-Sph was strongly decreased at 15°C, very similar to the effect found in intact cells. Most strikingly however, transport of t-GlcCer turned out not to be temperature dependent at all (Fig. 3A–D). These data imply that the luminally synthesized sphingolipid analog t-Sph is indeed transported in vesicles (Jeckel *et al.,* 1992).

Comparison with the biochemical criteria of transport of a bulk phase marker for the lumen of the ER (Wieland *et al.,* 1987) suggests a similar mechanism of vesicular transport from the ER to the plasma membrane (Helms *et al.,* 1990), but it remains to be established whether proteins and sphingolipids are transported within the same vesicles or whether different vesicular pathways exist that have in common the ability to obey the above-mentioned rules. The latter possibility is discussed by Urbani and Simoni (1990).

In contrast to t-Sph, t-GlcCer did not obey the criteria for vesicular transport. At this time, we do not take this to indicate that endogenous GlcCer is not transported within vesicles but rather speculate that this failure to follow the criteria of vesicular transport is due to the high

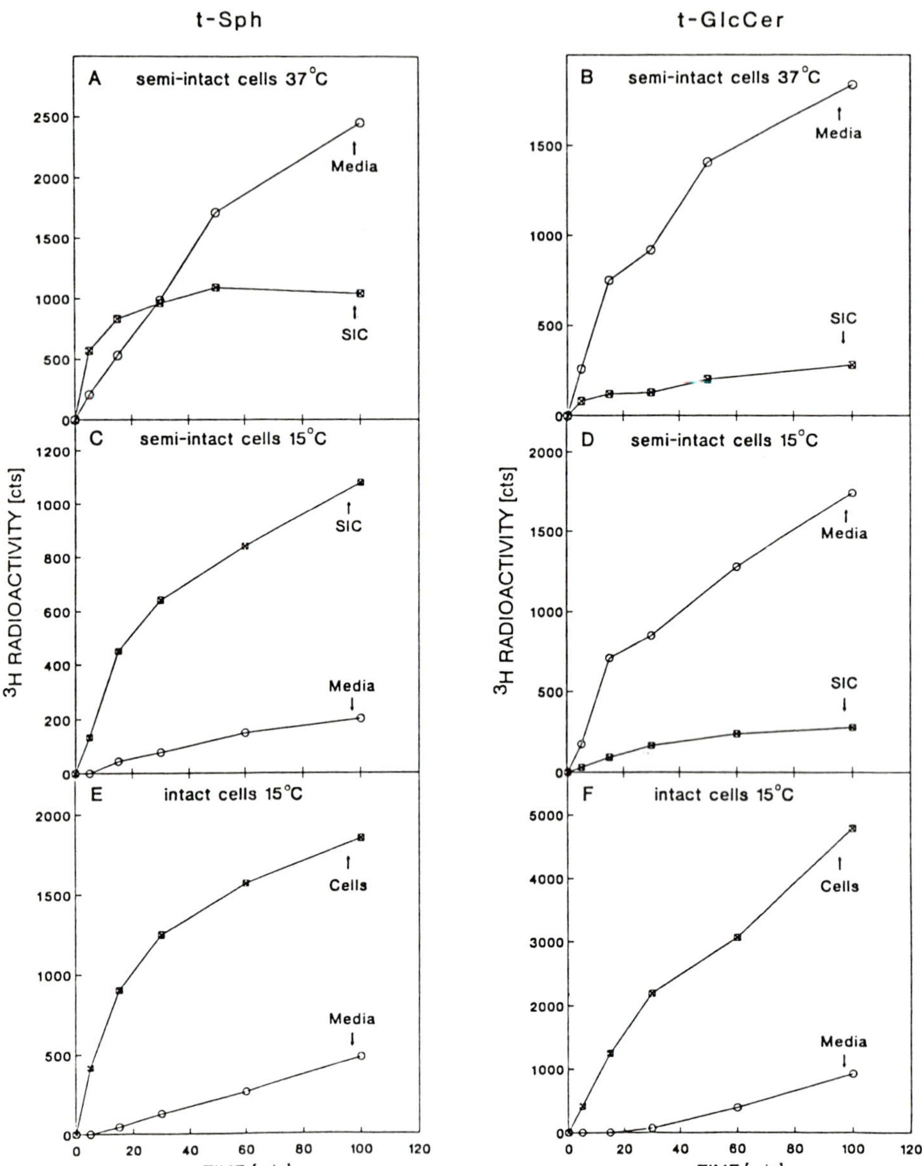

FIG. 3. t-GlcCer is not retained in semi-intact cells at 15°C. CHO-15B cells were permeabilized and aliquots of the resuspended semi-intact cells were incubated with t-[^{3}H]Cer at the temperatures and for the times indicated (A–D). As a control, intact CHO-15B cells were incubated with t-[^{3}H]Cer at 15°C, and truncated ^{3}H-sphingolipids were determined in the cells (□) and in the media (○) (E and F). SIC, Seni-intact cells.

solubility in water of this compound (see discussion below). In any case, this behavior provides strong additional evidence that it is the cytosolic surface of the various subsites of the Golgi where GlcCer is synthesized.

B. How Is Glucosylceramide Expressed at the Cell Surface?

Various results obtained with independent methods by four groups (Coste *et al.*, 1986; Futerman and Pagano, 1991; Trinchera *et al.*, 1991; Jeckel *et al.*, 1992) very strongly support the view that GlcCer is synthesized at the cytosolic surface of the Golgi. This raises two principal problems: first, how is GlcCer transported to the plasma membrane from its site of synthesis, and second, as GlcCer serves as the precursor for virtually all complex glycosphingolipids, and the subsequent reactions in their assembly are localized in the lumen of the Golgi, how is GlcCer translocated through the Golgi membrane?

One translocation step of GlcCer through a membrane is required for its expression on the cell surface. There are two possibilities as to how this may be achieved: either translocation takes place into the lumen of the Golgi and, subsequently, vesicular transport to the plasma membrane leads to the expression of GlcCer at the cell surface, or GlcCer reaches the plasma membrane by another mechanism (e.g., lipid transport proteins) and translocation is mediated by a plasma membrane flippase. A recent study in polarized epithelial cells (Kobayashi *et al.*, 1992) with fluorescent sphingolipid analogs provides some indication that the translocation step actually takes place in the Golgi membrane. Transport vesicles have been described that contain fluorescently labeled GlcCer which cannot be transferred to acceptors like serum albumin and therefore are regarded as staying in the luminal leaflet of the vesicular membrane. Why then is the t-GlcCer not translocated into the Golgi in semi-intact cells? This failure might be explained by the experimental conditions: in semi-intact cells, the volume of the medium surrounding the Golgi is infinitely larger when compared to the volume of the cytosol in intact cells. Therefore, with a lower affinity to a presumed Golgi GlcCer translocator than the endogenous glycolipid, the short-chain analog would be diluted out in semi-intact cells before it becomes translocated.

If GlcCer expression at the cell surface is due to vesicular transport and fusion of the vesicles with the plasma membrane, how then can two luminal pools of GlcCer in the Golgi be discriminated—GlcCer for glycosphingolipid synthesis and GlcCer for the expression at the plasma membrane as the monohexosyl lipid? A speculative explanation (Fig. 4) may be found in the fact that it is not a single subcompartment of the

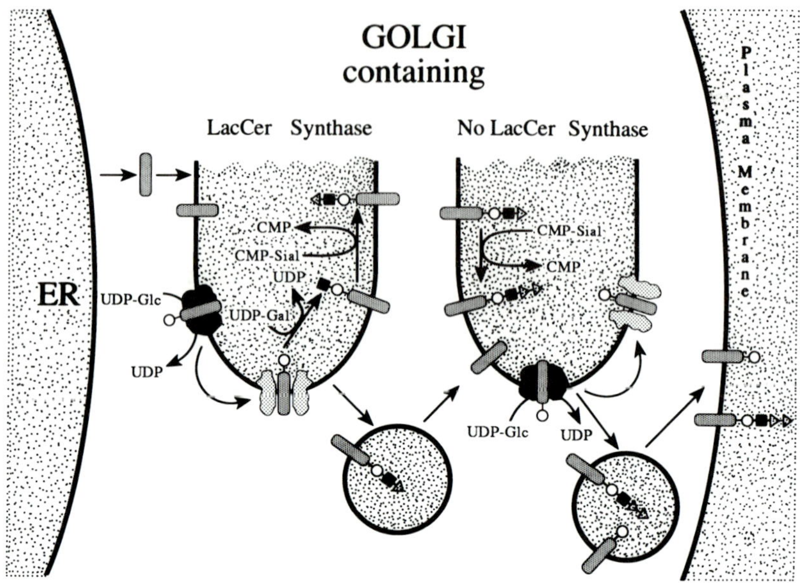

FIG. 4. Hypothetical scheme for a possible function of different Golgi subcompartments involved in GlcCer synthesis. In the Golgi subcompartment containing LacCer synthase, GlcCer synthesized at the cytosolic side of the Golgi is translocated into the lumen and converted to LacCer, the precursor for the synthesis of higher glycosphingolipids. In a subcompartment lacking LacCer synthesis, the cytosolically synthesized GlcCer is also translocated into the lumen, and then transported to the plasma membrane in vesicles. The possibility of cytosolically synthesized GlcCer reaching the membrane by diffusion is not shown. ▭, Ceramide; ○, glucose; ■, galactose; △, neuraminic acid (sial).

Golgi that is capable of GlcCer synthesis. Possibly, GlcCer synthesized on and translocated into the lumen of the early Golgi is converted to LacCer, and subsequent vesicular transport through the medial and distal Golgi gives rise to its conversion into higher glycosphingolipids. Any GlcCer synthesized downstream of the subcompartment active in LacCer synthesis would escape this fate and reach the plasma membrane as the monohexosyl lipid. Preliminary data indicate that LacCer synthase does indeed preferentially reside in the early Golgi (D. Jeckel and F. T. Wieland, unpublished). Trinchera *et al.*, (1991) have described LacCer synthase to be located at the cytosolic surface of the Golgi, like GlcCer synthase. Thus, our concept would have to be extended to include a putative LacCer translocator. With this translocator restricted to the early Golgi, our speculation of the mechanisms underlying the discrimination of two luminal Golgi pools of GlcCer would still hold. However, Brändii *et al.*, (1988) have demonstrated that LacCer synthesis is inhibited in a MDCK mutant cell line that is defective in UDP-Gal translocation into the Golgi. At this time, this conflict has not been resolved.

IV. Sphingolipid Biosynthesis and Transport in Intact CHO Cells

A. Kinetics of Sphingomyelin and Glucosylceramide Formation and Transport

According to the results discussed above, the t-Sph appears to be a promising marker for the vesicular transport from the cis-Golgi to the plasma membrane, and therefore the rate of its release into the medium was investigated in cultured cells. To this end, CHO wild-type cells in stirred suspension were incubated with t-[^{3}H]Cer and aliquots were withdrawn at various times. Cells and media were separated and their contents of truncated sphingolipids were analyzed by TLC and subsequent quantitation of the ^{3}H radioactivity in the individual spots. Autoradiographs of such chromatograms are depicted in Fig. 5, and a quantitative evaluation of such experiments performed at 37°C is shown in Fig. 6. Mean residence times for t-Sph and t-GlcCer were determined by linear extrapolation of the straight line of secretion to the time axis. Surprisingly, t-GlcCer appeared in the medium at a much faster rate (average $t_{1/2}$ = 3.3 minutes) than t-Sph, which averaged $t_{1/2}$ = 9.8 minutes (Karrenbauer *et al.*, 1990).

B. Comparison of the Rate of Bulk Flow from the ER with the Flow from the Cis-Golgi Compartment to the Plasma Membrane

Wieland *et al.* (1987) used *N*-octanoyl-Asn-[^{125}I]Tyr-Thr-NH_2 (^{125}I-OTP) as a membrane-permeable precursor for a marker to study the kinetics of constitutive secretion in CHO cells. The tripeptide contains the consensus sequence for *N*-glycosylation, a process that occurs in the lumen of the ER. The tripeptide is glycosylated in the ER and thereby restricted to the lumen of organelles due to the impermeability of membranes for complex carbohydrates. It is further processed in the Golgi complex, transported to the plasma membrane, and secreted into the medium.

In order to assess the rates of vesicular biosynthetic transport from the two intracellular sites to the plasma membrane relative to each other, we have measured the rates of transport from the ER to the plasma membrane (^{125}I-OTP) and from the proximal Golgi to the plasma membrane (t-

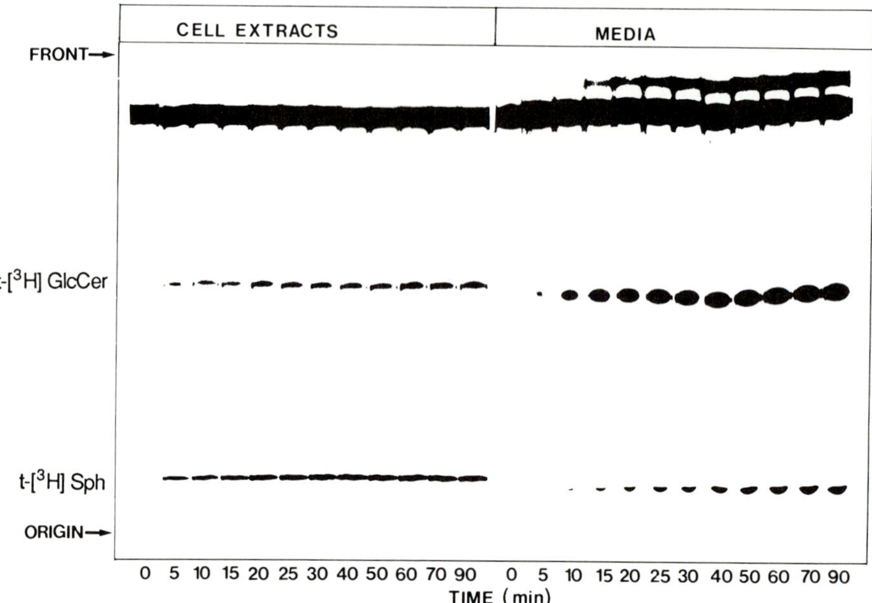

FIG. 5. Time course of synthesis and transport of t-[^{3}H]GlcCer and t-[^{3}H]Sph. After addition of t-[^{3}H]Cer to a stirred CHO cell suspension at time 0, aliquots were taken at the times indicated and centrifuged. The pellets were extracted and cell extracts and media analyzed by TLC. A corresponding fluorograph is shown.

[^{3}H]Sph) in one and the same batch of cells. This was performed at 30°C, because only slight differences in the lag times of the transport of ^{125}I-OTP and t-[^{3}H]Sph were observed at 37°C. At 30°C, vesicular flow from the ER to the plasma membrane in CHO cells has a lag time of about 26 minutes, whereas t-Sph shows a lag time of about 20 minutes. This corresponds to half-times of transport of 18 and 14 minutes, respectively. By subtraction we estimate the half-time for transit from the ER to the proximal site of sphingomyelin synthesis in the Golgi by bulk flow to be about 4 minutes at 30°C. This sets an upper limit on the transit time from the ER to the cis face of the Golgi stack (Karrenbauer *et al.*, 1990).

At this time, we cannot explain the unexpectedly short half-time of transport of the t-GlcCer (~3.3 minutes) as compared to t-Sph (~9.8 minutes) at 37°C. Two highly speculative views are considered. On the one hand, t-GlcCer synthesis takes place in a very late Golgi compartment. Vesicular transport would then include only one budding and fusion step for the marker to reach the plasma membrane. Given that similar mechanisms underly the budding and fusion events along the total secretory pathway, then a $t_{1/2}$ of 3.3 minutes, which is in close agreement with the 4 minutes we have calculated for the transport from ER to cis-Golgi, could be taken as an indication that release of t-GlcCer actually represents the last step of vesicular secretion. On the other hand, we cannot exclude at this time that t-GlcCer reaches the plasma membrane by cytosolic diffusion and is translocated to the extracellular space either by a GlcCer transloca-tor or by a more unspecific pump, similar to the multidrug resistance transporters.

C. Sphingomyelin Synthesis Is Increased by the Macrolide Antibiotic Brefeldin A

The antibiotic brefeldin A (BFA) has attracted the interest of biochemists and cell biologists because it acts on the equilibrium state of intracellular organelles (Fujiwara *et al.*, 1988; Doms *et al.*, 1989; Lippincott-Schwarz et al., 1989) and thus interferes with vesicular transport. A few minutes after addition of BFA, Golgi enzymes are translocated to the ER. These effects are fully reversible provided that ATP is not depleted in the cells. Brüning *et al.* (1992) used t-Cer as a tool for the quantitative determi-

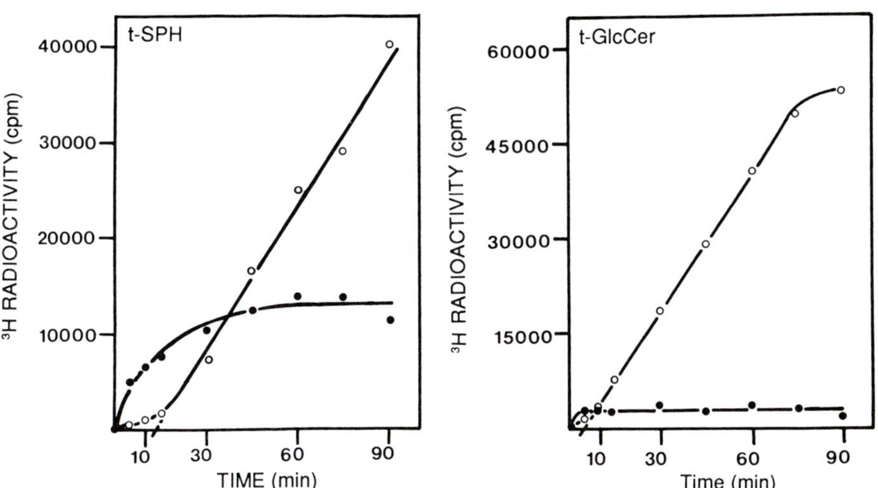

Fig. 6. Quantitative evaluation of an experiment such as is shown in Fig. 5. Spots corresponding to t-[^{3}H]Sph and t-[^{3}H]GlcCer were scraped out and ^{3}H radioactivity determined by scintillation counting. The time courses of t-[^{3}H]Sph and t-[^{3}H]GlcCer in cells (●) and media (○) are shown.

nation of this action of BFA. As mentioned above, biosynthesis of t-Sph occurs by transfer of a phosphorylcholine moiety from phophatidylcholine to ceramide. Sph synthase is located in the cis- or cis/medial-Golgi, whereas phosphatidylcholine, the second substrate, is synthesized in the ER, a membrane system of about 10 times the size of the Golgi apparatus with correspondingly more phosphatidylcholine present than in the Golgi. If the availability of this substrate is rate limiting under physiological conditions of Sph synthesis in the Golgi, then translocation of Sph synthase from the Golgi to the ER caused by BFA should lead to an increase of Sph. Indeed, a drastic increase of t-Sph synthesis is observed after addition of BFA to CHO cells, to a maximal cellular content of about fivefold as compared to untreated control cells after 30 minutes. In contrast, the level of *t*-GlcCer is not influenced by the drug. Incubation of isolated Golgi membranes with BFA and determination of t-Sph synthesis excluded influences that BFA could exert directly on the activity of Sph synthase. This finding suggests that the increase of Sph observed after addition of BFA *in vivo* reflects fusion of ER with Golgi membranes, i.e., translocation of Sph synthase from the Golgi to a mixed ER–Golgi organelle in which a large supply of phosphatidylcholine is available for increased Sph synthesis. This view is further substantiated by the findings that the BFA-induced increase in Sph is inhibited by drugs that inhibit BFA-induced ER-translocation of Golgi marker proteins (Brüning *et al.*, 1992).

V. Future Problems

In our view, the mechanisms of biosynthetic protein transport and biosynthetic lipid transport cannot be seen separately. A first indication has been reported that vesicular glycolipid transport and protein transport might be coupled functionally (Rosenwald *et al.*, 1992). The major problems that need to be solved in the near future include

1. purification to homogeneity and characterization of the key enzymes of sphingolipid biosynthesis, Sph synthase and GlcCer synthase. With antibodies against these enzymes, their unequivocal localization should be possible by immunoelectronmicroscopy;
2. an answer to the question whether proteins and lipids are transported in separate vesicles or use the same carrier. To this end, a thorough analysis of the lipid composition of coated secretory Golgi transport vesicles will be helpful;
3. the identification and characterization of putative glycolipid translocators;

4. the development of highly specific *in vivo* inhibitors of Sph and glycosphingolipid biosynthesis that will allow the study of the functions of this challenging class of biomolecules.

ACKNOWLEDGMENTS

We thank Dr. W. Just and the members of the Wieland/Jeckel laboratory for their helpful comments, Mathias Neufeld for his help with the figures, and I. Speckhard and B. Schröter for their excellent secretarial assistance. The work from our laboratory is supported by the Deutsche Forschungsgemeinschaft (SFB 352).

References

Allan, D., and Walkin, C. M. (1988). *Biochim. Biophys. Acta* **938,** 403–410.

Balch, W. E., Dunphy, W. G., Braell, W. A., and Rothman, J. E. (1984). *Cell (Cambridge, Mass.)* **39,** 405–416.

Balch, W. E., Elliot, M. M, and Keller, D. S. (1986). *J. Biol. Chem.* **261,** 14681–14689.

Beckers, C. J. M., Keller, D. S., and Balch, W. E. (1987). *Cell (Cambridge, Mass.)* **50,** 523–534.

Brändli, A. W., Hansson, G. C., Rodriguez-Boulan, E., and Simons, K. (1988). *J. Biol. Chem.* **263,** 16283–16290.

Brüning, A., Karrenbauer, A., Schnabel, E., and Wieland, F. T. (1992). *J. Biol. Chem.* **267,** 5052–5055.

Coste, H., Martel, M. B., and Got, R. (1986). *Biochim. Biophys. Acta* **858,** 6–12.

Doms, R. W., Russ, G., and Yewdell, J. W. (1989). *J. Cell Biol.* **109,** 61–72.

Fujiwara, T., Yokota, O. K., Takatsuki, A., and Ikehara, Y. (1988). *J. Biol. Chem.* **263,** 18545–18552.

Futerman, A. H., and Pagano, R. E. (1991). *Biochem. J.* **280,** 295–302.

Futerman, A. H., Stieger, B., Hubbard, A. L., and Pagano, R. E. (1990). *J. Biol. Chem.* **265,** 8650–8657.

Goud, B, Salminen, A., Walworth, N. C., and Novick, P. J. (1988). *Cell (Cambridge, Mass.)* **53,** 753–768.

Helms, J. B., Karrenbauer, A., Wirtz, K. W. A., Rothman, J. E., and Wieland, F. T. (1990). *J. Biol. Chem.* **265,** 20027–20032.

Jeckel, D., Karrenbauer, A., Birk, R., Schmidt, R. R., and Wieland, F. T. (1990). *FEBS Lett.* **261,** 155–157.

Jeckel, D., Karrenbauer, A., Birk, R., Schmidt, R. R., and Wieland, F. T. (1990). *FEBS Lett.* **261,** 155–157.

Karrenbauer, A., Jeckel, D., Just, W., Birk, R., Schmidt, R. R., Rothman, J. E., and Wieland, F. T. (1990). *Cell (Cambridge, Mass.)* **63,** 259–267.

Kobayashi, T., Pimplikar, S. W., Parton, R. G., Bhakdi, S., and Simons, K. (1992). *FEBS Lett.* **300,** 227–231.

Lippincott-Schwarz, J., Yuan, L. C., Bonifacino, J. S., and Klausner, R. D. (1989). *Cell (Cambridge, Mass.)* **56,** 801–813.

Lipsky, N. G., and Pagano, R. E. (1983). *Proc. Natl. Acad. Sci. U.S.A.* **80,** 2608–2612.

Lipsky, N. G., and Pagano, R. E. (1985). *J. Cell Biol.* **100,** 27–34.

Mandon, E. C., Ehses, I., Rother, J., van Echten, G., and Sandhoff, K. (1992). *J. Biol. Chem.* **267,** 11144–11148.

Matlin, K. S., and Simons, K. (1983). *Cell (Cambridge, Mass.)* **34,** 233–234.

Melancon, P., Glick, B. S., Malhotra, V., Weidman, P. J., Serafini, T., Gleason, M. L., Orci, L., and Rothman, J. E. (1987). *Cell (Cambridge, Mass.)* **51,** 1053–1062.

Pagano, R. E. (1990). *Biochem. Soc. Trans.* **18,** 361–366.

Randerath, K. (1970). *Anal. Biochem.* **34,** 188–205.

Rosenwald, A. G., Machamer, C. E., and Pagano, R. E. (1992). *Biochemistry* **31,** 3581–3590.

Rothman, J. E., and Orci, L. (1992). *Nature (London)* **355,** 409–415.

Simons, K., and Virta, H. (1987). *EMBO J.* **6,** 2241–2247.

Tabas, I., and Kornfeld, S. (1979). *J. Biol. Chem.* **254,** 11655–11663.

Trinchera, M., and Ghidoni, R. (1989). *J. Biol. Chem.* **264,** 15766–15769.

Trinchera, M., Fabbri, M., and Ghidoni, R. (1991). *J. Biol. Chem.* **266,** 20907–20912.

Urbani, L., and Simoni, R. D. (1990). *J. Biol. Chem.* **265,** 1919–1923.

van Meer, G. (1989). *Annu. Rev. Cell Biol.* **5,** 247–275.

Voelker, D. R., and Kennedy, E. P. (1982). *Biochemistry* **21,** 2753–2759.

Walter, V. P., Sweeny, K., and Morré, D. J. (1983). *Biochim. Biophys. Acta* **750,** 346–352.

Wattenberg, B. W. (1990). *J. Cell Biol.* **111,** 421–428.

Wieland, F. T., Gleason, M. L., Serafini, T. A., and Rothman, J. E. (1987). *Cell (Cambridge, Mass.)* **50,** 289–300.

ADVANCES IN LIPID RESEARCH, VOL. 26

Molecular Approaches to Studying the Intracellular Trafficking of Glycosphingolipids

WILLIAM W. YOUNG, JR.

Departments of Biological and Biophysical Sciences, and Biochemistry
Health Sciences Center
University of Louisville
Louisville, Kentucky 40292

I. Introduction

Newly made secretory and membrane proteins are transported from the endoplasmic reticulum (ER) through the Golgi apparatus prior to reaching their final destinations. At each step of this pathway the mechanism for forward movement is thought to consist of budding of lipid vesicles from a donor membrane and fusion with an acceptor membrane (Rothman and Orci, 1992). Hence, knowledge of the trafficking of lipid components in this pathway will be required to gain a complete understanding of the system. Unlike glycerophospholipids (Voelker, 1990) and cholesterol (Urbani and Simoni, 1990), which utilize non-Golgi as well as Golgi pathways, the transport of glycosphingolipids (GSL) through the Golgi takes place by similar if not identical mechanisms to those for glycoproteins (Wattenberg, 1990). Analysis of secretion mutants indicates that these similarities exist in yeast (Puoti *et al.*, 1991) as well as in higher eukaryotes. Hence, GSL are the preferred lipid class for studying lipid vesicle traffic in the ER–Golgi–plasma membrane pathway. The purpose of this review is to summarize available data regarding several features of the traffic of the GSL. The reviews of Rosenwald and Pagano and Sandhoff and van

Echten in this volume also cover many aspects of this topic. Therefore, the present review focuses on the locations of GSL biosynthetic enzymes, the kinetics of forward movement of GSL through the pathway, and the steady-state distributions of GSL. Studies concerning endogenous GSL in mammalian cells are emphasized; other articles in this volume are concerned with studies of exogenous GSL and the GSL of yeast cells. The present article also points out areas for future exploration, particularly with the tools of molecular biology. Other reviews have appeared recently on GSL traffic (Schwarzmann and Sandhoff, 1990) and on traffic of lipids in general (van Meer, 1989; Voelker, 1990).

II. Locations of Glycosphingolipid Biosynthetic Enzymes

A. Model of Glycosphingolipid Biosynthesis

The current model of GSL biosynthesis involves maturation of oligosaccharide chains by sequential addition of monosaccharides as the GSL proceed through successive Golgi cisternae. This model is based partly on the localization of the enzymes for terminal glycosylation of N-linked sugar chains in a roughly cis to trans direction that follows the sequence in which these enzymes act (Kornfeld and Kornfeld, 1985). With regard to this model, several considerations of Golgi structure should be kept in mind. First, the Golgi apparatus is primarily a carbohydrate assembly line. The fact that such an organized structure has survived evolution should provide solace for the glycobiologist concerned about the importance of protein and lipid glycosylation. Second, there is considerable variation between cell types with regard to the location of particular glycosyltransferases (reviewed by Mellman and Simons, 1992). Therefore, extrapolation of results to other cell types may be risky. Third, a given glycosyltransferase may have multiple isoforms, each of which may be targeted to different locations, such as the Golgi and plasma membrane forms of β1,4-galactosyltransferase (Lopez *et al.*, 1991). Fourth, the prevailing view of Golgi organization is that vesicular transport is directed forward from a cis donor cisterna to a trans acceptor compartment and that each cisterna differs in its content of enzymes from that of its neighboring compartments (Rothman and Orci, 1992). Mellman and Simons (1992) recently proposed an alternative view, which is that the Golgi may consist of only three regions, a cis-Golgi network responsible for entry into the Golgi from the ER and recycling back to the ER; a medial region which may be one functional entity, even though it appears to consist of morphologically separate cisternae; and a trans-Golgi network (TGN), in which components are sorted for passage to their ultimate destinations. Although EM immunolocalization and subcellular fractionation data have distinguished the locations of enzymes of the cis-Golgi from those residing in more trans compartments, the sub-Golgi locations of many N-linked processing enzymes have not been defined. Therefore, it is possible that many glycosyltransferases could reside in the same functional compartment (such as the medial Golgi).

Coresidence of several enzymes in the same compartment may not be a problem for the enzymes of N-linked chain glycosylation; in fact, for the synthesis of repeating determinants such as polylactosamine it may be a necessity. However, GSL synthesis is distinctly different from N-linked chain synthesis in that there are several early branchpoints for GSL synthesis at which a substrate may be used to form more than one product, each of which leads to a distinct pathway of GSL synthesis (Fig. 1). For example, some cell types, such as Madin–Darby canine kidney (MDCK) epithelial cells, produce both glucosylceramide (GlcCer) and galactosyltransferase (GalCer) from ceramide (Cer) (Nichols *et al.*, 1986; Hansson *et al.*, 1986). Similarly, LacCer is a major branchpoint from which at least four families of GSL arise (Fig. 1). What are some of the factors that will determine which pathway is taken? First, of course, the relevant GSL glycosyltransferases must be expressed. Second, a GSL substrate may be used by the cis-most glycosyltransferase while later enzymes go unused; i.e., a GSL substrate is glycosylated by the first enzyme that it encounters. In such a case, physical separation of enzymes would provide the means for pathway selection. Third, enzymes in the same compartment may simply compete for substrate. Smith *et al.* (1990) demonstrated competition for glycoprotein substrates between endogenous $\alpha 2 \rightarrow 3$-sialyltransferase and a transfected $\alpha 1 \rightarrow 3$-galactosyltransferase in CHO cells. Fourth, environmental factors such as pH could regulate GSL synthesis as well. Iber *et al.* (1990) tested the branchpoint at G_{M3} (Fig. 1) in cerebellar cells and found that, by lowering the pH of the medium, there was a shift in biosynthesis from "a"-series to "b"-series gangiosides. They also found that, in rat liver Golgi, the peak of activity for the first enzyme for synthesis of "b"-series gangliosides, sialyltransferase II (SAT-2), occurred at a lower pH than the peak of activity for GalNAc transferase, the first enzyme for synthesis of "a"-series gangliosides. The authors concluded that the lower pH in the medium may have led to a decreased Golgi pH, increasing the activity of SAT-2 and therefore increasing the GSL of the "b"- pathway. Studies with brefeldin A (see below) suggest that SAT-2 is located in a relatively cis compartment compared to the GalNAc transferase (van Echten *et al.*, 1990; Young *et al.*, 1990). Therefore, the lowered pH in effect caused more of the G_{M3} substrate to be

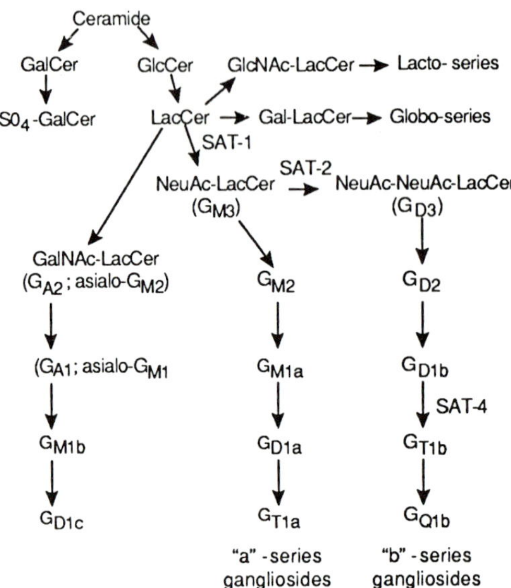

FIG. 1. Glycosphingolipid biosynthetic pathways. This flowchart shows the branchpoints at ceramide (Cer) and lactosylceramide (LacCer) at which the substrate can be used for several synthetic pathways. The steps of ganglioside synthesis are according to the model of Pohlentz *et al.* (1988). Gangliosides are abbreviated according to Svennerholm (1963). SAT, Sialyltransferase.

consumed in the earlier compartment, thus making less available for the GalNAc transferase. In conclusion, a complete understanding of the mechanisms responsible for choice of GSL pathway will only come when the locations and regulation of GSL glycosyltransferases become defined. In the interim, studies with drugs such as brefeldin A may provide insight (see below).

Unlike the early steps of GSL synthesis described above, many enzymes involved in late steps of GSL synthesis act sequentially only on their proper GSL substrate and therefore do not have to be separated into distinct compartments. As mentioned above, the polylactosamines, which are found in GSL as well as glycoproteins, may require colocalization of the Gal and GlcNAc transferases responsible for their synthesis. Based on *in vivo* labeling of rat brain gangliosides, Caputto *et al.* (1976) proposed that each type of ganglioside was synthesized by its own multienzyme complex. Similarly, Kijimoto-Ochiai *et al.* (1980) proposed a "baton pass" model in which Forssman glycolipid was synthesized from ceramide trihexoside without accumulation of the intermediate globoside by a complex of β- and α-GalNAc transferases in NIL-2K hamster fibroblasts.

B. Structural Basis for Glycosphingolipid Synthesis

What is our current understanding of the structural basis for GSL synthesis? The ceramide backbone is synthesized in the ER by a set of cytoplasmically oriented enzymes (Mandon *et al.*, 1992). Ceramide can be converted into either GalCer, from which sulfatide and GalGalCer are produced, or GlcCer, which is the precursor of the majority of GSL (Fig. 1).

1. *GalCer Synthesis*

Compared to GlcCer synthesis (described in the next paragraph), there is relatively little information available concerning the site and topology of GalCer synthesis. Both ceramide galactosyltransferase and cerebroside sulfotransferase (CST) have been identified in mouse brain as Golgi components (Siegrist *et al.*, 1979), but no sub-Golgi localization has been reported. The only data on the topology of this pathway indicate that GalCer is located on the outer half of the myelin membrane (Tennekoon *et al.*, 1983) and that CST was localized to the lumenal side of Golgi membranes (Tennekoon *et al.*, 1983).

2. *GlcCer Synthesis*

The location and topology of GlcCer synthesis have been of interest to several groups. In all of the following reports the donor of glucose is believed to be UDP-Glc, but the synthesis of GlcCer from Dol-P-Glc has been demonstrated as well (Suzuki *et al.*, 1984); the significance of this second pathway remains to be determined. Early work on the site of GlcCer synthesis localized this activity to either smooth microsomes or the Golgi apparatus (Sasaki, 1990). Coste *et al.*(1986) made the striking observation that both the enzyme responsible for synthesizing GlcCer and the glycolipid itself were oriented toward the cytoplasmic side of Golgi vesicles. Their data supporting cytoplasmic orientation of the enzyme were based on sensitivity of the enzyme to proteases present on the cytoplasmic side of intact Golgi vesicles. Such protease sensitivity experiments must be interpreted with caution. The glucosyltransferases responsible for adding glucose to the dolichol-P-P-GlcNAc$_2$Man$_9$ lipid precursor for N-linked chain synthesis are sensitive to proteases on the cytoplasmic side of microsomes, but other data strongly suggests that the catalytic portions of the enzymes are on the lumenal side (Abeijon and Hirschberg, 1992). Recently, two groups have reported on the location of GlcCer synthesis. Subfractionation of rat liver Golgi membranes by Futerman and Pagano (1991) revealed that the peak of specific activity of GlcCer synthesis in rat liver extended from the smooth ER through the intermediate or salvage compartment and into the cis-Golgi and therefore covered a range of membranes cis to the peak of sphingomyelin synthesis. In striking contrast, Jeckel *et al.* (1992) found by subfractionation of rat liver Golgi and HepG2 cells that, whereas sphingomyelin synthesis was restricted to the cis-Golgi, the synthesis of GlcCer was found in both cis- and trans-Golgi peaks. This apparent deviation from the pattern of early-acting enzymes being found in early compartments could explain why considerable GlcCer is never utilized as substrate for glycosylation; i.e., if it is synthesized beyond the point of LacCer in the forward flow of membrane transport, then there may not be a chance for enzyme and substrate to meet. Synthesis of GlcCer late in the pathway could also explain the very rapid kinetics of transport of truncated GlcCer to the plasma membrane (Karrenbauer *et al.*, 1990; see Section III below).

Recently, three groups have localized the site and topology of GlcCer synthesis (Futerman and Pagano, 1991; Trinchera *et al.*, 1991; Jeckel *et al.*, 1992). All three groups have confirmed the cytoplasmic orientation of GlcCer synthesis. Futerman and Pagano (1991) found that GlcCer synthase activity in intact Golgi vesicles from rat liver was sensitive to protease. In addition, they found that the transport of UDP-Glc into Golgi vesicles was insufficient to support the rate of GlcCer synthesis, indicating the need for synthesis on the cytoplasmic side. Jeckel *et al.* (1992) reported that GlcCer synthesis in intact Golgi membranes from rabbit liver was sensitive to protease and to the non-membrane-permeable reagent 4,4′-diisothiocyano-2,2′-stilbenedisulfonic acid (DIDS). They also found that GlcCer itself was cytoplasmically oriented based on the following data. First, truncated GlcCer, in which both acyl chains are C_8 and which had been synthesized from truncated Cer, was released from intact Golgi membranes of permeabilized CHO cells even at low temperature and in the presence of GTPγS, conditions that blocked the export of truncated sphingomyelin which was localized to the Golgi lumen. Second, newly synthesized NBD-GlcCer could be removed from Golgi membranes of HepG2 cells by back exchange with albumin, whereas NBD-SM could not. Trinchera *et al.* (1991) found that intact Golgi vesicles from rat liver could catalyze the formation of GlcCer when the ceramide substrate was either present in liposomes or attached to Sepharose beads, thus indicating the cytoplasmic orientation of this enzymatic activity. Surprisingly, they also found that LacCer synthesis was cytoplasmically oriented, whereas later acting glycosyltransferases were lumenally oriented. Cytoplasmically oriented synthesis of LacCer in all cells would be inconsistent with the existence of CHO and MDCK cell mutants which are deficient in UDP-Gal uptake into Golgi lumen and have markedly reduced levels of LacCer (Deutscher *et al.*, 1984; Stanley, 1980; Deutscher and Hirschberg, 1986; Brandli *et al.*, 1988).

The cytoplasmic orientation of GlcCer raises intriguing possibilities about the trafficking of this glycolipid which have been considered by Sasaki (1990). GlcCer may behave like phosphatidylcholine (PC) in certain respects. Following its synthesis on the cytoplasmic side of microsomes, a portion of PC is translocated to the inner leaflet by a "flippase" (Bishop and Bell, 1988) in order to equalize the phospholipid mass of the two monolayers. Other PC molecules move rapidly ($t_{1/2} \approx$ 1–2 minutes) from the microsomal cytoplasmic leaflet to the plasma membrane (Kaplan and Simoni, 1985). This transport continues at 15°C and is not sensitive to energy poisons. Because the PC transport kinetics are much faster than any reported vesicular transport (Wieland *et al.*, 1987), a nonvesicular mode is likely which could be mediated by cytosolic lipid transfer proteins or by transient collisions of membranes. Both flipping across membranes and transport between intracellular membranes may also occur for GlcCer. With the exception of the report of cytoplasmically oriented LacCer synthesis in rat liver (Trinchera *et al.*, 1991), all other GSL glycosylation steps apparently occur in the Golgi lumen. Therefore, GlcCer, in order to serve as substrate for further glycosylation, must flip so that the glucose moiety is in the Golgi lumen; the search for a GlcCer "flippase" is in progress (Jeckel *et al.*, 1992).

GlcCer may move to other intracellular membranes and the plasma membranes by non-Golgi pathways. Although the intermembrane movement of GlcCer monomers in artificial lipid bilayers is extremely slow ($t_{1/2} > 30$ days) (Correa-Freire *et al.*, 1992), this process can be accelerated greatly by a cytoplasmic glycolipid transfer protein (Wong *et al.*, 1984). In many cells, there is a large pool of GlcCer that is not utilized for further glycosylation. This lack of precursor utilization may indicate inefficiency of GSL glycosylation. However, in several cell types which have accumulated GlcCer, such as L5178Y lymphoma cells (Young *et al.*, 1990) and CHO cells (Young *et al.*, 1992), there is efficient use of LacCer such that it is nearly quantitatively converted to more complex products with minimal accumulation of LacCer. An alternative explanation for accumulation of GlcCer is that a portion moves from the cytoplasmic leaflet in which it is synthesized to other membranes including the plasma membrane and, therefore, can no longer serve as a substrate for further glycosylation. Another possibility is based on the finding of Jeckel *et al.* (1992) that GlcCer is synthesized in trans-Golgi membranes as well as cis membranes (described above). This trans-synthesized GlcCer may be beyond the site of LacCer synthesis, and therefore would be unable to form higher products.

A complete understanding of GlcCer trafficking is currently hampered by the need for two methods, one that can distinguish intracellular lipid from that in the plasma membrane and a second that can distinguish GlcCer located on the cytoplasmic face of the plasma membrane from that on the external face. The former problem may be overcome with a technique such as magnetic affinity chromatography (Becich and Baenziger, 1991; Becich *et al.*, 1991), which can produce plasma membrane preparations in high yield and purity. The latter question can be answered by using short-chain lipid derivatives that can be back-exchanged from the external leaflet but not the cytoplasmic leaflet of the plasma membrane. Studies with these lipid analogs have been very informative but must be interpreted with caution because the analogs function like their natural counterparts in only some but not all respects; e.g., NBD-ceramide is metabolized only to sphingomyelin and GlcCer derivatives in MDCK cells (van Meer *et al.*, 1987), which have a complex pattern of natural GSL (Nichols *et al.*, 1986; Hansson *et al.*, 1986). Surface-labeling methods, such as galactose oxidase/tritiated borohydride labeling of cell surface glycoconjugates having nonreducing terminal Gal and GalNAc (Lampio *et al.*, 1986) and periodate/tritiated borohydride labeling of gangliosides (Miller-Podraza and Fishman, 1982), are not of use for detecting GlcCer in the external leaflet of the plasma membrane; also, no antibodies or lectins have been described that could detect cell surface-exposed GlcCer.

3. Complex Glycosphingolipid Synthesis

Early studies to define the sites of synthesis of GSL more complicated than ceramide monohexosides featured use of the ionophore monensin which, among its diverse effects, blocks trafficking between cis- and trans-Golgi elements. Experiments from several laboratories (Schwarzmann and Sandhoff, 1990) all reported an increase in GlcCer synthesis and decrease in more complex GSL in the presence of the drug, suggesting that more complex GSL are synthesized in compartments that are separable from and more trans than the relatively earlier compartment(s) in which GlcCer was synthesized. These results supported the model in which GSL chain elongation occurred progressively during passage through the Golgi vesicular transport system.

Experimental support for this model has come from Wattenberg (1990), who used an *in vitro* biochemical complementation assay to show that glycoprotein and GSL transport through the Golgi are indistinguishable biochemically and kinetically. This assay utilized Chinese hamster ovary (CHO) cell glycosylation mutants developed by Pamela Stanley. Donor Golgi membranes were from the Lec2 mutant which, because of a block in uptake of CMP–sialic acid into the Golgi, cannot synthesize G_{M3} but instead accumulates LacCer. Acceptor membranes were from the Lec8 mutant which, because of a defect in UDP-Gal uptake into the Golgi, accumulates GlcCer but not LacCer. Synthesis of labeled product G_{M3} was then a measure of vesicular transport of the GSL substrate LacCer from the donor Lec2 membranes to the acceptor Lec8 membranes. By infecting donor membranes with vesicular stomatitis virus, both glycoprotein (namely, the VSV G protein) and GSL transport were measured. Mellman and Simons (1992) have questioned whether this type of *in vitro* Golgi transport assay is detecting cis-to-trans stack vesicular transport as opposed to possible cis-to-cis or medial-to-medial transport that could be mediated by direct fusion of stacks or vesicular transport. Regardless of the mechanism, Wattenberg's study clearly showed that glycoprotein and GSL transport were indistinguishable.

In yeast, the synthesis of GSL is, like glycoprotein synthesis, dependent on genes involved in Golgi vesicular transport (Puoti *et al.,* 1991). In that study, the synthesis of more complex GSL was blocked in a set of secretion mutants developed by Novick and Schekman (1979), indicating that GSL substrates are transported from the ER to the Golgi by vesicular transport.

The first direct evidence that ganglioside synthetic activities were in physically separable sub-Golgi compartments came from Trinchera and Ghidoni (1989), who separated SAT-1 and SAT-4 (Fig. 1) activities from rat liver on a sucrose density gradient. They found the peak of SAT-1 activity in cis-Golgi stacks and that of SAT-4 activity in trans-Golgi stacks, findings consistent with the model of maturation of GSL as vesicular traffic moved through the Golgi. In a later report, this group determined the activity profiles for several other ganglioside synthetic enzymes using the same sucrose gradient subfractionation of rat liver Golgi (Trinchera *et al.,* 1990). Although there was considerable overlap of activity profiles, the peaks of activity corresponded to the model of gangliosides maturing as vesicular flow proceeded in a cis to trans direction, i.e., moving from the heaviest to lightest fractions. The activity peaks were found in the order: heaviest, SAT-1; intermediate, GalNAc transferase for synthesizing G_{M2} and G_{D2} and Gal transferase(s) for synthesizing G_{M1} and G_{D1b}; and lightest, sialyltransferase(s) for synthesizing G_{Q1b}, G_{D1a}, and G_{T1b}. The activity of the sialyltransferase (SAT-2; Fig. 1) responsible for synthesizing G_{D3} from G_{M3} had a broad profile over the entire range of Golgi fractions.

Generally, subcellular fractionation of tissue culture cells has proved to be difficult (Howell *et al.* 1989), but Iber *et al.* (1992) were able to fractionate primary cultured neurons using sucrose gradients into four fractions, two of which contained Golgi enzyme markers. The patterns of activities were similar to those described for rat liver in the preceding paragraph in that SAT-1 (Fig. 1) was enriched in the heavier Golgi fraction, SAT-4 in the lighter fraction, and SAT-2 activity was equivalent in the two fractions. The antibiotic brefeldin A (BFA) is a fungal metabolite that has profound effects on intracellular trafficking (Klausner *et al.,* 1992). BFA blocks protein secretion due to a disassembly of the Golgi apparatus and mixing of the Golgi elements with the ER. Whereas cis, medial, and trans stacks are redistributed into the ER, the trans-Golgi network (TGN) remains separate. BFA acts in this striking manner by preventing the reversible binding of the cytoplasmic protein β-COP110 to the Golgi, which leads to the fusion of the Golgi stacks via tubular connections. This BFA-induced cleavage of the secretory pathway between the trans-Golgi and the TGN has been used to map the diverse activities responsible for N-linked chain modification of glycoproteins (Sampath *et al.,* 1992). Analysis of glycoproteins labeled with [^{3}H]mannose in the presence and absence of BFA indicated that certain early steps occurred prior to the point of the BFA block, some activities apparently straddled the BFA block, and several late activities were beyond the point of the BFA block, including polylactosamine chain addition, outer-branch fucosylation, and addition of terminal α-Gal residues. Similarly, two groups found that BFA blocked synthesis of ganglio series GSL (van Echten *et al.,* 1990; Young *et al.,* 1990). In both reports, BFA caused the accumulation of the simpler GSL GlcCer, LacCer, G_{M3}, and G_{D3}. Van Echten *et al.* found that the synthesis of more complex gangliosides of the "a"- and "b"- series was strongly inhibited in murine cerebellar cells. Young *et al.* tested the effect of BFA on several cell types and localized the block to the GalNAc transferase activity(ies) responsible for synthesizing G_{A2}, G_{M2}, and G_{D2}; i.e., BFA caused the accumulation of the substrates LacCer, G_{M3}, and G_{D3} while also drastically decreasing the synthesis of the products of the GalNAc transferase(s), G_{A2}, G_{M2}, and G_{D2}. Thus, the GalNAc transferase(s) was localized to a late compartment, possibly the TGN, which remains functionally distinct from the ER in the presence of BFA. BFA results must be interpreted with caution because they are based on several assumptions: (1) the environment of the redistributed ER/Golgi complex will allow the enzymatic activities being tested to function, if in fact they are present; (2) the respective nucleotide–sugar transporters can function in the redistributed membranes; and (3) BFA does not specifically inhibit the glycosyltransferase(s) being tested.

C. Future Studies

Clearly, there is a great deal of information to be learned about the locations of GSL synthetic enzymes. One informative approach is to generate specific anti-glycosyltransferase antibodies which could be used for EM immunoperoxidase or immunogold cytochemistry. Because of the difficulty of biochemical purification of glycosyltransferases, a method of choice for studying these enzymes is expression cloning of the respective genes (see Section V below).

Additional studies with BFA may be useful to determine whether enzymes of other GSL synthetic pathways (Fig. 1) are sensitive to this drug. Of particular interest are the following questions: (1) Is the branchpoint at which Cer is converted into either GlcCer or GalCer determined by a trans location of the Gal transferase that synthesized GalCer? (2) Is sulfation of GalCer sensitive to BFA, as is the sulfation of N-linked chains (Sampath *et al.*, 1992)? Similarly, is GSL fucosylation sensitive to BFA, as is the terminal but not the core fucosylation of glycoproteins? (3) Are enzymes of the globo and lacto series (Fig. 1) sensitive to BFA, as is the GalNAc transferase of the ganglio series? Each new cell type must be tested for the sensitivity of its Golgi apparatus to BFA, as cells from the kangaroo rat (Ktistakis *et al.*, 1991) and dog (Hunziker *et al.*, 1991) have been shown to be resistant. Studies with BFA will also be informative about the pathway the different GSL take to the plasma membrane. Whereas protein transport is blocked by BFA, cholesterol (Urbani and Simoni, 1990) and phosphatidylethanolamine (Vance *et al.*, 1991) passage to the plasma membrane is unaffected by BFA. Thus, if a significant portion of GlcCer moves to the plasma membrane by non-Golgi pathways, then, in the presence of BFA, GlcCer will still move to the plasma membrane, while the movement of more complex GSL will be blocked because they utilize the Golgi route exclusively.

III. Kinetics of Forward Movement of Glycosphingolipids through the Golgi Pathway

The "bulk flow" hypothesis (Pfeffer and Rothman, 1987) states that forward transport of passenger proteins through the Golgi system occurs by default; i.e., retention signals are required for retention in compartments along the pathway but no special signal is required for forward transport. Support for this model has come from the elegant demonstration by Pelham's group that retention in the ER of resident soluble proteins is mediated by the C-terminal sequence KDEL (Lewis and Pelham, 1992). The rate of bulk flow was estimated by Rothman and co-workers using exogenous markers. Tripeptides that were designed to cross membranes readily and become *N*-glycosylated in the ER moved from the Golgi to the cell surface of CHO cells after a transit time of about 20 minutes (Wieland *et al.*, 1987). A truncated analog of ceramide was taken up into CHO cells and converted to truncated GlcCer and truncated sphingomyelin, which were secreted after transit times of 5 and 14 minutes, respectively (Karrenbauer *et al.*, 1990). Because these rates were as fast as or faster than the fastest reported rates for protein secretion, these results support the proposal that no special signal is required for forward transport. The very rapid secretion of truncated GlcCer is not consistent with this GSL being synthesized in an early compartment (Futerman and Pagano, 1991). However, as described above (Section II,B,2), synthesis of truncated GlcCer was detected recently in distal as well as proximal compartments (Jeckel *et al.*, 1992). Therefore, it seems likely that this rapid transit time represents secretion from the most distal compartment; in fact, this time is similar to the 5–6 minutes required for G_{M3} to reach the plasma membrane of CHO cells (Young *et al.*, 1992), as described in the following paragraph.

Young *et al.* (1992) addressed a crucial prediction of the bulk flow model, which is that the lipid components of the transport vesicles would reach the plasma membrane at the rapid rate of bulk flow. Because of the extensive non-Golgi traffic of cholesterol (Urbani and Simoni, 1990) and glycerophospholipids (Voelker, 1990), the GSL are the preferred lipid class for this analysis. CHO cell monolayers were labeled with tritiated palmitate and the times at which each glycolipid species became labeled were determined. Label was incorporated into ceramide and GlcCer at time zero without a detectable lag. These kinetic data indicate that these two lipids are synthesized in the same compartment or compartments that are kinetically coupled, which is consistent with the data of Futerman and Pagano (1991) that GlcCer is synthesized in rat liver in a very early compartment. In contrast to the immediate labeling of GlcCer, labeling of LacCer did not begin until 5–6 minutes after applying the labeled precursor. This time interval represented the time required for GlcCer to flip to the inner leaflet of the membrane in which it is synthesized and move to the site of LacCer synthesis. An additional 5–6 minutes transpired before label appeared in G_{M3} ganglioside. Using a periodate cell surface-labeling technique to distinguish cell surface gangliosides, labeled G_{M3} was detected at the plasma membrane after an additional 5–6 minutes had transpired following G_{M3} synthesis. Overall, the transit time from the start of labeling of the ceramide precursor in the ER until labeled G_{M3} reached the plasma membrane was 17–18 minutes. These results are consistent with the 20-minute transit time described above for model tripeptides (Wieland *et al.*, 1987) and indicate that endogenous GSL move from the ER to the plasma membrane at a rate consistent with bulk flow estimates.

As indicated above, truncated sphingomyelin moved from the Golgi to the cell surface of CHO with a transit time of 14 minutes (Karrenbauer *et al.*, 1990). In rat liver, sphingomyelin is synthesized in a later Golgi compartment than GlcCer (Futerman and Pagano, 1991). If this same compartmentation of synthetic activities holds true in CHO cells, then the transit time of 17–18 minutes from synthesis of GlcCer until the product G_{M3} reaches the plasma membrane is consistent with the transit time for truncated sphingomyelin. The transit time for gangliosides metabolically labeled with [^{3}H]galactose to reach the plasma membrane of neuroblastoma and glioma cells was 20–33 minutes (Miller-Podraza and Fishman, 1982), a figure much longer than the 5–6 minutes required for labeled G_{M3} to reach the plasma membrane in CHO cells, and actually longer than the entire time from labeling of Cer until G_{M3} reached the CHO plasma membrane (Young *et al.*, 1992). This discrepancy of ganglioside transport times may reflect differences between the large specialized neuronal cells and CHO cells. Fluorescent C_6-NBD sphingomyelin and C_6-NBD GlcCer required a half-time of 20–30 minutes for transport from the Golgi to the cell surface of MDCK epithelial cells (van Meer *et al.*, 1987). Again, this value is much longer than the transit times determined in CHO cells for truncated sphingomyelin (Karrenbauer *et al.*, 1990) and for GSL (Young *et al.*, 1992) and may reflect cell type differences, in this case between the polarized epithelial cells and CHO cells.

One approach to clarifying these discrepancies in data obtained from different cell types is to construct CHO cells that express other glycolipid pathways so that all analyses can be performed in the same cell type. Such constructs can be obtained by expression cloning as described in Section V.

IV. Locations of Glycosphingolipids

A part of the dogma that has developed in the GSL field is that GSL are found primarily in the external leaflet of the plasma membrane. Surprisingly, there are minimal data to support this widely held view. One of the few clear determinations of cell surface GSL was from Miller-Podraza *et al.* (1982), who showed by cholera toxin binding to neuroblastoma cells that 78–87% of G_{M1} was accessible on the cell surface. Subcellular fractionation of rat liver indicated that, while the plasma membrane contained 80–90% of G_{M1}, G_{D1b}, and G_{Q1b}, there was only 14% of G_{D1a} and 28% of G_{D3} in that fraction (Matyas and Morré, 1987). Their data indicated that 11% of G_{D1a} and 5% of G_{D3} were present in the ER, which contained only trace levels of G_{M1}, G_{D1b}, and G_{Q1b}. Several other reports have described the presence of GSL in intracellular locations. Symington *et al.* (1987) found the majority of LacCer in human neutrophils in granules, with less than 25% being in the plasma membrane. In MDCK epithelial cells, approximately two-thirds of the Forssman glycolipid has been found on intracellular membranes (Hansson *et al.*, 1986; Van Genderen *et al.*, 1991; Butor *et al.*, 1991). Although Forssman was initially detected by immunofluorescence in large intracellular vacuoles (Hansson *et al.*, 1986), recent EM immunostaining localized intracellular Forssman to the nuclear envelope, endosomes, and lysosomes (Van Genderen *et al.*, 1991).

Two groups have described association of GSL with cytoskeletal elements. Sakakibara *et al.* (1981) found that antibodies against GalCer stained a colchicine-sensitive microtubule-like cytoskeletal structure in several cultured cell lines. Gillard *et al.* (1991) reported that in human endothelial cells staining with antibodies against globoside and G_{M3} colocalized with vimentin intermediate filaments but not with tubulin or actin. To date, there are no data to indicate if these cytoskeletal-associated GSL are housed in lipid vesicles.

V. Use of Molecular Biology to Study Glycosphingolipid Trafficking

A more complete understanding of GSL trafficking will come when cloned glycosyltransferases are available, so that their regulation and locations can be determined. To date, the genes for a few glycosyltransferases involved in the terminal steps of glycosylation of both glycoproteins and GSL have been isolated from cDNA libraries using antibodies specific for the purified enzyme (Shaper *et al.*, 1986; Weinstein *et al.*, 1987; Yamamoto *et al.*, 1990). Despite a common overall structure (Paulson and Colley, 1989), none of the genes cloned to date show sequence homology with each other, so cloning of additional enzymes based on homology has not been possible. Because biochemical purification of glycosyltransferases is a difficult and lengthy process, expression cloning has become the method of choice for isolating these genes. The genes for several enzymes responsible for terminal glycosylation of glycoproteins and GSL have been isolated in this manner (Kumar *et al.*, 1991; Potvin *et al.*, 1990; Ernst *et al.*, 1989; Rajan *et al.*, 1989; Joziasse *et al.*, 1989; Larsen *et al.*, 1989). To date, the only glycosyltransferase gene that has been cloned which is specific for GSL substrates is the GalNAc transferase for synthesizing G_{M2} and G_{D2} (Nagata *et al.*, 1992).

A. Source of DNA to Be Used for Transfection

To achieve expression of exogenous glycosyltransferases, the DNA to be used for transfection should be derived from cells expressing high levels of the desired GSL product. A cDNA library from such cells can be constructed in a mammalian expression vector such as pCDM8. Genomic DNA can be used as an alternative to a cDNA library, but the procedure will succeed only if the gene of interest is small enough to be contained in single pieces of genomic DNA which typically range from 50–200 kb. The total length of the $\alpha 2 \rightarrow 6$-sialyltransferase gene is approximately 80 kb (Wen *et al.*, 1992), and in fact several glycosyltransferase genes have been cloned from genomic DNA (Smith *et al.*, 1990; Kumar *et al.*, 1990, 1991). A major difficulty of transfecting with genomic DNA is the problem of rescuing the transfected DNA. Rescue using species-specific repetitive sequences, like human Alu sequences, has been used for cloning glycosyltransferases (Kumar *et al.*, 1990, 1991). Another problem of genomic DNA transfection is the possibility that the transfection procedure will activate an endogenous host gene, as occurred for $\alpha(1,3)$-fucosyltransferase (Potvin *et al.*, 1990).

B. Recipient Host Cells to Be Transfected

Expression cloning using cDNA libraries requires a recipient host cell that expresses the early SV40 genes so that the shuttle vector, such as CDM8, will be autonomously replicated by the SV40 origin of replication. Although COS cells have been used frequently for cloning with CDM8, the need for a recipient cell with an appropriate glycosylation pattern (see below) led to the construction of large T antigen-expressing variants of CHO cells (Heffernan and Dennis, 1991) and B16 melanoma cells (Nagata *et al.*, 1992). Whether transfection is accomplished with genomic DNA or a cDNA library, the recipient host cell must satisfy several criteria from a GSL standpoint. First, the recipient cells must be able to transfer the respective nucleotide sugar into the Golgi and must also synthesize the precursor GSL substrate to which the transfected enzyme can add the respective sugar. Transfection of $\alpha 1 \rightarrow 3$-galactosyltransferase into CHO cells resulted in expression of this determinant on glycoproteins but not glycolipids because the appropriate GSL substrate was not produced by the cells (Smith *et al.*, 1990). Second, the recipient cells must not make the desired GSL product prior to transfection. A related problem is that the cells must not synthesize the same determinant on glycoproteins prior to transfection. Third, a less obvious requirement is that the cells must not express the enzymes for synthesizing GSL beyond the desired product in a pathway; e.g., the Gal transferase for synthesizing globotriosylceramide should not be cloned into cells that express the GalNAc transferases for synthesizing globoside and Forssman. Fourth, to maximize the yield of a given GSL product, recipient cells should be chosen in which competition for GSL substrate will be minimized, e.g., cloning of the glycosyltransferases responsible for any of the other pathways that use LacCer as substrate might be best accomplished in cells in which LacCer is an end product.

C. Screening Procedures

For expression cloning, a specific antibody or lectin must be available in relatively large quantities in order to screen for transfected cells. Large panels of anti-carbohydrate antibodies (Magnani, 1986, 1987; Spitalnik, 1987) and plant lectins (Sharon and Lis, 1990) include reagents specific for many but not all GSL determinants. The most widely used screening procedures consist of multiple rounds of panning on antibody- or lectin-coated dishes interspersed with sorting with a fluorescence-activated cell sorter (Ernst *et al.*, 1989) or rosetting of transfected cells using antibody-coated erythrocytes (Potvin *et al.*, 1990).

D. Uses for Cloned Glycosyltransferase Genes

First, understanding the regulation of GSL trafficking will require a prior knowledge of the regulation of each of the component glycosyltransferases. The complexity of glycosyltransferase regulation has been demonstrated for the $\alpha 2 \rightarrow 6$-sialyltransferase gene which utilizes not only alternative splicing but possibly alternative promoters to achieve tissue-specific forms of this enzyme (Wen *et al.*, 1992). Second, cells can be constructed in which competition for branchpoint substrates such as LacCer (Fig. 1) can be established. By regulating the expression of each inserted gene, the factors that govern the choice of a GSL pathway can be determined. Third, expression of exogenous glycosyltranserases in a single cell line like CHO will allow for the kinetics of traffic for the various GSL pathways to be compared against a common background. Fourth, analysis of additional glycosyltransferases should provide a clearer definition of the signals required for retention of proteins in the various compartments of the Golgi pathway (Bendiak, 1990; Colley *et al.*, 1992; Russo *et al.*, 1992; Tang *et al.*, 1992; Wong *et al.*, 1992; Aoki *et al.*, 1992). Finally, analysis of substrate specificities of cloned glycosyltransferases should provide more definitive data than analysis of whole cells or tissues in which multiple transferases may be expressed.

Acknowledgment

WWY is supported by National Institutes of Health Grant GM42698 and Faculty Research Award 358 from the American Cancer Society.

References

Abeijon, C., and Hirschberg, C. B. (1992). *Trends Biochem. Sci.* **17,** 32–36.

Aoki, D., Lee, N., Yamaguchi, N., Dubois, C., and Fukuda, M. N. (1992). *Proc. Natl. Acad. Sci. U.S.A.* **89,** 4319–4323.

Becich, M. J., and Baenziger, J. U. (1991). *Eur. J. Cell Biol.* **55,** 71–82.

Becich, M. J., Mahklouf, S., and Baenziger, J. U. (1991). *Eur. J. Cell Biol.* **55,** 83–93.

Bendiak, B. (1990). *Biochem. Biophys. Res. Commun.* **170,** 879–882.

Bishop, W. R., and Bell, R. M. (1988). *Annu. Rev. Cell Biol.* **4,** 579–610.

Brandli, A. W., Hansson, G. C., Rodriguez-Boulan, E., and Simons, K. (1988). *J. Biol. Chem.* **263,** 16283–16290.

Butor, C., Stelzer, E. H. K., Sonnenberg, A., and Davoust, J. (1991). *Eur. J. Cell Biol.* **56,** 269–285.

Caputto, R., Maccioni, H. J., Arce, A., and Cumar, F. A. (1976). *Adv. Exp. Med. Biol.* **71,** 24–44.

Colley, K. J., Lee, E. U., and Paulson, J. C. (1992). *J. Biol. Chem.* **267,** 7784–7793.

Correa-Freire, M. C., Barenholz, Y., and Thompson, T. E. (1992). *Biochemistry* **21,** 1244–1248.

Coste, H., Martel, M. B., and Got, R. (1986). *Biochim. Biophys. Acta* **858,** 6–12.

Deutscher, S. L., and Hirschberg, C. B. (1986). *J. Biol. Chem.* **261,** 96–100.

Deutscher, S. L., Nuwayhid, N., Stanley, P., Briles, E. I. B., and Hirschberg, C. B. (1984). *Cell (Cambridge, Mass.)* **39,** 295–299.

Ernst, L. K., Rajan, V. P., Larsen, R. D., Ruff, M. M., and Lowe, J. B. (1989). *J. Biol. Chem.* **264,** 3436–3447.

Futerman, A. H., and Pagano, R. E. (1991). *Biochem. J.* **280,** 295–302.

Gillard, B. K., Heath, J. P., Thurmon, L. T., and Marcus, D. M. (1991). *Exp. Cell Res.* **192,** 433–444.

Hansson, G. C., Simons, K., and van Meer, G. (1986). *EMBO J.* **5,** 483–489.

Heffernan, M., and Dennis, J. W. (1991). *Nucleic Acids Res.* **19,** 85–92.

Howell, K. E., Devaney, E., and Gruenberg, J. (1989). *Trends Biochem. Sci.* **14,** 44–47.

Hunziker, W., Whitney, J. A., and Mellman, I. (1991). *Cell (Cambridge, Mass.)* **67,** 617–627.

Iber, H., van Echten, G., Klein, R. A., and Sandhoff, K. (1990). *Eur. J. Cell Biol.* **52,** 236–240.

Iber, H., van Echten, G., and Sandhoff, K. (1992). *J. Neurochem.* **58,** 1533–1537.

Jeckel, D., Karrenbauer, A., Burger, K. N. J., van Meer, G, and Wieland, F. T. (1992). *J. Cell Biol.* **117,** 259–267.

Joziasse, D. H., Shaper, J. H., Van den Eijnden, D. H., Van Tunen, A. J., and Shaper, N. L. (1989). *J. Biol. Chem.* **264,** 14290–14297.

Kaplan, M. R., and Simoni, R. D. (1985). *J. Cell Biol.* **101,** 441–445.

Karrenbauer, A., Jeckel, D., Just, W., Birk, R., Schmidt, R. R., Rothman, J. E., and Wieland, F. T. (1990). *Cell (Cambridge, Mass.)* **63,** 259–267.

Kijimoto-Ochiai, S., Yokosawa, N., and Makita, A. (1980). *J. Biol. Chem.* **255,** 9037–9040.

Klausner, R. D., Donaldson, J. G., and Lippincott-Schwartz, J. (1992). *J. Cell Biol.* **116,** 1071–1080.

Kornfeld, S., and Kornfeld, R. (1985). *Annu. Rev. Biochem.* **54,** 631–664.

Ktistakis, N. T., Roth, M. G., and Bloom, G. S. (1991). *J. Cell Biol.* **113,** 1009–1024.

Kumar, R., Yang, J., Larsen, R. D., and Stanley, P. (1990). *Proc. Natl. Acad. Sci. U.S.A.* **87,** 9948–9952.

Kumar, R., Potvin, B., Muller, W. A., and Stanley, P. (1991). *J. Biol. Chem.* **266,** 21777–21783.

Lampio, A., Rauvala, H., and Gahmberg, C. G. (1986). *Eur. J. Biochem.* **157,** 611–616.

Larsen, R. D., Rajan, V. P., Ruff, M. M., Kukowska Latallo, J., Cummings, R. D., and Lowe, J. B. (1989). *Proc. Natl. Acad. Sci. U.S.A.* **86,** 8227–8231.

Lewis, M. J., and Pelham, H. R. B. (1992). *Cell (Cambridge, Mass.)* **68,** 353–364.

Lopez, L. C., Youakim, A., Evans, S. C., and Shur, B. D. (1991). *J. Biol. Chem.* **266,** 15984–15991.

Magnani, J. L. (1986). *Chem. Phys. Lipids* **42,** 65–74.

Magnani, J. L. (1987). *In* "Methods in Enzymology" (V. Ginsburg, ed.), vol. 138, pp. 484–491. Academic Press, Orlando, FL.

Mandon, E. C., Ehses, I., Rother, J., van Echten, G., and Sandhoff, K. (1992). *J. Biol. Chem.* **267,** 11144–11148.

Matyas, G. R., and Morré, D. J. (1987). *Biochim. Biophys. Acta* **921,** 599–614.

Mellman, I., and Simons, K. (1992). *Cell (Cambridge, Mass.)* **68,** 829–840.

Miller-Podraza, H., and Fishman, P. H. (1982). *Biochemistry* **21,** 3265–3270.

Miller-Podraza, H., Bradley, R. M., and Fishman, P. H. (1982). *Biochemistry* **21,** 3260–3265.

Nagata, Y., Yamashiro, S., Yodoi, J., Lloyd, K. O., and Furukawa, K. (1992). *J. Biol. Chem.* **267,** 12082–12089.

Nichols, G. E., Lovejoy, J. C., Borgman, C. A., Sanders, J. M., and Young, W. W., Jr. (1986). *Biochim. Biophys. Acta* **887,** 1–12.

Novick, P., and Schekman, R. (1979). *Proc. Natl. Acad. Sci. U.S.A.* **76,** 1858–1862.

Paulson, J. C., and Colley, K. J. (1989). *J. Biol. Chem.* **264,** 17615–17618.

Pfeffer, S. R., and Rothman, J. E. (1987). *Annu. Rev. Biochem.* **56,** 829–852.

Pohlentz, G., Klein, D., Schwarzmann, G., Schmitz, D., and Sandhoff, K. (1988). *Proc. Natl. Acad. Sci. U.S.A.* **85,** 7044–7048.

Potvin, B., Kumar, R., Howard, D. R., and Stanley, P. (1990). *J. Biol. Chem.* **265,** 1615–1622.

Puoti, A., Desponds, C., and Conzelmann, A. (1991). *J. Cell Biol.* **113,** 515–526.

Rajan, V. P., Larsen, R. D., Ajmera, S., Ernst, L. K., and Lowe, J. B. (1989). *J. Biol. Chem.* **264,** 11158–11167.

Rothman, J. E., and Orci, L. (1992). *Nature (London)* **355,** 409–415.

Russo, R. N., Shaper, N. L., Tattjes, D. J., and Shaper, J. H. (1992). *J. Biol. Chem.* **267,** 9241–9247.

Sakakibara, K., Momoi, T., Uchida, T., and Nagai, Y. (1981). *Nature (London)* **293,** 76–79.

Sampath, D., Varki, A., and Freeze, H. H. (1992). *J. Biol. Chem.* **267,** 4440–4455.

Sasaki, T. (1990). *Experientia* **46,** 611–616.

Schwarzmann, G., and Sandhoff, K. (1990). *Biochemistry* **29,** 10865–10871.

Shaper, N. L., Shaper, J. H., Meuth, J. L., Fox, L., Chang, H., Kirsch, I. R., and Hollis, G. F. (1986). *Proc. Natl. Acad. Sci. U.S.A.* **83,** 1573–1577.

Sharon, N., and Lis, H. (1990). *FASEB J.* **4,** 3198–3208.

Siegrist, H. P., Burkart, T., Wiesmann, U. N., Herschkowitz, N. N., and Spycher, M. A. (1979). *J. Neurochem.* **33,** 497–504.

Smith, D. F., Larsen, R. D., Mattox, S., Lowe, J. B., and Cummings, R. D. (1990). *J. Biol. Chem.* **265,** 6225–6234.

Spitalnik, S. L. (1987). *In* "Methods in Enzymology" (V. Ginsburg, ed.), Vol. 138, pp. 492–503. Academic Press, Orlando, FL.

Stanley, P. (1980). *ACS Symp. Ser.* **128,** 213–221.

Suzuki, Y., Ecker, C. P., and Blough, H. A. (1984). *Eur. J. Biochem.* **143,** 447–453.

Svennerholm, L. (1963). *J. Neurochem.* **10,** 613–623.

Symington, F. W., Murray, W. A., Bearman, S. I., and Hakomori, S.-I. (1987). *J. Biol. Chem.* **262,** 11356–11363.

Tang, B. L., Wong, S. H., Low, S. H., and Hong, W. (1992). *J. Biol. Chem.* **267,** 10122–10126.
Tennekoon, G., Zaruba, M., and Wolinsky, J. (1983). *J. Cell Biol.* **97,** 1107–1112.
Trinchera, M., and Ghidoni, R. (1989). *J. Biol. Chem.* **264,** 15766–15769.
Trinchera, M., Pirovano, B., and Ghidoni, R. (1990). *J. Biol. Chem.* **265,** 18242–18247.
Trinchera, M., Fabbri, M., and Ghidoni, R. (1991). *J. Biol. Chem.* **266,** 20907–20912.
Urbani, L., and Simoni, R. D. (1990). *J. Biol. Chem.* **265,** 1919–1923.
Vance, J. E., Aasman, E. J., and Szarka, R. (1991). *J. Biol. Chem.* **266,** 8241–8247.
van Echten, G., Iber, H., Stotz, H., Takatsuki, A., and Sandhoff, K. (1990). *Eur. J. Cell Biol.* **51,** 135–139.
Van Genderen, I. L., van Meer, G., Slot, J. W., Geuze, H. J., and Voorhout, W. F. (1991). *J. Cell Biol.* **115,** 1009–1019.
van Meer, G. (1989). *Annu. Rev. Cell Biol.* **5,** 247–275.
van Meer, G., Stelzer, E. H., Wijnaendts van Resandt, R. W., and Simons, K. (1987). *J. Cell Biol.* **105,** 1623–1635.
Voelker, D. R. (1990). *Experientia* **46,** 569–579.
Wattenberg, B. W. (1990). *J. Cell Biol.* **111,** 421–428.
Weinstein, J., Ujita, E., McEntee, K., Lai, P.-H., and Paulson, J. C. (1987). *J. Biol. Chem.* **262,** 17735–17743.
Wen, D. X., Svensson, E. C., and Paulson, J. C. (1992). *J. Biol. Chem.* **267,** 2512–2518.
Wieland, F. T., Gleason, M. L., Serafini, T. A., and Rothman, J. E. (1987). *Cell (Cambridge, Mass.)* **50,** 289–300.
Wong, M., Brown, R. E., Barenholz, Y., and Thompson, T. E. (1984). *Biochemistry* **23,** 6498–6505.
Wong, S. H., Low, S. H., and Hong, W. (1992). *J. Cell Biol.* **117,** 245–258.
Yamamoto, F., Marken, J., Tsuji, T., White, T., Clausen, H., and Hakomori, S. (1990). *J. Biol. Chem.* **265,** 1146–1151.
Young, W. W., Jr., Lutz, M. S., Mills, S. E., and Lechler-Osborn, S. (1990). *Proc. Natl. Acad. Sci. U.S.A.* **87,** 6838–6842.
Young, W. W., Jr., Lutz, M. S., and Blackburn, W. A. (1992). *J. Biol. Chem.* **267,** 12011–12015.

Part III

FUNCTIONAL AND DYSFUNCTIONAL EFFECTS OF SPHINGOLIPID METABOLISM

ADVANCES IN LIPID RESEARCH, VOL. 26

Metabolic Effects of Inhibiting Glucosylceramide Synthesis with PDMP and Other Substances

NORMAN S. RADIN,[*,†] JAMES A. SHAYMAN,[*]
AND JIN-ICHI INOKUCHI[‡]

[*]Department of Internal Medicine
and [†]Mental Health Research Institute
University of Michigan Medical Center
Ann Arbor, Michigan 48109
[‡]Seikagaku Corporation
Tokyo Research Institute
Higashiyamato
Tokyo 207, Japan

I. Introduction

Nearly all of the hundreds of glucosphingolipids (GSLs) found in nature are derived from glucosylceramide (GlcCer). Thus, any factor that affects the rate of GlcCer synthesis or hydrolysis could have an important effect on the levels of many GSLs as well as on the levels of GlcCer precursors, particularly ceramide and the sphingols (long-chain bases of sphingoid bases). Because of the many different kinds of known enzymatic involvement, physiological roles, and effects of these compounds, interference in GlcCer levels will produce many effects. Despite this complexity, studies of this sort have yielded interesting findings, described below.

II. Chemistry of PDMP

In 1977, Radin initiated a search for an inhibitor of GlcCer synthase (ceramide: UDP-glucose glucosyltransferase) in order to help patients with Gaucher's disease. In this genetic disorder, GlcCer accumulates over time because of a lack of GlcCer β-glucosidase. A keto amine (Fig. 1) proved to be a fairly good inhibitor of GlcCer synthase: 300 μM produced complete inactivation of the enzyme within 1 hour. Substantial inhibitory activity was also obtained with compounds in which the morpholine ring was replaced with other amines (1). When the keto DL-morpholino compound was injected i.p. into mice, there was a rapid decrease in the concentration of GlcCer in liver (2). Within 2 hours it had dropped 48%; by 24 hours the concentration was now 19% *above* the controls. The latter effect was presumably a rebound phenomenon, in which the liver responded to a lack of glucosyltransferase by synthesizing more than enough new enzyme molecules. A considerable increase in ceramide levels was observed, evidently due to lack of a negative feedback effect on the fatty acyltransferase that makes ceramide. (This lack of feedback control is seen also in Farber's disease, where ceramide accumulates due to insufficient ceramidase activity.) Assay of hepatic GlcCer synthase *in vitro* showed that 28% of the enzyme activity had been lost by 2 hours after ketone injection.

The injected mice demonstrated behavioral abnormalities suggestive of hyperirritability. Since the morpholino ketone was also found to inhibit

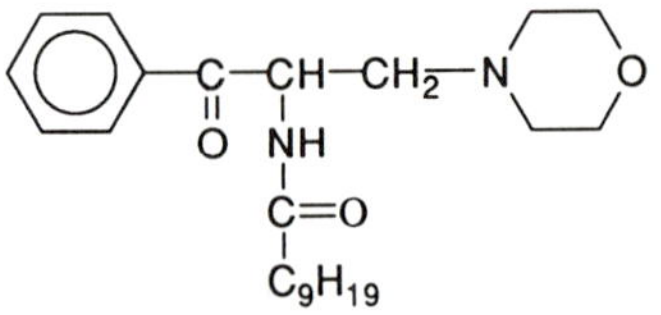

FIG. 1. DL-2-Decanoylamino-3-morpholino-propiophenone-1.

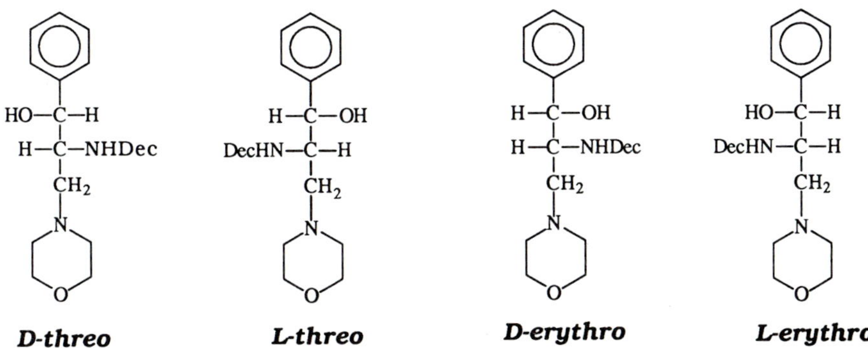

FIG. 2. Formation of PDMP isomers by reduction of the ketone. (Dec is the decanoyl group.) By way of comparison: sphingosine has the D-*erythro* configuration.

monoamine oxidase (3), it seemed possible that it was elevating CNS amine levels and producing the observed neurological changes. We therefore tried administering a neuroleptic agent, chlorpromazine (CPZ). This worked behaviorally as expected, but the surprising result was that CPZ itself, just like the ketone, lowered the level of GlcCer. The effect was additive: the combination lowered GlcCer concentration by 67% in 2 hours. Both CPZ and the ketone also lowered the body temperature of the mice, by as much as 7°C. The cooling phenomenon is known to occur in patients who take CPZ.

Chemical reduction of the DL-morpholino ketone with $NaBH_4$ yielded the morpholino alcohol, PDMP (1-phenyl-2-decanoylamino-3-morpholino-1-propanol; Fig. 2). As the figure shows, four isomeric products were formed because two asymmetric centers are present. The mixture of isomers proved to be much more effective than the ketone in inhibiting GlcCer cynthase *in vitro* (1). Luckily, PDMP did not inhibit monoamine oxidase. The inhibition was not time dependent, suggesting it acted as a competitive inhibitor. When assayed with lyophilized mouse brain as the source of GlcCer synthase, 38 μM PDMP produced 84% inhibition. The reduced piperidino ketone seemed to be significantly more effective than PDMP.[1]

Rosenwald *et al.* removed the decanoyl group of PDMP by heating it

[1] Table VI in Ref. 1 erroneously describes the piperidino analog of PDMP, RV 471, as being PDMP (RV 378).

with 0.8 *M* methanolic KOH under nitrogen for 18 hours at 100°C and extracting the free base with ether (4). This was acylated with a fluorescent acyl ester of *N*-hydroxysuccinimide. This sequence of reactions would also be usable for attaching a radioactive fatty acid. The free base is currently available commercially from Matreya (Pleasant Gap, PA), as are the DL-decanoyl and DL-palmitoyl amides.

A review of recent findings with PDMP has appeared (5).

III. Using PDMP

Most of the experiments have been done with the hydrochloride salt, but early *in vivo* mouse experiments were done with other salts. These salts were made by partitioning ion exchange, using a "Folch" solvent ratio, C : M : W 8 : 4 : 3. Ammonium hydroxide (1 *N*) in the water yielded the free base in the lower layer, which could then be converted to any other salt by addition of a stoichiometric amount of acid. All the salts were readily soluble in isopropanol or ethanol, but the latter may be undesirable because of its tendency to form acetaldehyde on storage.

The acetate salt could be dissolved at a reasonable concentration (1%) on heating to ~50°C. Because crystal formation does not begin immediately on cooling, one can inject the solution at body temperature. For convenience, we now use the hydrochloride the same way, frequently with inclusion of a nonionic detergent, Myrj 52, which seems to help keep the inhibitor in solution (6). The concentrated solution of PDMP.HCl (for animal injection) is surprisingly acidic and we now buffer it to some extent by using 8 mg of NaOAc, 8 mg of PDMP.HCl, and 12 mg of Myrj per milliliter of saline. Recent work showed that Myrj produces some interference with DNA synthesis in cultured cells at a low cell density (7), as well as deposits in mouse liver.

PDMP.HCl can also be used in culture media by evaporating the stock isopropanol solution in a sterile flask with sterile nitrogen and sonicating it in a water bath with sterile medium (8). It can also be prepared as an aqueous 4 m*M* solution and sterile-filtered (9). PDMP.HCl seems quite stable in aqueous solution, as evaluated by thin-layer chromatography (TLC) and a autoradiogram of [^{3}H]PDMP. However, there have been some tentative indications that the solution in isopropanol loses its effectiveness even when stored in a freezer, so further study is needed.

Aqueous PDMP is adsorbed to some extent by the plastic used in cell incubations (7). Thus, it is more active in cells than predicted from its initial concentration in the medium.

IV. The Enzymatically Active Component of PDMP

Thin-layer chromatography of the mixture of PDMP isomers (obtained by borohydride reduction of the ketone) yielded two spots of similar intensity, evidently the DL-*threo* and DL-*erythro* forms (1, 10). The two forms could be separated by crystallizing the hydrochlorides from chloroform and diethyl ether (11). Only the material corresponding to the slower migrating spot showed inhibitory power.

The optical enantiomers of the active product were resolved by crystallization of the salts formed with optically active dibenzoyltartaric acid, followed by recovery of the free base and reconversion to the hydrochloride salt. Only one of the enantiomers was enzymatically active (11). This was identified as D-*threo*-PDMP, although the structural proof was not as secure as one might desire. In the rest of this review we refer to this compound as PDMP, except where related isomers were used.

Liver microsomes, tested at 5 μM PDMP, exhibited 50% inhibition of GlcCer synthase. From kinetic evaluations, it seemed that PDMP is an uncompetitive inhibitor with respect to UDP-glucose and a mixed-type competitive inhibitor against ceramide.

V. Comparison of PDMP Homologs

A. Metabolism of PDMP and the Myristoyl Homolog (PMMP)

[1-^{3}H]PDMP was made from the intermediate ketone (Fig. 1) by reduction with labeled $NaBH_4$ (12). The mixture of four isomers (Fig. 2) was converted to the (−)-camphanic acid esters and separated with a mixed mode (cation/C_8) HPLC column. Fortunately, the desired enantiomer eluted ahead of the other three isomers, which could not be separated from each other. Thus, it was unnecessary to first separate the *threo* and *erythro* isomers. The camphanate group was readily removed by alkaline methanolysis and solvent partitioning.

Initial trials with mice injected with [^{3}H]PDMP at a typical dose (80 mg/kg) showed that much of the radioactive material in liver and kidney consisted of polar products that partitioned into the upper "Folch phase." Presumably, these were the products of hydroxylation and conjugation reactions. Brain and spleen did not contain these metabolites but urine contained them almost exclusively. It seems likely that the products originated in liver and kidney and were rapidly excreted. It was thus necessary to isolate the unmetabolized PDMP by TLC in order to determine the true concentrations of the inhibitor.

PDMP was found to enter the blood and organs very rapidly, brain showing a slight lag. The maximal concentration in liver and kidney was ~170 nmol/g of tissue, which corresponds to ~218 μM. Spleen demonstrated a lower level, ~140 nmol/g, and brain showed the lowest maximum, ~65 nmol/g. Even in brain, the glucosyltransferase must have been completely inhibited.

The PDMP concentrations in the mice dropped to near zero in 3 hours. Thus, whatever effects PDMP produced, were produced in a short time. We investigated the possibility of blocking the degradation of PDMP, which is presumably due to one or more forms of cytochrome *P*-450. Piperonyl butoxide has been used commercially in insect sprays for many years to block the *P*-450(s) in insects, in order to block their inactivation of insecticides. We injected the mice with 600 mg/kg of piperonyl butoxide and then, 4 hours later, injected them with 80 mg/kg of PDMP.HCl in Myrj/saline. This yielded much higher levels of PDMP in all four organs studied, with a maximum of 540 nmol/g of kidney. High levels were maintained for at least 16 hours, but all of the drug was gone by 24 hours. It would seem advisable to test PDMP together with piperonyl butoxide in future animal studies.

An interesting observation in our drug metabolism study (12) is that a small portion of the [^{3}H]PDMP, ~1%, may have been oxidized back to the original keto amine, forming labeled water. The keto amine apparently inactivated part of the GlcCer synthase, which was below normal at the 3-hour and 8-hour points; by 24 hours the specific activity of the enzyme was above normal. Thus, PDMP appears to act by two mechanisms: by reversible inhibition and by covalent inhibition of the synthase. A third mechanism, observed with DL-PDMP, involves elevation of GlcCer β-glucosidase activity (see Section XII). This mechanism, if it does indeed function in intact cells, acts to lower the level of GlcCer more rapidly than one would expect from normal hydrolytic breakdown.

A fluorescent analog of PDMP was found to be taken up by CHO cells, most of it going to the lysosomes (4). A small portion was associated with Golgi membranes, where part or most of GlcCer synthesis occurs. Thus (assuming the fluorescent analog behaves like the aliphatic decanoyl compound), much of the amine is wasted, in the sense that it does not come into contact with the synthase. In intact animals, which metabolize and excrete the inhibitor, this sequestration may act as a buffering process in which the inhibitor gradually dissociates from the lysosomes and prolongs the duration of contact with the enzyme.

The sequestration in lysosomes may explain why mice injected with D-

PDMP exhibited significant increases in the levels of several lysosomal sphingolipid hydrolases in kidneys (see Section XII).

PDMP enters and leaves cells very rapidly. The study with cultured CHO cells (4) found that marked reduction in GlcCer synthesis occurred in just 2.5 minutes with 10 μM PDMP. This supports the belief that PDMP itself, not some metabolite, is the primary inhibitor of GlcCer synthase, at least initially. Inokuchi *et al.* (13) found partial recovery of GSL concentrations in B16 melanoma cells in just 2 hours after replacing PDMP medium with fresh medium.

B. Efficacy of PDMP Homologs

Homologs of PDMP, differing in acyl chain length, were synthesized and resolved into the enantiomers as described in Section V,A (7). The purity of each enantiomer was checked with an Opti-Pak chiral column. Surprisingly, the D-*threo* homologs—C_8 to C_{18}—did not differ greatly in their ability to inhibit glucosyltransferase activities in subcellular preparations. The C_6 homolog, however, was quite low in activity. When the homologs were tested at various concentrations by incubation with intact MDCK (Madin–Darby canine kidney) cells for 24 hours, marked differences were seen in their ability to slow cell growth. For example, at 8 μM, the hexanoyl derivative reduced thymidine conversion to DNA by only 15%, while PDMP reduced it by 42% and the stearoyl derivative blocked it almost completely. The C_{14} and C_{16} homologs were more effective than the C_{18}. Similar differences were seen when cell protein was measured. Thus, it seemed that these anti-growth effects were not simply related to the ability of the different homologs to inhibit GlcCer synthase.

Some clarification of the discrepancy was obtained by labeling both PDMP and PMMP (the tetradecanoyl homolog) with tritium and incubating them at 2 μM with MDCK cells. The uptake of the longer chain inhibitor was 20 times that of PDMP after 30 minutes and 46 times as high after 24 hours. Whether the difference is due to differences in permeation or binding affinity (by lysosomes?) is unknown. Whatever the explanation, it was demonstrated by TLC that GlcCer synthesis by intact cells was much more sensitive to the myristoyl derivative. A similar finding about the effectiveness of the different homologs was reported by Paul *et al.* (14).

These differences in sensitivity to PDMP homologs were seen also in the ability of the compounds to inhibit the growth of Ehrlich ascites cells in mice (15).

An important observation with the MDCK cells was that GlcCer synthesis could be blocked very effectively at low concentrations of inhibitor, yet inhibition of DNA synthesis required higher concentrations. One possible explanation is that the two effects were unrelated. Another explanation is that normal cells have more GlcCer than they need for normal cell growth—it is only when the level of GlcCer goes below a critical level that growth is slowed. This idea is consistent with studies showing that a GSL present at low concentrations in the plasma membrane may not react with an antibody to the GSL, but reaction does occur when a critical concentration is exceeded (see Section IX).

VI. Effects on Phospholipids

A. Inositides in MDCK Cells

Cells typically respond to agonists (hormones, neurotransmitters, etc.) by activation of a receptor, resulting in phosphorylation of certain proteins. Bradykinin is such a controller for MDCK cells. It rapidly stimulates the hydrolysis of phosphatidylinositol bisphosphate by a phospholipase C, forming diacylglycerol (DAG) and inositol 1,4,5-trisphosphate ($InsP_3$). The former helps to activate PKC and the latter acts to increase cytosolic Ca^{2+} concentration. DAG is of interest in sphingolipid metabolism because it is formed from lecithin by a phosphocholine transferase when ceramide is converted to sphingomyelin (SM).

When MDCK cells were treated with PDMP for 2 days, the last day in contact with a labeled precursor, the usual depletion of GSLs was seen (16). The incorporation of thymidine into DNA was reduced, showing in another way the interference in cell proliferation produced by PDMP. Ceramide and SM radioactivity were increased, as observed with other cells. Exposure of the control cells to bradykinin for only 15 seconds produced a rise in the level of $InsP_3$ but the rise in PDMP-treated cells was much larger. From this observation, it was concluded that GlcCer or a related GSL exerts a *negative* modulatory action on the phospholipase.

The ability of PDMP to produce the increased sensitivity to the peptide hormone could be prevented almost completely by including GlcCer–lecithin liposomes in the medium, but galactosylceramide (GalCer) liposomes had no effect. This is further evidence that PDMP acts via its depleting effect on GlcCer or a higher anabolic product. If the effect was due to ceramide accumulation and subsequent formation of sphingols by ceramidase action, we would expect the extra GlcCer or GalCer to *increase* the effect, not counteract it. However, the effect of exogenous GlcCer on ceramide and sphingol levels was not determined. This question clearly needs further study.

Guanosine trisphosphate plays a vital role in the response to bradykinin.

In the cells studied above (16), GTPγS (the stable thio ether analog of GTP) increased the formation of $InsP_3$ to a small, not significant extent in control cells but the PDMP-treated cells were very responsive to the addition of GTPγS with bradykinin. The size of the PDMP effect was much more impressive when the degradation of $InsP_3$ was blocked by the addition of 2,3-bisphosphoglycerate, an inhibitor of the $InsP_3$ 5-phosphatase (S. Mahdiyoun and J. Shayman, unpublished work). This supports the idea that the depleted cells really had a more active phospholipase C rather than slower hydrolysis of $InsP_3$.

Presumably the PDMP-treated cells, because of their higher basal level of $InsP_3$, had a higher level of internal Ca^{2+}. This cation is an activator of GlcCer synthase (17), so we see the possibility of a stabilizing feedback loop.

Further support for GlcCer or a related GSL as a negative modulator of inositide phospholipase comes from a study of the effects of conduritol B epoxide (18). MDCK cells grown in the presence of this inhibitor of GlcCer glucosidase exhibited time- and concentration-dependent increases in GlcCer levels and, conversely, decreases in the bradykinin effect. The inclusion of GlcCer in the medium with the epoxide enhanced the reduction in $InsP_3$ production. These data strongly suggest that GlcCer itself, or an anabolic product, rather than a catabolic product (ceramide?), is the primary modulator of the activity of this important phospholipase.

It is interesting to examine reports of other modifications of second messenger phenomena and consider whether GSL changes might be involved. For example, Hirata *et al.* (19) found that oxidized low-density lipoprotein (LDL) inhibited the production of $InsP_3$ by bradykinin in aortic endothelial cells. From this, one could conclude the LDL (which typically contains oxidized LDL) stimulates GSL uptake or synthesis. Such an effect has indeed been found (20): LDL stimulated lactosylceramide (LacCer) formation in proximal tubular cells. It would be interesting to know whether GlcCer or ganglioside G_{M3} (the closest metabolites) also increase under such conditions. These questions are very relevant to the characterization of atherosclerosis, which is increasingly being considered a disease of excessive proliferation. Other work also points to a mitogenic action of LacCer but here, too, the possibility that LacCer acts via its conversion to GlcCer remains to be determined (21).

B. General Phospholipid Changes

When rabbit skin fibroblasts were incubated with 25 μM DL-PDMP for 7 days, several effects were observed (22). The number of cells per dish was about one-third lower, while the level of phosphatidylethanolamine (per milligram of protein) was markedly higher. Phosphatidylcholine was also elevated, but not as much. The level of phosphatidylserine was half of normal. Unfortunately, the inositides were not determined.

The GSLs, including gangliosides G_{M3} and G_{D3}, were lowered by the inhibitor, and the level of ceramide was greatly elevated, as seen in other cell types. An interesting point is that the loss of GSLs could be demonstrated in the cell surface, using galactose oxidase/borotritide.

It may be possible to explain the increase in lecithin by noting the report that exogenous gangliosides seem to stimulate the synthesis of methylated compounds, such as lecithin (23). If gangliosides inhibit GlcCer synthesis, like PDMP (see Section XI), the mechanism of this phenomenon may involve GSL metabolism. The report is consistent with the observation that PDMP increases the formation of methylated sphingosine (Section VII). Thus, while PDMP causes ceramide accumulation and a decrease in lecithin by SM synthase, it also stimulates lecithin synthesis via transmethylation. It could be useful to see if exogenous gangliosides cause an increase in dimethylsphingosine.

C. Sphingomyelin

MDCK and 3T3 cells incubated with PDMP and [^{3}H]palmitic acid have shown somewhat enhanced labeling of SM (16, 24). This was attributed to the greater labeling and mass accumulation of ceramide, arising from blockade of the glucosylation pathway. However, in the case of intact mice injected with DL-PDMP (6), there was no distinct change in the SM content of kidney, although there was a rapid decrease in GlcCer.

This apparent discrepancy finds an explanation in the observation that mice treated with PDMP exhibited significant elevations in their kidney levels of sphingomyelinase (SMase) (6). Young male mice injected with DL-PDMP at time zero and again after 210 minutes, then killed 210 minutes afterward, had 33% more SMase. Thus, it appears that an increased level of ceramide gives rise to faster synthesis of SM, but the increase in SMase induced by PDMP speeds SM breakdown and maintains the level constant. We have a hint here of a possible control system that maintains the homeostasis of SM. It would be of interest to see if ceramidase also responds to PDMP treatment. (See also Section IX,B for comments on SM synthesis in PDMP-treated choline-labeled cells.)

DL-PDMP, at relatively high concentrations (50–100 μM), has been found to slow SM synthesis in a variety of cell types (4). This was detected by incubating the cells with [^{14}C]hexanoylsphingosine, which was converted to labeled SM. However, the slowing of SM synthesis may have been only apparent, since the unlabeled ceramide formed by PDMP's blockade of GlcCer synthesis must have competed with the labeled ceramide for reaction with lecithin.

VII. Effect of PDMP on Ceramides, Sphingols, Diacylglycerols, and Protein Kinase C

Analysis of MDCK cells after exposure to 20 μM PDMP for 24 hours showed that there was an increase in the concentrations of ceramide (+160%), sphingosine (+167%), and DAG (+500%) (25). Higher PDMP levels produced greater increases. An increased level of ceramide was also indicated by following the incorporation of labeled palmitic acid into 3T3 cells (24) and MDCK cells (16).

Significant changes in the level of ceramide were also demonstrated by mass analysis in the kidneys of mice injected with DL-PDMP (6). Here, however, the changes were generally in the opposite direction. Five hours after injecting 100 mg/kg of PDMP.HCl, we found that the concentration of ceramide dropped 9% to 96 μg/g wet kidney. (This concentration applies only to the less polar ceramides; the hydroxy acid and phytosphingosine ceramides were not measured, but visibly followed the changes in the measured ceramides.) The decrease was greater (−21%) if L-cycloserine was also injected, at 40 mg/kg. This effect is to be expected since cycloserine blocks sphingol synthesis. Mice injected twice a day for 4 days showed an even larger decrease in ceramide (−23% with PDMP alone and −47% with the mixture of inhibitors). In one experiment with mice injected with D-PDMP (not the racemic mixture) at 80 mg/kg, there was a small but significant increase (9%) in ceramide after 7 hours. This may mean that the L-enantiomer of PDMP interferes with ceramide synthesis.

The ability of PDMP to elevate free sphingol levels was noted first with murine IL-2-dependent T lymphocytes (CTLL) grown with 10 μM PDMP for periods of up to 4 days (26). The cells were grown in the presence of labeled serine, a relatively specific precursor of the sphingols. A radioactive band corresponding to sphingosine appeared in the inhibited cells, as well as a band corresponding to *N,N*-dimethylsphingosine. This was the first evidence that the sphingol could undergo *N*-methylation; the enzyme reaction was confirmed in a later study.

Another study with human tumor cells, *in vitro* and in nude mice, showed protective action of exogenous dimethylated sphingol (27). This suggests that one mode of PDMP's anti-cancer action is due to its sphingol-accumulating effect, as well as depletion of the cancer cells' GSLs.

Many studies with exogenous sphingols have shown that they inhibit protein kinase C (PKC), so endogenous sphingols may therefore control growth and the many other processes that seem to need PKC. Direct measurement of PKC in MDCK cells treated with 20 μM PDMP has shown that there is indeed a loss of activity (25). By 24 hours of exposure to inhibitor, the PKC specific activity in both cytosol and membranes was distinctly lower. Thus, the effect was not simply on translocation of the kinase. Complicating the conclusion from this observation is that there was also an even greater elevation in cellular DAG level. Perhaps we have an example of downregulation of PKC due to prolonged DAG elevation. Experiments with exogenous dioctanoyl glycerol, which activates PKC, may not be comparable because the exogenous diester is probably hydrolyzed quickly; in the PDMP cells, DAG is constantly being formed.

Further evidence that endogenous sphingols do not affect PKC activity comes from new unpublished work with MDCK cells (J. Shayman, S. Mahdiyoun, and N. Radin). This showed that L-cycloserine, which lowers sphingol levels, does not increase PKC activity. Instead, it blocks the decrease in PKC induced by PDMP, evidently through its depleting action on ceramide. These findings strongly support the idea that much of the DAG that controls PKC in cells arises from the enzymatic reaction between lecithin and ceramide, which forms SM and DAG.

The β-glucosidase inhibitor conduritol B epoxide was also administered to MDCK cells (25). This inhibitor causes an *increase* in the level of GlcCer. To some extent, it produced effects opposite to those of PDMP: DNA synthesis was stimulated, PKC activity increased, and sphingosine levels dropped. However, ceramide levels were not affected, although the levels of DAG increased. Perhaps, under these conditions, the ceramide:lecithin phosphocholinetransferase reaction relied on accelerated formation of ceramide *de novo*.

The report that CPZ inhibits PKC (28) may explain why it lowered the level of GlcCer in mouse liver (Section II). PKC apparently stimulates the synthesis of many substances, including GlcCer; this effect is stimulated by phorbol ester (29). Thus, it looks as though any inhibitor of PKC should lower GlcCer synthesis, like PDMP, and slow cell growth.

VIII. Depletion of Glycosphingolipids

In all cell types and animals thus far tested, PDMP acted to block GlcCer synthesis and cause depletion of the GSLs derived from GlcCer. In cultured cell experiments, fetal serum (if present in the medium) acts as a supplier of GSLs, slowing the depletion. In mouse experiments, piperonyl butoxide markedly enhanced the effectiveness of PDMP, producing rapid decreases in the concentrations of kidney GlcCer and LacCer

(12) (see Section V,A). The speed of depletion suggests that the lower GSLs (and possibly the higher ones) normally undergo rapid turnover.

This phenomenon indicates that PDMP can be useful in distinguishing whether a cell's particular antigenic reaction or other type of binding reaction is due to its surface GSL or to its glycoprotein. The value of this technique has been pointed out by Paul *et al.* (14). Of course, such a comparison would also benefit from the use of an endoglycoceramidase.

PDMP does slow protein synthesis, as shown by measuring the amount of total protein per incubation dish; this seems to be due to interference with cell proliferation since the number of cells also decreases. Barbour *et al.* (30) found that the amount of protein per cell was unaffected by 10 μM PDMP. In the case of mice injected with PDMP, there was a distinct slowing of growth or even loss of weight but this could also be due to some indirect effect.

A puzzling observation is that cells can survive even when highly depleted of GSLs. Fish embryos were incubated in control medium or medium containing 20 μM PDMP and allowed to hatch (9). Once hatched they were placed in normal medium and allowed to develop further. GSL depletion had relatively little effect on the time required to hatch (~10 days) and the matured young fish behaved normally, presumably due to newly synthesized GSLs during growth in normal medium. The GSLs of the depleted fish, before hatching, were examined by immunostaining of sectioned fry. While the antibodies to three classes of GSLs stained beautifully in the normal fish, the PDMP-treated fish showed virtually no staining.

Other studies with individual cells have also shown that PDMP could produce very large depletion of the GSLs, given enough time and not too high a PDMP level.

Immunostaining, as noted in Section IX, can give a falsely high picture of the extent of GSL depletion if there is sufficient depletion of GlcCer. This kind of test is not sufficient to prove that nearly all the GSLs have disappeared. In the fish study there was no mass analysis of individual GSLs, although autoradiograms showed a great decrease in incorporation of labeled sugars into GSLs. The latter test is fairly convincing, yet there is the possibility that the depletion—by some unknown mechanism of feedback adjustment—causes the radioactive metabolites to be diluted with nonradioactive metabolites or shunted to a different pathway, such as enhanced synthesis of glycoproteins. This question ought to be examined in isotope incorporation studies by analyzing several classes of cell components (glycoproteins, other lipids, etc.) to see if there is indeed such a diversion of sugars.

Perhaps, if the depletion rate is slow enough, cells can adapt by accumulating other lipids (glyceroglycolipids?) or glycoproteins to take over some functions of the lost GSLs. It is also possible that much of the GSL present in cells (particularly GlcCer and LacCer) is there primarily for storage purposes, like glycogen or fat. The amount needed for life functions may be quite low.

IX. Changes in Cell Membranes

A. Changes in Cell Surface Antibody Reactivity

Several observations support the idea that the GSLs in the plasma membrane exist as islands, or aggregates. First, there are fluorescence micrographs that show nonuniform or punctate distribution. In the case of GalCer, which behaved very differently from GlcCer in several PDMP studies, the distribution is relatively uniform (31).

Several studies have shown that antibodies directed against specific GSLs do not react with cells having a relatively low concentration of the GSL. When the cells exceed a critical concentration, they now bind the antibodies. It is possible to explain the phenomenon by assuming that the binding affinity of the antibodies is insufficiently strong to bind noticeably to monomeric GSL molecules, such as one might see in cells having a low GSL concentration. The monomeric GSL molecules might be bound to specific proteins and thus incapable of reacting with the antibody. In cells possessing aggregated GSL molecules, *both* arms of the antibody are able to bind to the aggregate, forming a much stronger linkage. The aggregated molecules may contain relatively little protein and thus be able to deform as needed to form the Ab–Ag complex.

The aggregation hypothesis explains findings of this sort in melanoma cells that had been depleted of GlcCer and LacCer with 10 μM PDMP for 20 hours (13). The cells were then fixed with formalin and treated with antibody M2590, which reacts quite specifically with ganglioside G_{M3}. Fluorescent antibody to M2590 was then added to define the sites of attachment. The control cells exhibited the usual punctate distribution of G_{M3}, together with paler uniform areas, but the cells that had been depleted with 10μM PDMP showed only the paler, uniformly fluorescent areas. Cells treated with 25 μM PDMP showed virtually no fluorescence at all. Chemical analysis of the cells revealed that the level of G_{M3} had not decreased at all at the lower inhibitor level, although the GlcCer level fell by two-thirds. The cells treated with the higher inhibitor level did indeed show a loss of G_{M3} but only by one-third (the GlcCer level here was less than one-sixth of normal). These findings can be interpreted to mean that most of the G_{M3} is normally associated with GlcCer aggregates. When the size of GlcCer aggregates decreases below a certain point, the associated G_{M3} molecules diffuse away and fail to react with antibody.

B. Mobility of Plasma Membrane Components

The technique of fluorescence photobleaching was applied to human A431 cells that had been depleted by GSLs by 5 μM PDMP for 6 days (30). In this procedure, the cells are allowed to incorporate fluorescent substances, then a very small beam of light is used to bleach a small region in the plasma membrane. As the adjacent fluorescent molecules diffuse into the bleached area, fluorescence measurements of the area are made, yielding data on the rate of lateral diffusion in the membrane and the fraction of the total material that is actually diffusible (the "mobile fraction").

The rates of diffusion of four labeled substances—ganglioside G_{M1}, lecithin, and two proteins—were unaffected by GSL depletion. However, the mobile fraction of the two lipids was significantly higher in the depleted cells. From this finding, and the aggregate hypothesis described in the above section, one could conclude that GlcCer-depleted cells have a higher proportion of their lipids (lecithin as well as GSLs) in nonaggregated, rapidly diffusible forms.

The effect of PDMP on the lecithin mobile fraction was reversed by including 10 μM ganglioside G_{M3} in the medium. The reversal may be the result of hydrolysis to GlcCer and LacCer. This reversal supports the belief that PDMP's effects were due primarily to loss of GlcCer, rather than some secondary change.

At the low PDMP concentration used in this study, the GSL levels were dramatically reduced, yet the cells survived. This point is discussed in Section VIII. Using 10 μM PDMP for 3 days, Barbour *et al.* (30) found no significant changes in the amount of radioactive choline or palmitate incorporation into lecithin, SM, or triglycerides. However, ceramide radioactivity, as noted with other cells treated with PDMP, was significantly elevated. The failure of labeled choline to reveal enhanced synthesis of lecithin and SM may be due to enhanced biosynthesis of unlabeled choline by PDMP-treated cells (see Section VI,B on the possibility of increased transmethylation). The specific activity of labeled exogenous choline may thus be lowered by PDMP so that increased mass synthesis of choline lipids would not be evident.

C. Adhesion to Matrix Proteins and Cell Migration

Several laboratories have observed marked changes in the appearance of cells treated with PDMP (22, 24, 30). Perhaps there is no surprise, therefore, in learning that the cells changed in their ability to bind to surfaces and to migrate. The first observations were made with B16 melanoma cells, exposed to 10 or 25 μM PDMP for 20 hours (13). The cells were suspended in plastic wells precoated with laminin, collagen IV, fibronectin, or bovine serum albumin (BSA). Binding to BSA was negligible with control or treated cells. The binding to fibronectin was not affected by GSL depletion, despite reports that GSLs are needed for cell adherence to this protein. Perhaps these particular lipids have a relatively slow turnover rate, and a longer period of PDMP treatment would be necessary to lower their concentrations. In contrast, binding to laminin and collagen was markedly inhibited; the effect was greater with cells that had been depleted more extensively. Cells treated with L-PDMP were unaffected, although they exhibited marked increases in LacCer and G_{M3} levels. Even with exposure to 25 μM PDMP there were still some cells that could bind, and it is an interesting question whether further depletion of GSLs would have produced complete loss of adhesion.

The inhibition of laminin and collagen binding was blocked by including lecithin–GlcCer liposomes in the medium. Lecithin alone did not affect the inhibitory action of PDMP. This is further evidence for the specificity of PDMP effects.

Normal B16 cells that were attached to laminin and collagen prior to GSL depletion rounded up on exposure to PDMP and some of them became detached. Thus, the binding phenomenon seems to be reversible, and dependent on the concentration of GSL in the plasma membrane. Evidently, the presence of the binding regions near the GSLs does not protect the latter against hydrolytic breakdown.

The PDMP-treated cells, which had lost most of their GlcCer but only one-third of their G_{M3}, had lost 70% of their binding ability toward laminin. This suggests that the GlcCer content is crucial for the phenomenon and that the binding to matrix proteins takes place primarily with the aggregated GlcCer. This interpretation is supported by the finding that addition to the cell medium of G_{M3} or antibody against G_{M3} did not interfere with the binding of normal cells. Hakomori has proposed that cell–cell binding also occurs via GSL aggregates in the two cell surfaces (32).

A similar study with PDMP and murine Lewis lung carcinoma 3LL cells yielded similar specificity of action (8). These cells, which do not bind to collagen (when it is applied to the dishes at 10 μg/ml), were tested against laminin and fibronectin. Depletion with 5 μM PDMP for 3 or 6 days yielded cells with reduced ability to bind to laminin, while fibronectin binding was normal. L-PDMP elevated the level of LacCer, globoside, and G_{M3} but did not affect binding to the matrix proteins. This makes it unlikely that the latter three lipids are involved in the binding reaction.

Another aspect of matrix binding is the migration of cells through a filter toward a matrix protein. The GSL-depleted 3LL cells migrated more slowly than control cells when laminin was present in the lower chamber of a Boyden-like cell (8). Thus, one would expect that PDMP treatment would slow metatasis of cancer cells. The inhibitory effect was corrected by incubating the depleted cells in normal medium for 24 hours even though the GSL levels were still not back to normal.

Spreading of cells on substrata is another test of adherence. PDMP-treated A431 cells, after 3 days of GSL depletion, were tested for spreading on fibronectin-coated surfaces (30). Only 40% of the cells showed the spreading phenomenon, compared to 76% of the controls. This difference was not seen if G_{M3} (20 μg/ml) was included with the PDMP, perhaps due to hydrolysis to GlcCer. L-PDMP had no effect. After 10 days of exposure to 10 μM PDMP, only 4% of the cells were able to spread.

In all but the last of these adhesion studies, PDMP failed to block adhesion *completely*. Studies of the matrix proteins have shown that more than one type of binding site is present, so it would appear that GSL is needed for only one of the binding sites. In a preliminary study (33), renal mesangial cells were treated with PDMP and tested for adhesion to a variety of matrix proteins. In this instance, adhesion to all proteins tested—laminin, collagen IV, fibronectin, vitronectin, fibrinogen, and thrombospondin—was substantially but incompletely impaired. Immunoassays demonstrated that the integrins were *in*creased, rather than depleted. While the increased reactivity with integrin antibodies could be attributed to depletion of obscuring GSLs from the plasma membrane surface, it nevertheless suggests that the integrins do not need GSLs for *some* degree of binding. It is also possible that the depleted cells respond to the problem of poor adherence by synthesizing additional integrin molecules.

A recent study of the adherence of 3LL cells to type IV collagen, laminin, and fibronectin (34) showed that exogenously added sphingosine interfered with binding to the collagen (used in a thick layer here) and laminin, but not to fibronectin. Thus, this finding is parallel to the abovementioned effect of PDMP and, since PDMP probably raised the sphingosine level in these cells (see Section VII), one can attribute the PDMP effect to its sphingosine effect. The effect of sphingol was attributed to inhibition of PKC and, indeed, phorbol ester counteracted the sphingol effect. However, one can point to the report that phorbol ester stimulates ceramide and GSL synthesis (29), which would tend to decrease the level of cellular sphingosine. This could explain the restoration of cell–matrix protein binding by phorbol ester.

X. The L-Enantiomer of PDMP

A study of 3LL cells treated with 5 μM PDMP showed that the cells lost most of their GSLs in 6 days without losing viability (8). The L-isomer of PDMP at the same concentration produced no distinct effect by 3 days but a *higher* concentration of all the GSLs was observed after 6 days. The largest increases were in LacCer and globoside.

Another unexpected observation was that the L-isomer of PDMP produced a marked increase in the level of LacCer in B16 melanoma cells, as well as a 50% increase in the level of its anabolic product, G_{M3} (13). One could conclude that the accumulation was due to inhibition of either or both galactosidases that act on LacCer, but tests with L-PDMP *in vitro* showed that it did not inhibit either enzyme.

Another study, with human A431 cells, showed some reduction in GSLs (including GlcCer and LacCer) with L-PDMP (30). Incubation for 3 days with 10 μM D- or L-PDMP lowered the rates of incorporation of labeled galactose or glucosamine into six different GSLs (comparing equal numbers of cells). The decreases were two or four times greater with the D-enantiomer.

Similar inhibition by L-PDMP was seen with murine CTLL cells [a lymphocyte line that is dependent on interleukin-2 (IL-2)] (26). Incubation for 2 days with 5 μM L-PDMP lowered the incorporation of [^{3}H]galactose and glucosamine into GSLs by ~50%, while D-PDMP lowered it much more. Palmitate incorporation into ceramide was stimulated by D-PDMP, as observed in the other studies, but L-PDMP had no effect. This suggests that the L-enantiomer acts further along the sequence of GSL synthesis, not on the metabolism of ceramide. Choline incorporation into SM was unaffected by either enantiomer.

This study also showed that the proliferative response of CTLL cells to IL-2 was blocked almost completely by depletion with 5 μM D-PDMP but only slightly by L-PDMP. Apparently, a relatively strong depletion of GSLs is necessary for markedly blocking the IL-2 effect. No changes in the receptor for IL-2 could be found.

The formation of *N,N*-dimethylsphingosine from labeled serine was stimulated over 300% by D-PDMP but only <50% by L-PDMP. The observation that L-PDMP could cause *any* accumulation of ceramide and also lower precursor sugar incorporation into GlcCer seems to contradict the *in vitro* demonstration that it does not inhibit GlcCer synthase (11).

Stimulation of LacCer synthesis by L-PDMP was demonstrated in smooth muscle cells of the aorta (35). These cells responded to exogenous LacCer by proliferating rapidly. This effect could be blocked by antibodies to LacCer, and the antibodies, without exogenous LacCer, decreased cell proliferation below normal. Incubating the cells with 10 μM L-PDMP for 24 hours raised the number of cells by 42% and the incorporation of [^{3}H]thymidine by 261%. The galactosyltransferase that makes LacCer from GlcCer, when assayed in broken cells, was also increased, but only by 18%. It is possible that L-PDMP in the intact cells exerted a direct stimulatory effect on the transferase but the effect was weakened on washing the cells and diluting the cell contents prior to assay.

These findings agree with the findings in melanoma cells and Lewis lung tumor cells, where the actual mass of LacCer was measured after incubation with L-PDMP (8, 13). However, no such effect was seen in kidneys of intact mice (8). Chatterjee (35) found that exogenous LacCer did not stimulate, and even inhibited, the proliferation of two other cell types, so it is likely that only certain cells respond positively to LacCer and only certain cell types respond to L-PDMP by increasing their LacCer concentration. This topic is clearly worth additional study.

In the case of aortic smooth muscle cells, as Chatterjee and Kwiterovich have pointed out, there is a marked possibility that their response to LacCer is a causative factor in atherosclerosis.

XI. Gangliosides, Equivalence to PDMP

Several papers have reported that the addition of gangliosides to culture media can block cell growth and produce differentiation. It is possible that the exogenous sialolipids act by lowering the level of cellular GlcCer. It has already been shown that gangliosides can stimulate GlcCer glucosidase (36); thus, if they are present near the lysosomal enzyme, they may act to lower GlcCer levels. When gangliosides were tested *in vitro* with murine brain microsomes in an assay for GlcCer synthase, rather effective inhibition was observed (37). The more highly glycosylated gangliosides were the most effective, with G_{D1a} producing 55% inhibition at 40 μM. The effect was reduced by ganglioside-binding material present in cells (i.e., use of a higher amount of tissue per incubation tube lowered the inhibition). It was shown that the enzyme, after treating the membranes with ganglioside, was inhibited even after centrifugal removal of excess, soluble ganglioside; this indicates that tight binding occurs.

From the above considerations, it seems that gangliosides can act to control GlcCer levels by two mechanisms, synthesis inhibition and hydrolysis stimulation. In rat brain, the assayed level of glucosyltransferase drops sharply as the level of gangliosides in brain reaches a maximum and cell division slows down (38). This possible causal relationship was examined by injecting a mixture of brain gangliosides i.p. into mice, at 50 mg/kg daily for 5 days (37). Significantly lower glucosyltransferase activities were found in brain (−11%), liver (−27%), and kidney (−9%). Presumably, the decrease was due to tight binding of the injected ganglioside by the enzyme, so that the inhibition could be detected even when the tissues were homogenized and diluted for the assay procedure.

Gangliosides are now being evaluated for therapy of people suffering traumatic damage to nerves. Damage to the central nervous system normally does not result in nerve regeneration (in higher animals), possibly because of excessive growth by oligodendroglia or astroglia, which may block the regrowth of adjacent axons. Glia apparently cannot synthesize GlcCer (39, 40) and presumably get it, or the higher GSLs, from adjacent neurons. If the exogenous gangliosides act by blocking GlcCer synthesis in neurons, the glia will be deprived of the primary GSL and fail to proliferate or grow. If this explanation for the healing effect of gangliosides is correct, it seems likely that administering PDMP would yield similar results or enhance the effects of gangliosides.

XII. Miscellaneous Effects

Injection of DL-PDMP into mice, twice in a 7-hour period, led to distinct increases in the specific activities of three kidney enzymes (6), GlcCer glucosidase (+25%), GalCer galactosidase (+34%), and sphingomyelinase (+33%). Activity of glucuronidase, a non-sphingolipid hydrolase, *de*creased significantly (−11%). Cycloserine injection under the same conditions, which lowered the kidney level of ceramide by half, did not affect the enzyme activities in either control or PDMP-treated mice, so the effects seem to be due to loss of GlcCer, not to loss of sphingols or ceramides. Brain and liver glucosidase, oddly enough, did not respond to PDMP. Curiously, L-PDMP also produced elevations in the three sphingolipid hydrolases. Perhaps both enantiomers accumulated in the lysosomes (the sites of the sphingolipid hydrolases) and somehow slowed the catabolism of the enzymes.

If these changes are due to loss of GlcCer, one could expect to find opposite changes in animals with elevated levels of GlcCer. Mice injected i.p. with GlcCer took up much of the lipid in the liver, where the level of β-glucosidase rose (41). This finding supports the idea that PDMP acts directly on lysosomes, not via its effect on GlcCer.

3T3 cells that were exposed to DL-PDMP for some time showed enhanced glucose uptake, as measured with [^{3}H]deoxyglucose (24). The effect could be prevented by including trisialoganglioside in the cell medium, possibly because it was converted in part to GlcCer. The enhanced glucose uptake capability may be due to a compensatory system that "seeks" to increase GlcCer synthesis by enhancing the synthesis of UDP-glucose.

Parathyroid hormone secretion by the parathyroid gland is suppressed by elevation of the extracellular Ca^{2+} level. In a recent study, McKay *et al.* (42) incubated dispersed cells from the bovine gland with elevated levels of Ca^{2+} and found that this produced an elevation in ceramide level. This suggests that the gland controls its hormone production by changing its ceramide content. When the parathyroid cells were incubated with 15 μM PDMP for 24 hours, the ceramide level in the treated cells rose to 31% above normal and the secretion of extracellular parathyroid decreased 25%. The PDMP may have worked in two ways in this system: it caused ceramide accumulation by blocking glucosylation, and it raised intracellular Ca^{2+} by raising the rate of inositide hydrolysis and $InsP_3$ production (see Section VI,A).

At relatively high levels in cell media (50 μM), DL-PDMP was found to slow the synthesis of SM (4). In addition, the transport of newly formed viral protein, GlcCer, and SM from the Golgi membranes to the cell periphery of CHO cells was also slowed. (The lipids were followed by incubating the cells with fluorescent ceramide before being treated with PDMP.) The authors point out that these reductions in transport occurred so quickly on exposure to inhibitor that a reduction in total GSL could not be involved, but rather that the reduction in lipid synthesis and transport are functionally related. However, it is possible that the relatively few sphingolipid molecules in the transport system are the ones that control transport rates and that they undergo very rapid turnover. Thus, the level of these lipids could drop rapidly very soon after PDMP reaches the synthases.

At 25 μM or higher concentration of DL-PDMP, there was a significant reduction in the sialylation of viral protein in these CHO cells (4). This raises the possibility that D- or L-PDMP directly inhibits sialyltransferases that act on proteins and, perhaps, on GSLs too. However, in two other cell systems, L-PDMP produced an *increase* in ganglioside G_{M3}, the sialylation product of LacCer (8, 13).

Slight but interesting changes in the appearance of the mitochondria and Golgi stacks were also observed in this detailed study. Very densely staining bodies have been detected by electron microscopy in PDMP-treated MDCK cells (J. Shayman and N. Radin, unpublished work).

The differentiation of cloned rat skeletal muscle cells has been under study (43). PDMP (25 μM) greatly inhibited the normal cell fusion that should lead to formation of myotubes. However, the synthesis of protein, glycoprotein, phospholipids, and nucleic acids was unaffected. The lack of an effect on growth is unusual. The blockage of fusion could be partially prevented by including LacCer or G_{M3} in the medium. These GSLs, when added to normal, uninhibited cells, had no effect on cell fusion. Myotube formation presumably involves a type of intercellular binding, a phenomenon based on binding between specific GSLs on the surface of both cells, as well as the mutual binding between the surface matrix proteins and their receptors on adjacent cells.

One of the curious features in the GSL literature is the finding that certain GSLs stimulate growth while others inhibit growth and lead to differentiation (indicated by morphological changes). Thus, PDMP, by lowering the level of all GSLs, should block both growth *and* differentiation. Supplementation of cells with GlcCer should lead to increases in all GSLs, including G_{M3}, which appears to counteract the growth effect. The recent work of Ogura and Sweeley with skin fibroblasts points to LacCer as a proliferation stimulator and G_{M3}, its anabolic product, as a proliferation inhibitor (21). Yet, G_{M3} reversed the inhibitory effect of PDMP in several studies. No doubt other factors control which one of these two effects is dominant at a particular stage.

Fragments of goldfish retinas, cultured in a simple medium, produce neuron-like outgrowths called neurites. This growth process was readily inhibited by D- or L-PDMP (A. Heacock, B. Agranoff, and N. Radin, unpublished work), but the addition of GlcCer to the medium failed to reverse the inhibition. Possibly the retinal explants, which are several cell layers thick, are too impervious to the lipid. A similar study with murine neuroblastoma cells (44) also showed blockage of outgrowth by PDMP. Ganglioside G_{M1} partially reversed the inhibition. Curiously, cell division was inhibited by PDMP only in dilute cultures. Perhaps this is due to a slowing of GlcCer synthesis as cultured cells approach confluence; alternatively, it is possible that more PDMP is absorbed by each cell when there are fewer competing cells. The content of LacCer and G_{M2} actually rose a little when DL-PDMP was used. This seems to be another example of the ability of L-PDMP to induce an elevation in LacCer level.

XIII. PDMP as an Anticancer Drug

The first study with the optically active form of PDMP was done in mice that were inoculated i.p. with Ehrlich ascites carcinoma cells (15, 45). The rationale for trying this was based on a host of published reports, discussed in Radin and Inokuchi (46). These included reports on the occurrence of unusual GSLs in tumors, use of immunotherapy with anti-ganglioside antibodies, the shedding of GSLs by tumors and their possible blockage of immunoresistance, the changes in GSL composition that take place as a cell becomes transformed or more malignant, the elevated levels of serum GSLs in cancer patients, the abnormal levels of GSL-metabolizing enzymes in tumors, and the many reports of GSL involvement in growth processes. Hakomori has suggested that many forms of cancer are the result of aberrant expression of GSLs (32).

The control mice inoculated with Ehrlich cells died in ~24 days. When PDMP was injected i.p. once a day for 10 days or twice a day for 5 days, starting 1 day after the inoculation, marked protection against development of the tumor cells was seen. About 70% of the mice died eventually, somewhat later than the untreated mice, and ~30% *never* showed any growth of the tumor cells. This promising observation was supplemented by the observation that reinoculating the survivors with new Ehrlich cells failed to yield tumor growth in all the mice. This suggests that the treated, surviving mice had developed antibodies to the tumor and that PDMP does not interfere with the normal process of immunorejection (shown also by the normal blood cell counts).

Several studies have shown that tumors shed GSLs that block the normal activity of immunorejecting cells (47). This helps explain why so few cancer patients are able to generate antibodies to their tumors. Presumably, the tumor cells in the mice treated with PDMP could not protect themselves against immunological attack.

In two of the animals, a small solid tumor formed despite the PDMP treatment but this was resorbed later. Presumably, these mice developed enough antibodies to lyse the tumor eventually. It seems as though the blockage effect on glucolipid synthesis has to slow tumor growth long enough to give the animal time to develop its antibody defense.

In experiments in which the administration of PDMP was delayed, to allow the Ehrlich cells to grow to a high cell density, PDMP could only retard their growth during the period of treatment. GlcCer, injected i.p. into untreated tumor-bearing mice, stimulated cell growth ~50%, which suggests that the tumor cells could not synthesize GlcCer fast enough to achieve their full growth potential.

Preliminary tests in this system suggested that the palmitoyl and stearoyl homologs of PDMP were more effective in curing the inoculated mice. (See Section V,B for a discussion of homologs.) The C_{18} homolog cured 75% of the inoculated mice.

Tests with L-PDMP showed that it too had distinct therapeutic activity. At no time was ascitic fluid visible in the treated mice but solid tumors did develop ultimately. We now know that L-PDMP, in some systems, has a fair amount of ability to inhibit GSL synthesis (see Section X), so the antitumor effect is consistent with the GSL–cancer hypothesis.

Two "long-term" toxicity tests were made with normal mice. In one experiment, they were injected with D-PDMP at 100 mg/day for 10 days, then killed 5 hours later. The treated mice were 8% lighter and several organs were significantly smaller: liver (−18%), kidney (−17%), and brain (−5%). When these decreases were corrected for body weight, only liver and kidney showed significant decreases. In the other test, the PDMP was given 12 times and the mice were allowed to recover for 40 hours. Here, only the kidneys were significantly smaller (−11%). From our tracer study with labeled PDMP (12), we learned that the kidney has a particular affinity for the inhibitor, which explains why the kidneys took longer to recover from the drug. Perhaps PDMP would be especially useful in cases of renal cell carcinoma.

The dosages used here were relatively high, compared to a typical commercial chemotherapeutic agent. We now know that most of the injected drug was metabolized by a *P*-450 system and that this can be somewhat blocked by a *P*-450 inhibitor (12).

The studies described above have been criticized because Ehrlich ascites cells are genetically somewhat different from Sprague-Dawley mice. They are thus more likely to elicit antibodies than the autologous tumors formed in patients *de novo*. While this factor is no doubt important, people have in some instances been shown to produce antibodies to their own tumors. Moreover, exogenous antibodies to gangliosides have shown promise in treating melanoma patients.

PDMP was tested with C_6 glial tumor cell line, with the cells stereotactically implanted into the striatum of rat brains (J. Olson and N. Radin, unpublished work). Starting one day after implantation, the rats were injected i.p. with DL-PDMP four times a day for 5 days. The untreated and PDMP-treated rats were killed 5 hours after the last injection and the brains were sectioned and stained with a radioactive glia-specific reagent, PK 11195. Every fifth section was autoradiographed and the area of each tumor-positive section was determined with a video-computer. The tumor volumes in the treated rats were 71% smaller, showing, as with the [^{3}H]PDMP study (Section V,A), that PDMP readily enters brain and specifically slows the growth of tumor cells, even if they are solid rather than free-floating in ascites fluid.

Another tumor cell type, the mouse Lewis lung carcinoma 3LL cell, was studied by a different approach (8). These cells were first depleted of GSLs by growth in medium containing 5 μM PDMP for 6 days. While the growth rate was somewhat slowed, the cells showed good trypan blue rejection and could replete their GSLs when placed in PDMP-free media. (This is another study that showed cells could survive a large loss of GSLs.)

When the cells were injected i.v. into mice, colonies of the tumor formed in the lungs. By the end of a 20-day period, undepleted cells had formed 28 colonies per mouse, whereas depleted cells had formed only 8.5 colonies. The former cells also formed extrapulmonary tumors but the latter cells formed none. It seems likely that the depleted cells, after inoculation into host mice, were able to resynthesize their GSLs or absorb GSLs from the host, but were poorly resistant to the mouse's immunoprotective system.

XIV. Discussion

A. Questions of Specificity

Considering the structure of PDMP and its L-enantiomer, one would anticipate that any enzyme dealing with ceramide or a GSL might be affected. We did report slight inhibition of GalCer synthase in a microsomal preparation by 100 μM D-PDMP (11), and Rosenwald *et al.* (4) also reported some inhibition of SM synthesis in cultured cells by 50–100 μM PDMP, so the inhibition of GlcCer synthase is not absolutely specific. However, the latter enzyme is blocked very strongly in media containing only 5 μM inhibitor and growth inhibition has been detected even at 1 μM. A test of the α-galactosidase that acts on GalGalCer and GalGalGlcCer showed no effect of L- or D-PDMP (A. Abe and N. Radin, unpublished work). Cultured CHO cells incubated with as much as 100 μM PDMP showed no changes in the appearance and reactivity of the endoplasmic reticulum and microtubules, or in lysosomal uptake or endocytosis (4). The concentration of protein within cells was unaffected.

A possible factor for any lipoidal amine (such as PDMP or chloroquine or sphingosine) is that the amine may be taken up by lysosomes, where it might induce an increase in pH and slow the action or turnover of lysosomal hydrolases. Such uptake might also slow the breakdown of the

GSLs, but in fact the rate of disappearance of the GSLs from PDMP-treated cells seemed to be unexpectedly fast. As mentioned in Section V,A, PDMP is indeed largely localized in the lysosomes. Moreover, it was found (4) that other lipoidal amines did not affect the transport of fluorescent GlcCer and SM from the Golgi to the plasma membrane, nor did stearylamine block GlcCer synthesis.

On the other hand, stearylamine, DL-PDMP, sphinganine, and the antidepressant drug imipramine were all found to inhibit the transport of cholesterol out of lysosomes (48). They produced a model form of Niemann-Pick disease, type C, a genetic disorder in which cholesterol, GlcCer, and SM accumulate. Sphingol accumulation in this disorder was proposed as the primary genetic lesion. This would seem to bypass PKC as a factor in sphingol effects.

In several studies, the effect of PDMP was partially or completely blocked by the addition of exogenous GlcCer or trisialoganglioside or G_{M3}. (The latter two are presumably hydrolyzed to GlcCer.) This is a good test of specificity of action: if PDMP's sole effect is on GlcCer synthesis, then replacement of needed GlcCer from exogenous sources (if uptake is fast enough) should block the effects of PDMP. However, one must consider the possible metabolic roles of the exogenous lipids; these can be different in different cell types or intact animals. GalCer should fail to reverse the effects of PDMP, an observation we made in a few instances and which suggests also that the ceramide and sphingols formed from GalCer were not important.

There is a technical problem with getting good uptake of an insoluble lipid: when a solution of GlcCer in ethanol or dimethyl sulfoxide is added to cell medium, the lipid precipitates. The precipitate may not be noticeable for a day if the GlcCer "solution" is dilute. Presumably, the aggregates settle in the incubation dish, with some cells receiving a particle and other cells not. Then the corrective or other effect of the exogenous lipid may appear to be incomplete. This phenomenon can give rise to excessive interexperimental variability, depending on the size of the aggregates in each preparation. Some authors have stated that they could complex GlcCer with low-fatty acid BSA, but in our experience the lipid eventually precipitated. We have used a nonionic detergent, Myrj 52, to emulsify GlcCer, by coevaporating solutions of the two and sonicating the residue with a small volume of water. However, we later found that the Myrj adversely affected MDCK cell growth, particularly when the number of cells in the dish was low (7).

In recent experiments with MDCK cells (48a), we resorted to a mixture of synthetic lecithins (dioleoyl and stearoyl–palmitoyl), dissolved in warm ethanol with the GlcCer, injected into culture media. To make the dispersion more stable, we utilized GlcCer containing octanoic acid as the fatty acid moiety. However, the abnormal GlcCer led to rapid conversion to octanoyl sphingosine and sphingosine, with concomitant greatly elevated formation of *natural* ceramide and GlcCer. In addition, there was a decrease in PKC, growth rate, LacCer, and SM. The use of unnatural lipids is clearly fraught with danger.

B. What Does PDMP Actually Do to Cells?

1. PDMP clearly lowers the levels of all the glucosphingolipids. It seems unlikely that some GSLs do not undergo hydrolysis.

2. PDMP makes the phospholipase C that acts on phosphatidylinositol diphosphate more active and more sensitive to hormonal activation. This must affect the distribution of Ca^{2+}. Perhaps other messenger reactions are similarly slowed by the presence of GSLs? Phosphatidylinositol bisphosphate activates PKC (49), so a loss of the lipid in PDMP-treated cells should lower the level of PKC, in agreement with our observations.

3. PDMP, in cultured cells, causes ceramide, sphingosine, and dimethylsphingosine to accumulate. In some experiments, PDMP also caused a considerable increase in DAG, probably due to the enzymatic reaction between lecithin and ceramide. Analysis of the fatty acids in the DAGs appearing in PDMP-treated cells showed that most of the diglyceride came from lecithin (50). Some DAGs, in normal cells, must also come from the action of phosphatase on phosphatidic acid, which could be formed from lecithin by phospholipase D. However, sphingosine, which is elevated in PDMP-treated cells, is said to block this phosphatase, so this route may be a minor one. DAG, of course, is a well-known stimulator of PKC.

Sphingosine, added to parotid acinar cells, produced increased synthesis of $InsP_3$ (51). This offers an additional mechanism whereby PDMP acts (by increasing ceramide and sphingol levels).

Ceramide is an effective activator of a newly described protein kinase that acts on the epithelial growth factor receptor and other proteins (52). Thus, PDMP can be expected to stimulate phosphorylation of EGF receptor. Phosphorylation of the receptor is said to stimulate the hydrolysis of phosphatidylinositol bisphosphate (53), an observation like the one made in PDMP-treated MDCK cells (16). Thus, it is possible that PDMP stimulated the formation of inositol trisphosphate by causing accumulation of ceramide rather than by lowering the level of GlcCer.

But which of these effects are significant in PDMP-treated cells? Mice accumulated little ceramide when they were injected with D-PDMP, so there may be a major difference in PDMP mechanisms *in vivo* and "in plastico."

4. PDMP inhibits cell growth (protein synthesis) and DNA synthesis (thymidine incorporation). Conduritol B epoxide, which causes GlcCer accumulation instead of depletion, produces the opposite effect. Yet, the growth inhibition by PDMP, if produced slowly by low concentrations, does not produce cell death even when most of the GSLs have disappeared. There is a need to analyze separately cells that remain adherent to the culture dish and cells that float off.

5. The activities of several sphingolipid hydrolases are increased by PDMP treatment. Presumably this speeds the disappearance of the sphingolipids. Perhaps the presence of PDMP in lysosomes stabilizes the enzymes against autolysis. Do other lipoidal amines do this?

6. PDMP-treated cells lose much of their ability to bind to matrix proteins. Binding between cells is, in some cases, dependent on the interaction between specific GSLs in the different cells (54); this means that PDMP should cause such cells to disaggregate. The binding of viruses, toxins, and bacteria to certain cells is dependent on the specific GSLs in those cells. This kind of binding should also be reduced by PDMP treatment; so should the metastasis of cancer cells.

C. Possible Therapeutic Uses of PDMP

The original motivation for developing a GlcCer synthase inhibitor was to identify a potential therapeutic modality for treating Gaucher's disease and related genetic disorders. Fabry's disease, a disorder arising from defective α-galactosidase activity, results in renal failure, in addition to neurovascular complications. Both α-galactosyl-GalCer and α-galactosyl-LacCer accumulate in these patients. The tendency of PDMP to localize in the kidney makes this drug a potentially useful agent for preventing accumulation of the trihexoside, but a new inhibitor would be needed to block synthesis of the dihexoside.

PDMP might prove useful in treating infections. Bacterial adhesins, for example, have been observed to bind specifically to the lactosyl moieties of GSLs, particularly in the urinary tract. PDMP administration should lower the cellular content of GSLs and block the binding. Preliminary tests *in vitro* with several transformed epithelial cells lines (C. Svanborg and R. Lindstedt, unpublished work) showed that binding of *Escherichia coli* cells was greatly reduced by depleting the epithelial cells with PDMP. In individuals who are already infected, do GSLs in the plasma membrane undergo hydrolysis even if they are *already* bound to a microorganism? If they are stabilized by the adhesin, PDMP would lower only the level of adjacent GSLs that are still free. At least such an effect would prevent binding by additional bacteria.

The potential antineoplastic effect of PDMP has been described in Section XIII. Hakomori (32) has endorsed the idea of further testing of PDMP, calling it a mode of orthosignaling therapy.

Several other diseases are associated with a disordered growth response and may be amenable to treatment with a GlcCer synthase inhibitor. Many collagen vascular diseases are associated with scarring due to uncontrolled proliferation of fibroblasts. Similarly, the proliferative response of keratinocytes associated with dermatologic disorders, such as psoriasis, may respond to PDMP. Recent reports have demonstrated the requirement of ceramides for the maintenance of dermal integrity (55). Because of the tendency of PDMP to cause ceramide accumulation, it could conceivably prove helpful in repairing damaged skin. (So too might GlcCer, which is converted to ceramide in the skin.)

The human lens has been found to accumulate a Lewis^x pentahexosylceramide with age. The cataract that typically appears with age has been attributed to this localized "sphingolipidosis," possibly due to the ability of the GSL to enhance cell–cell binding (56). PDMP should stop formation of the lipid and allow the lens glycosidases (if present) to destroy it.

Hypertrophic responses, in addition to hyperproliferative responses, may also be remedied by blockade of GlcCer synthesis. Recently, we have observed that the renal hypertrophy of early diabetes mellitus in rats is associated with increased levels of GlcCer and G_{M3} (due to the glucose-induced elevation in UDP-Glc level). Treatment of diabetic rats with PDMP and piperonyl butoxide for 4 days reversed the increase in kidney size, as measured by renal weights and glomerular volumes. These changes were observed in spite of the fact that the rats remained hyperglycemic (57).

Acknowledgments

Preparation of this manuscript was supported by USPHS grants DK-39255 and DK-41487, a Merit Review Award from the Dept. of Veterans Affairs (JAS), the Glycolipid Research Fund, and a grant from the Uehara Memorial Foundation (JI).

References

1. Vunnam, R. R., and Radin, N. S. (1980). *Chem. Phys. Lipids* **26,** 265–278.
2. Hospattankar, A. V., Vunnam, R. R., and Radin, N. S. (1982). *Lipids* **17,** 538–543.
3. Vunnam, R. R., Bond, D., Schatz, R. A., Radin, N. S., and Narasimhachari, N. (1980). *J. Neurochem.* **34,** 410–416.
4. Rosenwald, A. G., Machamer, C. E., and Pagano, R. E. (1992). *Biochemistry* **31,** 3581–3590.

5. Radin, N. S., and Inokuchi, J. (1991). *Trends Glycosci. Glycotechnol.* **3,** 200–213.
6. Shukla, G., Shukla, A., Inokuchi, J., and Radin, N. S. (1991). *Biochim. Biophys. Acta* **1083,** 101–108.
7. Abe, A., Inokuchi, J., Jimbo, M., Shimeno, H., Nagamatsu, A., Shayman, J. A., Shukla, G. S., and Radin, N. S. (1992). *J. Biochem.* (*Tokyo*) **111,** 191–196.
8. Inokuchi, J., Jimbo, M., Momosaki, K., Shimeno, H., Nagamatsu, A., and Radin, N. S. (1990). *Cancer Res.* **50,** 6731–6737.
9. Fenderson, B. A., Ostrander, G. K., Hausken, Z., Radin, N. S., and Hakomori, S.-I. (1992). *Exp. Cell Res.* **198,** 362–366.
10. Radin, N. S., and Vunnam, R. R. (1981). *In* "Methods in Enzymology" (J. Lowenstein, ed.), vol. 72, pp. 673–684. Academic Press, New York.
11. Inokuchi, J., and Radin, N. S. (1987). *J. Lipid Res.* **28,** 565–571.
12. Shukla, A., and Radin, N. S. (1991). *J. Lipid Res.* **32,** 713–722.
13. Inokuchi, J., Momosaki, K., Shimeno, H., Nagamatsu, A., and Radin, N. S. (1989). *J. Cell. Physiol.* **141,** 573–583.
14. Paul, P., Bordmann, A., Rosenfelder, G., and Towbin, H. (1992). *Anal. Biochem.* **204,** 265–272.
15. Inokuchi, J., Mason, I., and Radin, N. S. (1987). *Cancer Lett.* **38,** 23–30.
16. Shayman, J. A., Mahdiyoun, S., Deshmukh, G., Barcelon, F., Inokuchi, J., and Radin, N. S. (1990). *J. Biol. Chem.* **265,** 12135–12138.
17. Shukla, G. S., and Radin, N. S. (1990). *Arch. Biochem. Biophys.* **283,** 372–378.
18. Mahdiyoun, S., Deshmukh, G., Abe, A., Radin, N. S., and Shayman, J. A. (1992). *Arch. Biochem. Biophys.* **292,** 506–511.
19. Hirata, K., Akita, H., and Yokoyama, M. (1991). *FEBS Lett.* **287,** 181–184.
20. Chatterjee, S., Clarke, K. S., and Kwiterovich, P. O., Jr. (1986). *J. Biol. Chem.* **261,** 13474–13479.
21. Ogura, K., and Sweeley, C. C. (1992). *Exp. Cell Res.* **199,** 169–173.
22. Uemura, K., Sugiyama, E., Tamai, C., Hara, A., Taketomi, T., and Radin, N. S. (1990). *J. Biochem.* (*Tokyo*) **108,** 525–530.
23. Ferret, B., Hubsch, A., Dreyfus, H., and Massarelli, R. (1991). *Neurochem. Res.* **16,** 137–144.
24. Okada, Y., Radin, N. S., and Hakomori, S.-I. (1988). *FEBS Lett.* **235,** 25–29.
25. Shayman, J. A., Deshmukh, G., Mahdiyoun, S., Thomas, T. P., Wu, D., Barcelon, F. S., and Radin, N. S. (1991). *J. Biol. Chem.* **266,** 22968–22974.
26. Felding-Habermann, B., Igarashi, Y., Fenderson, B. A., Park, L. S., Radin, N. S., Inokuchi, J., Strassmann, G., Handa, K., and Hakomori, S.-I. (1990). *Biochemistry* **29,** 6314–6322.
27. Endo, K., Igarashi, Y., Nisar, M., Zhou, Q., and Hakomori, S.-I. (1991). *Cancer Res.* **51,** 1613–1618.
28. Mori, T., Takai, Y., Minakuchi, R., Yu, B., and Nishizuka, Y. (1980). *J. Biol. Chem.* **255,** 8378–8380.
29. Kiguchi, K., Henning-Chubb, C., and Huberman, E. (1986). *Cancer Res.* **46,** 3027–3033.
30. Barbour, S., Edidin, M., Felding-Habermann, B., Taylor-Norton, J., Radin, N. S., and Fenderson, B. A. (1992). *J. Cell. Physiol.* **150,** 610–619.
31. Dyer, C. A., and Benjamins, J. A. (1989). *J. Neurosci. Res.* **24,** 212–221.
32. Hakomori, S.-I. (1991). *Cancer Cells* **3,** 461–470.
33. Benjamin, K., Deshmukh, G., Liebert, M., Radin, N. S., and Shayman, J. A. (1991). *J. Am. Soc. Nephrol.* **2,** 571.
34. Inokuchi, J., Kumamoto, Y., Jimbo, M., Shimeno, H., and Nagamatsu, A. (1991). *FEBS Lett.* **286,** 39–43.

35. Chatterjee, S. (1991). *Biochem. Biophys. Res. Commun.* **181,** 554–561.
36. Mueller, O. T., and Rosenberg, A. (1979). *J. Biol. Chem.* **254,** 3521–3525.
37. Shukla, G. S., Shukla, A., and Radin, N. S. (1991). *J. Neurochem.* **56,** 2125–2132.
38. Brenkert, A., and Radin, N. S. (1982). *Brain Res.* **36,** 183–193.
39. Radin, N. S., Brenkert, A., Arora, R. C., Sellinger, O. Z., and Flangas, A. L. (1972). *Brain Res.* **39,** 163–169.
40. Byrne, M. C., Farooq, M., Sbaschnig-Agler, M., Norton, W. T., and Ledeen, R. W. (1988). *Brain Res.* **461,** 87–97.
41. Datta, S. C., and Radin, N. S. (1986). *Lipids* **21,** 702–709.
42. McKay, C. P., Mei, Z-Y., Radin, N. S., and Shayman, J. A. (1992). *Clin. Res.* **40,** 263A.
43. Cambron, L. D., and Leskawa, K. C. (1991). *Trans. Am. Soc. Neurochem.* **22,** 154.
44. Uemura, K., Sugiyama, E., and Taketomi, T. (1991). *J. Biochem.* (*Tokyo*) **110,** 96–102.
45. Radin, N. S., and Inokuchi, J. (1991). U. S. Patent 5,041,441.
46. Radin, N. S., and Inokuchi, J. (1988). *Biochem. Pharmacol.* **37,** 2879–2886.
47. Li, R., and Ladisch, S. (1991). *Biochim. Biophys. Acta* **1083,** 57–64.
48. Roff, C. F., Goldin, E., Comley, M. E., Cooney, A., Brown, A., Vanier, M. T., Miller, S. P. F., Brady, R. O., and Pentchev, P. G. (1991). *Dev. Neurosci.* **13,** 315–319.
48a. Abe, A., Wu, D., Shayman, J. A., and Radin, N. S. (1992). *Eur. J. Biochem.* **210,** 765–773.
49. Chauhan, A., Brockerhoff, H., Wisniewski, H. M., and Chauhan, V. P. S. (1991). *Arch. Biochem. Biophys.* **287,** 283–287.
50. Shayman, J. A., Sculley, D., Lee, C., Hajra, A. K., and Radin, N. S. (1992). *Clin. Res.* **40,** 216A.
51. Sugiya, H., and Furuyama, S. (1990). *Cell Calcium* **11,** 469–475.
52. Goldkorn, T., Dressler, K. A., Muindi, J., Radin, N. S., Mendelson, J., Menaldino, D., Liotta, D., and Kolesnick, R. N. (1991). *J. Biol. Chem.* **266,** 16092–16097.
53. Hepler, J. R., Jeffs, R. A., Huckle, W. R., Outlaw, H. E., Rhee, S. G., Earp, H. S., and Harden, T. K. (1990). *Biochem. J.* **270,** 337–344.
54. Kojima, N., and Hakomori, S. (1991). *J. Biol. Chem.* **266,** 17552–17558.
55. Holleran, W. M. (1991). *Adv. Lipid Res.* **24,** 119–139.
56. Ogiso, M., Irie, A., Kubo, H., Hishi, M., and Komoto, M. (1992). *J. Biol. Chem.* **267,** 6467–6470.
57. Zador, I. Z., Deshmukh, G. D., Kunkel, R., Johnson, K., Radin, N. S., and Shayman, J. A. (1993). *J. Clin. Invest.* **91,** 797–803.

Fumonisins and Other Inhibitors of *de Novo* Sphingolipid Biosynthesis

ALFRED H. MERRILL, JR.,* ELAINE WANG,*
DAVID G. GILCHRIST,† AND RONALD T. RILEY‡

**Department of Biochemistry*
Rollins Research Center
Emory University School of Medicine
Atlanta, Georgia 30322
†Department of Plant Pathology
University of California
Davis, California 95616
‡Toxicology and Mycotoxins Research Unit, USDA-ARS
Athens, Georgia 30613

I. Introduction

Fusarium moniliforme (Sheldon) is one of the most prevalent molds on corn, sorghum, and other grains throughout the world (Marasas *et al.*, 1984a). *F. moniliforme* has been shown to be toxic and carcinogenic for animals both as a contaminant of grains and as a culture isolate (Kellerman *et al.*, 1972; Kriek *et al.*, 1981; Marasas *et al.*, 1984b; Voss *et al.*, 1990; Gelderblom *et al.*, 1988). Consumption of *F. moniliforme*-contaminated corn has been correlated with human esophageal cancer in areas of southern Africa, China, and other countries (Li *et al.*, 1980; Yang, 1980; Marasas, 1982; Sydenham *et al.*, 1990a).

This mold contains a number of toxins, a group of which (termed fumonisins) (Bezuidenhout *et al.*, 1988; Sydenham *et al.*, 1990a,b) are thought to be mainly responsible for these diseases. Purified fumonisin B_1 has been shown to cause equine leukoencephalomalacia (ELEM) (Marasas *et al.*, 1988; Kellerman *et al.*, 1990; Thiel *et al.*, 1992), porcine pulmonary edema (Harrison *et al.*, 1990), and hepatotoxicity and liver tumors in rats (Gelderblom *et al.*, 1988, 1991).

The fumonisins bear a remarkable structural similarity to sphinganine (Fig. 1). This led us to hypothesize that the mechanism of action of these mycotoxins may be via disruption of sphingolipid metabolism or function (Wang *et al.*, 1991). As is described in this review, there is now considerable evidence in support of this hypothesis. Studies of the mechanism of action of these compounds are providing leads that will help us to understand the diseases associated with consumption of fumonisins as well as to develop methods that will be useful in monitoring fumonisin consumption.

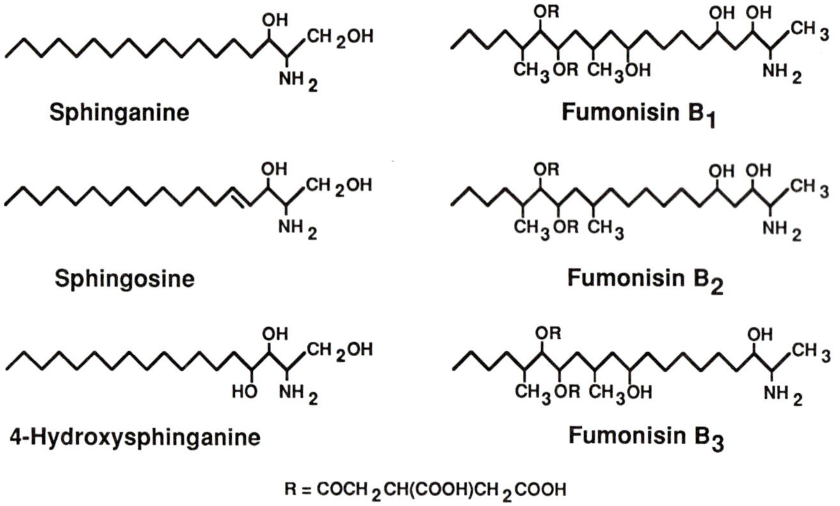

Sphinganine

Fumonisin B_1

Sphingosine

Fumonisin B_2

4-Hydroxysphinganine

Fumonisin B_3

R = $COCH_2CH(COOH)CH_2COOH$

FIG. 1. Structures of fumonisins and comparison to long-chain (sphingoid) bases. Shown on the left are the structures of the major long-chain bases of mammalian sphingolipids: most have alkyl chain lengths of 18 carbon atoms, but smaller amounts of other homologs are usually also found in mammalian sphingolipids. Other frequently used terms for these compounds are sphingenine (for sphingosine), dihydrosphingosine (for sphinganine), and phytosphingosine (for 4-D-hydroxysphinganine). On the right are the major fumonisins produced by *F. moniliforme*. A related series of fumonisins are acetylated on the amino group. These are designated fumonisin A1, A2, etc. and appear to have a much lower toxicity than the compounds with the free amino group (Gelderblom *et al.*, 1992).

II. Discovery of Fumonisins as Inhibitors of *de Novo* Synthesis of Sphingosine by Rat Liver Hepatocytes

Rat hepatocytes were selected as a model for studies of fumonisins because *F. moniliforme* culture materials and fumonisin B_1 are hepatotoxic and hepatocarcinogenic in the rat (Gelderblom *et al.*, 1991; Marasas *et al.*, 1984b; Voss *et al.*, 1990). Furthermore, sphingosine biosynthesis *de novo* by rat hepatocytes is relatively easy to follow using [^{14}C]serine (Messmer *et al.*, 1989), and appears to proceed via the reactions shown in Fig. 2, like in other cells (Merrill and Jones, 1990; Merrill and Wang, 1986). In this scheme, serine is first converted to 3-ketosphinganine, followed by reduction to sphinganine and acylation to *N*-acylsphinganines (also termed dihydroceramides). The 4,5-*trans* double bond of the sphingosine backbone of more complex sphingolipids is added at a later, as-yet-uncharacterized step in the pathway.

Fumonisin B_1 caused an almost complete inhibition of [^{14}C]sphingosine formation by hepatocytes (Fig. 3A), with an IC_{50} of approximately 0.1 μM for fumonisin B_1 (Fig. 3B) (Wang *et al.*, 1991). A similar degree of inhibition was obtained with fumonisin B_2, another mycotoxin often produced in substantial amounts by *F. moniliforme* (Voss *et al.*, 1989; Cawood *et al.*, 1991). In contrast, there was no reduction in the radiolabeling of fatty acids (Fig. 3A), phosphatidylserine, phosphatidylethanolamine, or phosphatidylcholine from [^{14}C]serine, nor in the mass of these phospholipids. Therefore, the inhibition of sphingolipid biosynthesis does not appear to be due to the inability of the cells to take up [^{14}C]serine and to incorporate it into lipids in general.

Initially, there was no significant reduction in the overall mass of sphingolipids when hepatocytes were incubated with fumonisin B_1; however, when the cells were incubated with 1 μM fumonisin B_1 for 4 days, the level of total sphingolipids was reduced to about half of the controls (i.e., from 4.6 $\pm$ 0.7 nmol/dish to 2.1 $\pm$ 0.1 nmol/dish (Wang *et al.*, 1991). This reduction was not due to cell death; fumonisins do not appear to be particularly cytotoxic to cultured hepatocytes, despite their hepatotoxicity *in vivo*.

III. Identification of Ceramide Synthase as the Target of Action of Fumonisins

There are several potential sites at which fumonisins might block the incorporation of [^{14}C]serine into sphingosine (Fig. 2). The primary target appears to be ceramide synthase, as depicted in the figure. This was

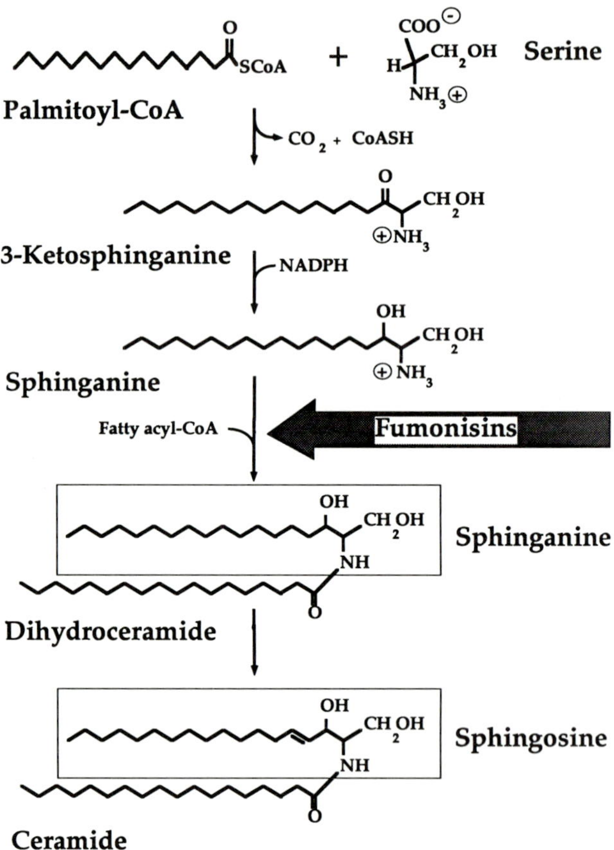

FIG. 2. Pathway of sphingolipid biosynthesis and site of inhibition by fumonisins. For the major chain length homologs (containing 18 carbon atoms), the pathway begins with palmitoyl-CoA as shown. Under usual conditions, little 3-ketosphinganine or sphinganine accumulates, whereas there is a somewhat slower conversion of dihydroceramides to ceramides. The precise step at which the 4-*trans* double bond of sphingosine is added is unclear, except that it appears to occur after dihydroceramide formation, as shown. Fumonisins inhibit sphinganine (sphingosine) *N*-acyltransferase (also called ceramide synthase), which results in an accumulation of sphinganine and reduced synthesis of more complex sphingolipids.

established by several analyses (Wang *et al.*, 1991): (1) The amount of radiolabel in [^{14}C]sphinganine increased when hepatocytes were treated with fumonisin B_1, whereas [^{14}C]sphingosine decreased. (2) Assays of ceramide synthase showed inhibition by fumonisin B_1 with apparent IC_{50} of approximately 0.1 μM. (3) Fumonisin B_1 inhibited the conversion of

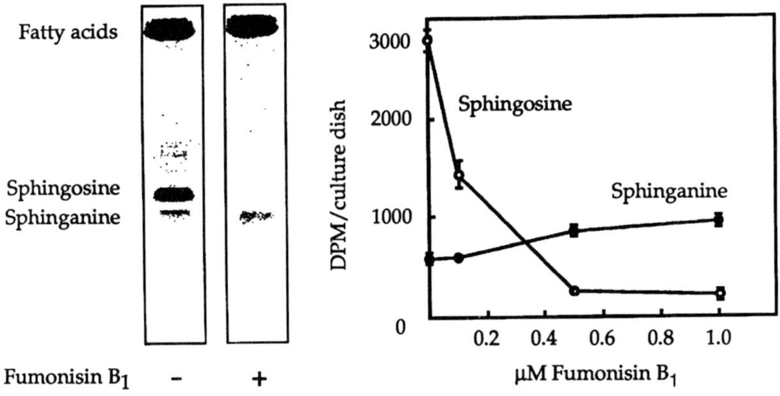

FIG. 3. Inhibition of sphingosine formation in hepatocytes by fumonisin B_1. (A) Isolated rat hepatocytes were incubated with [^{14}C]serine for 24 hours to label the long-chain base backbone of sphingolipids, which were extracted and acid hydrolyzed to liberate the sphinganine and sphingosine. Chromatography on silica gel plates developed with $CHCl_3$: Methanol: 2 N NH_4OH and detection of the radiolabel by autoradiography reveals that most of the label is in sphingosine and a lesser amount in sphinganine (a lot of label is also incorporated into fatty acids). In the presence of 1 μM fumonisin B_1, there is no reduction in the labeling of sphinganine or fatty acids, but the sphingosine band is completely abolished. (B) A dose response for the inhibition. Data redrawn from Wang *et al.* (1991).

[^{3}H]sphingosine to [^{3}H]ceramide by intact cells (Wang *et al.*, 1991, and Fig. 4). When [^{3}H]sphingosine was added to hepatocytes, most of the radiolabel was recovered in [^{3}H]sphingosine and [^{3}H]ceramides; however, a small amount of radiolabel was noted at the solvent front, where the cleavage products of [^{3}H]sphingosine catabolism (trans-4-hexadecenal and fatty acids derived from subsequent metabolism) migrate (no additional label was detected in the region where [^{3}H]sphingosine-1-phosphate migrated). In the presence of fumonisin B_1, there was both a reduction in the amount of radiolabel in [^{3}H]ceramide and an increase in the amount in such cleavage products (Fig. 4). This suggests that inhibition of the acylation of [^{3}H]sphingosine increases the amount that is degraded.

Despite degradation of long-chain bases in the presence of fumonisin B_1, the mass of sphinganine rose significantly in treated hepatocytes (Wang *et al.*, 1991). Increases in sphinganine were evident within hours after adding fumonisin B_1; after incubation with 1 μM fumonisin B_1 for 4 days, the mass of sphinganine increased 110-fold (i.e., to 1499 $\pm$ 18 pmol/dish compared to 13.6 $\pm$ 0.4 pmol/dish for the control). In contrast, there was a small reduction in the amount of free sphingosine within the first day

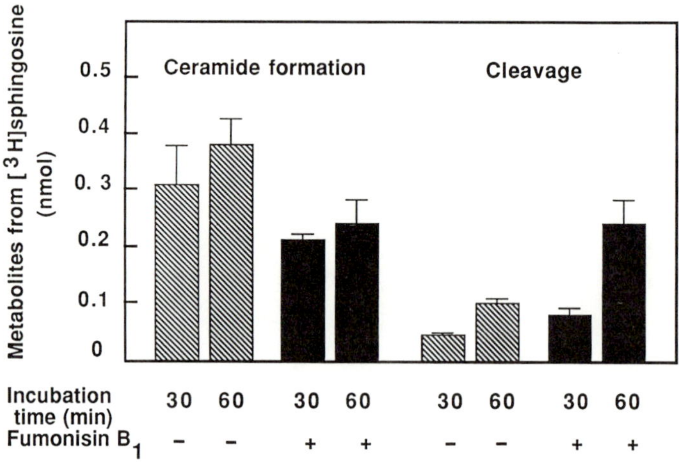

FIG. 4. Inhibition of [^{3}H]sphingosine incorporation into ceramide. Isolated hepatocytes were incubated with 1 μM [^{3}H]sphingosine with and without 1 μM fumonisin B_1 for the times shown, then the lipids were extracted and analyzed (using the thin-layer chromatographic system in Fig. 3) for the amount of radiolabel in [^{3}H]ceramides and at the solvent front (where fatty aldehydes, fatty alcohols, and fatty acids–products of cleavage of sphingosine—migrate). At these short times, no label was detected in more complex sphingolipids nor in the region of sphingosine 1-phosphate.

and, after 4 days of treatment, free sphingosine decreased from 233 ± 14 to 52 ± 1 pmol/dish.

These findings provide further evidence that, as shown in Fig. 2, addition of the 4-*trans* double bond of sphingosine is added after formation of *N*-acylsphinganines (Ong and Brady, 1973; Merrill and Wang, 1986; Messmer *et al.*, 1989)—a reaction that has proved difficult to demonstrate *in vitro* (Merrill and Jones, 1990). They are also consistent with the view that the free sphingosine found in hepatocytes is not an intermediate of the *de novo* biosynthetic pathway but arises from the turnover of complex sphingolipids (Slife *et al.*, 1989) and would, therefore, be most affected when there is a decrease in total sphingolipids due to longer term exposure to these compounds.

IV. Correlation of Inhibition of Sphingolipid Synthesis and Cellular Death in LLC-PK1 Cells

Fumonisins do not appear to be toxic to primary rat hepatocytes over several days of exposure (Wang *et al.*, 1991; Norred *et al.*, 1992), and longer term studies are made difficult by the usual loss of viability of hepatocytes when maintained in culture. Because the proximal tubules of the kidney are another target of fumonisins, we examined a pig kidney renal epithelial cell line (LLC-PK1) for the effects of fumonisin B_1 on cytotoxicity and sphingolipid metabolism (Yoo *et al.*, 1992).

Fumonisin B_1 inhibited proliferation of LLC-PK1 cells at concentrations $\geq 10\ \mu M$ (concentrations greater than 35 μM were cytotoxic and the ED_{50} for effects on cell growth was between 20 and 30 μM) (Fig. 5A). There was a lag period of approximately 24 to 48 hours preceding the inhibition of proliferation, during which the cells appeared normal. Afterward, some cells began to develop a fibroblast-like appearance, with loss of cell–cell contact and an elongated, spindle shape. There was no evidence of blebbing. If fumonisin was removed, the cells that survived resumed growth and had a normal epithelial morphology. Fumonisins B_1 and B_2 had similar effects on the cells (Yoo *et al.*, 1992).

The concentration dependence for inhibition of sphingosine biosynthesis (as reflected in reduced incorporation of [^{3}H]serine into sphingosine and as an accumulation of sphinganine) was approximately 35 μM (Fig. 5B). This probably underestimates the inhibitory potential of fumonisin B_1 in LLC-PK1 cells because the shorter labeling period was not long enough for the [^{3}H]serine to reach a steady state. Nonetheless, the IC_{50} for inhibition of sphinganine metabolism to sphingosine is very close to the EC_{50} for

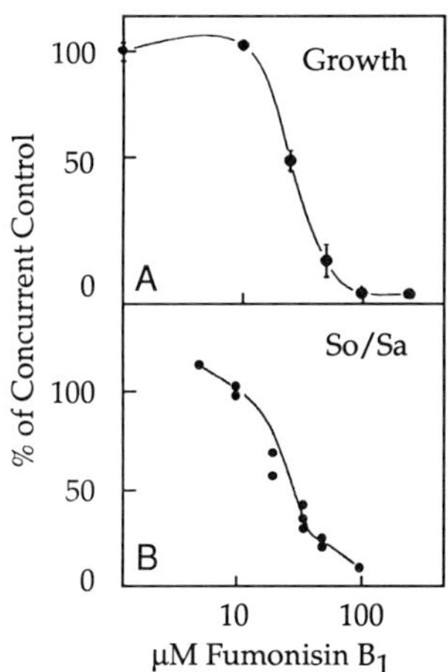

FIG. 5. Comparison of the dose response of (A) growth inhibition for LLC-PK1 cells by fumonisin B_1 and (B) the effects of this compound on the ratio of sphingosine to sphinganine in the cells. For each study, the cells were placed in medium that contains fumonisin B_1 at the concentration shown, then analyzed after 24 hours for the cell number and the amounts of sphingosine and sphinganine. These data were redrawn from Yoo *et al.* (1992).

cytotoxicity (Fig. 5) and strongly suggests that disruption of sphingolipid metabolism leads to cell death.

V. Demonstration of Disruption of Sphingolipid Metabolism on Consumption of Fumonisins by Ponies and Other Animals

Because inhibition at this step blocks the formation of complex sphingolipids while leading to accumulation of sphinganine, we hypothesized that exposure of animals to fumonisins could be detected by analyses of serum sphingolipids. This hypothesis has been proven by recent analyses of serum from ponies given fumonisin-contaminated feed (Wang *et al.*, 1992) and by similar analyses of serum and tissues of pigs (Riley *et al.*, 1993).

Figure 6 shows typical HPLC profiles for the free long-chain bases in serum from these animals using a straightforward method for analyzing long-chain bases (Merrill *et al.*, 1988). Before the ponies were intentionally given fumonisin-contaminated feed, serum contained 65 ± 29 nmol sphingosine/liter and 19.5 ± 6.5 nmol sphinganine/liter (mean ± SD, n = 12). The ratio of sphinganine to sphingosine was somewhat more constant for the animals (0.31 ± 0.12, in analyses of 4 ponies over a 2 to 9 month period). On consumption of fumonisins (Fig. 6, right profile), there was a large increase in sphinganine, a variable increase in sphingosine (see discussion below for more information about this discrepancy), and vari-

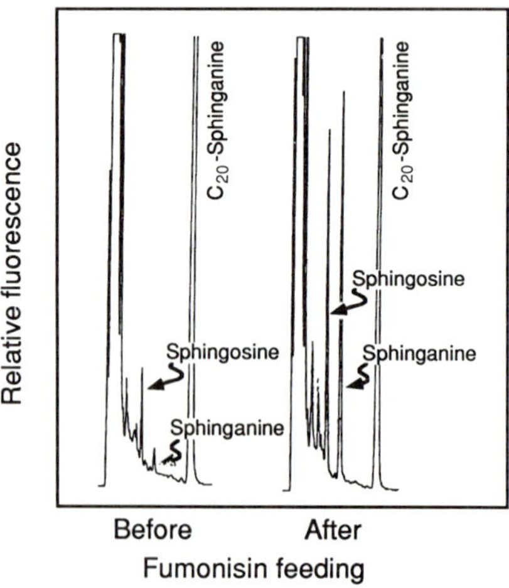

FIG. 6. HPLC profiles for serum from ponies before and after consumption of 44 ppm fumonisin.

able increases in other compounds with earlier (see Fig. 6) and later (although these are not seen in the sample shown in Fig. 6) retention times than sphinganine and sphingosine. These have not yet been structurally characterized, but may be chain length homologs or variously unsaturated long-chain bases (Karlsson, 1970).

A time course for one of the ponies is shown in Fig. 7. This pony consumed a diet containing 44 ppm fumonisin B_1, developed equine leukoencephalomalacia and was euthanized after 45 days because of the severity of the symptoms (Wang *et al.*, 1992). There was a 2.7-fold increase in sphinganine by day 2; by day 7, sphinganine had increased 4.5-fold, to about the same concentration as sphingosine (the sphinganine/sphingosine ratio was 0.95); and, by day 10, there were elevations in sphingosine (2-fold) and sphinganine (13-fold). On day 10, there was a 7-fold increase in serum transaminase activities (not shown), which indicated that the pony had significant cellular injury.

The pony refused to eat the contaminated feed between days 8 and 21, which allowed us to determine if the levels of sphinganine and sphingosine remained elevated after removal of the mycotoxin. The amounts of the long-chain bases were higher for several days after the animal reduced its consumption of the contaminated diet, then began to return to normal. Upon resumption of the fumonisin-contaminated diet, there was another increase in sphinganine and the sphinganine/sphingosine ratio, and a small

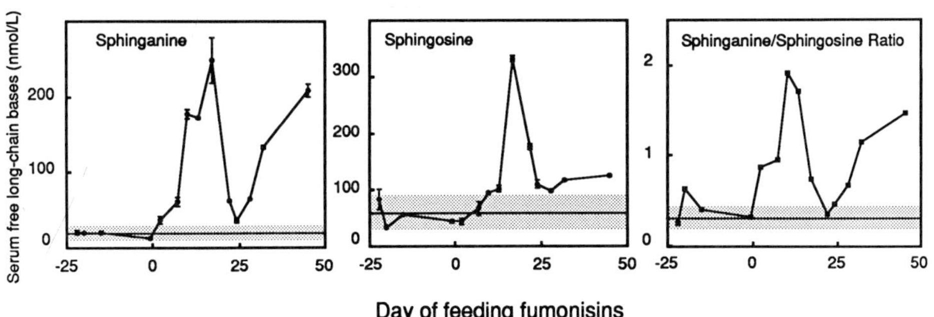

FIG. 7. Time course of the increases in sphinganine and sphingosine in serum from a pony consuming 44 ppm fumonisins. The third panel gives the ratio of sphinganine to sphingosine. From days −25 to 0 the pony was fed diets that contained no detectable fumonisins (<1 ppm). The data for this period, combined with control data from other ponies, are shown by the line and shaded area for the mean ± SD. The pony stopped eating feed contaminated with 44 ppm fumonisins on day 1; between days 8 and 21, the pony did not consume the contaminated diet which accounts for the return to a more normal concentration. On day 45, the pony was euthanized because of the severity of the symptoms of equine leukoencephalomalacia.

change in sphingosine. However, there was no change in serum transaminase activities or other standard parameters of tissue injury (not shown), even though the animal began to show signs of equine leukoencephalomalacia and required euthanasia on day 45.

Similar responses were seen with the other ponies in this study (Wang *et al.*, 1992). The results establish that a ration containing fumonisin B_1 at 22 to 44 ppm increases serum sphingosine and sphinganine levels (and most especially their ratio) after fumonisin exposure, and that these return to more normal levels on removal of (or more variable consumption of) contaminated feed, and increase again when the animals are reexposed to fumonisins. The ratio appears to change within days of exposure, and is significantly elevated before other signs of clinical symptoms are detectable. These changes occurred at fumonisin B_1 concentrations in feed that have been correlated with equine leucoencephalomalacia field cases (Wilson *et al.*, 1990, 1992; Ross *et al.*, 1991a,b), which may mean that analyses of serum long-chain bases might be useful as an early indicator of exposure of animals to fumonisins or related mycotoxins.

There was also a reduction in the amounts of complex sphingolipids in serum (Wang *et al.*, 1992), as would be expected if the elevation in sphinganine is due to inhibition of *de novo* sphingolipid biosynthesis. The concentrations of total (complex) sphingolipids were reduced by 50 to 95% during the early times after feeding fumonisin, but increased again when there was also an elevation in hepatic enzyme activities. One explanation for these changes might be that fumonisin B_1 is blocking the secretion of sphingolipids with lipoproteins, since liver is a major site of synthesis of sphingolipids (Messmer *et al.*, 1989; Merrill and Jones, 1990) and secretes sphingolipids with very-low-density lipoproteins (A. H. Merrill, Jr., S. Lingrell, and D. E. Vance, unpublished data). At later times, when cytotoxicity is evident, the increase in complex sphingolipids in serum might be due to cell death.

These observations are most likely explained by the model shown in Fig. 8. That is, inhibition of the acylation reaction should result in sphinganine accumulation in cells. Some sphinganine should appear in blood because long-chain bases can readily diffuse across cell membranes (Hope and Cullis, 1987; Merrill *et al.*, 1989; Merrill, 1991) and bind to albumin (Merrill *et al.*, 1989). Free sphingosine, on the other hand, would be more likely to arise from inhibition of the reacylation of the sphingosine that arises from the turnover of complex sphingolipids. This is suggested by the time course for the increase in sphingosine (Fig. 7), which increases after changes in sphinganine, perhaps because it involves a significant contribution from the turnover of sphingolipids on cell injury/death. It is not known which tissues are responsible for the elevation in free long-chain bases in

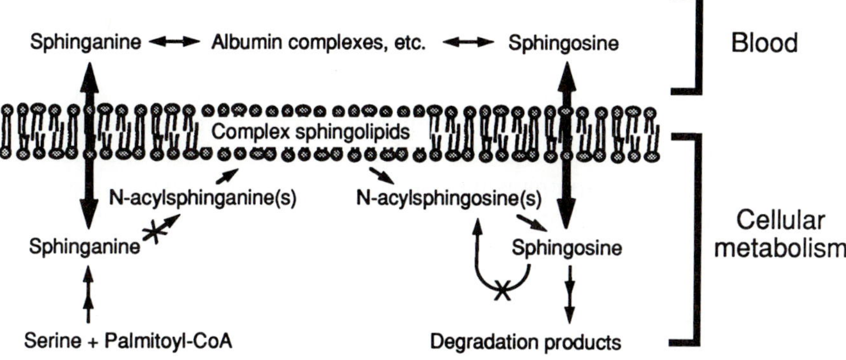

FIG. 8. A scheme to account for the changes in serum sphinganine and sphingosine on consumption of fumonisins. Inhibition of the *de novo* biosynthetic pathway causes accumulation of sphinganine, which can exit the cells and bind to albumin. Fumonisins also block the reacylation of sphingosine released by the turnover of complex sphingolipids and this, too, can appear in serum.

serum because most, if not all, tissues appear to have the capacity for long-chain biosynthesis *de novo* (Merrill *et al.*, 1985).

In a subsequent study (Riley *et al.*, 1993) of pigs fed corn screenings containing 0 to 175 ppm fumonisins (fumonisin B_1 plus B_2), free sphinganine was elevated in liver, lung, and kidney at fumonisin concentrations of 23 ppm or greater. At 175 ppm fumonisins, the free sphinganine concentrations in these organs were elevated 96-, 17-, and 126-fold, whereas sphingosine levels were only increased 5-, 3-, and 5-fold, respectively. At every fumonisin dose, the free long-chain base concentration was much greater in lung than for liver or kidney. For example, at 101 ppm fumonisin, the amounts of sphinganine plus sphingosine were 275, 75, and 100 nmol/g wet weight, respectively. The amount of total complex sphingolipids was also reduced by fumonisins. These changes were accompanied by signs of respiratory distress due to severe pulmonary interstitial edema with pleural effusion for three of the five pigs fed 175 ppm fumonisins. Hepatic injury was observed in most of the pigs fed >39 ppm fumonisins, as characterized by hepatocyte disorganization, single cell necrosis, and mild inflammation.

Free sphinganine and sphingosine in serum paralleled that in tissues, and statistically significant increases in the ratio of these long-chain bases were seen at levels as low as 5 ppm. The concentrations of free sphingosine showed no clear dose–response relationship; however, there was a statistically significant relationship between the fumonisin level in the diet and the ratio of sphinganine to sphingosine. Serum ratios were elevated in pigs that did not yet exhibit observable gross or microscopic lesions in liver, lung, or kidney.

Similar studies have been conducted with chickens and turkeys (Wiebking *et al.*, 1993) and also find that fumonisins increase the levels of serum sphinganine (and the sphinganine/sphingosine ratio). Thus, the increase in the sphinganine to sphingosine ratio appears to be a sensitive indicator of fumonisin consumption. Changes in the ratio are seen much sooner than detectable tissue lesions, which lends further evidence for this being a useful early indicator of fumonisin exposure before irreversible tissue injury has occurred.

VI. Other Naturally Occurring Inhibitors of Sphingolipid Synthesis

The structural basis for this inhibition is unknown; however, one can speculate that similarities between the fumonisins and long-chain (sphingoid) bases (Fig. 1) allow them to be recognized as substrate (or transition state or product) analogs by ceramide synthase. The absence of a hydroxyl group at carbon 1 may alter their orientation in the active site of this enzyme and preclude acylation or, if acylated, result in an inhibitory ceramide that cannot be removed by addition of a sphingolipid headgroup at that position. These compounds are also unable to follow the usual pathway of long-chain base catabolism, which proceeds via phosphorylation at position 1 followed by cleavage to ethanolamine phosphate and hexadecanal (from sphinganine 1-phosphate) or *trans*-2-hexadecanal (from sphingosine-1-phosphate), and this may contribute to the persistence of the inhibition. The 5-hydroxyl group is not critical because fumonisin B_3, which is structurally similar to fumonisin B_1 but lacks the hydroxyl group at position 5 (Fig. 1) has an IC_{50} of approximately 1 μM (Fig. 9A). Thus far, all of the fumonisins with a free amino group have been inhibitory.

One can envision an additional mode of inhibition wherein the negatively charged tricarbyllic acid groups bind to ceramide synthase in the site that normally interacts with the coenzyme A moiety of the cosubstrate fatty acyl-CoA. This may occur, however, the tricarbyllic acid groups are not critical for inhibition because their removal (which yields a compound termed fumonisin B_{1H}) still leaves a compound that inhibits [^{14}C]serine incorporation into [^{14}C]sphingosine by rat hepatocytes with a low IC_{50} (1 μM) (Fig. 9B). Furthermore, we have found that fumonisin B_{1H} is still cytotoxic to LLC-PK_1 cells (Riley *et al.*, 1993).

The lack of a hydroxymethyl group at position 1 is shared by a number

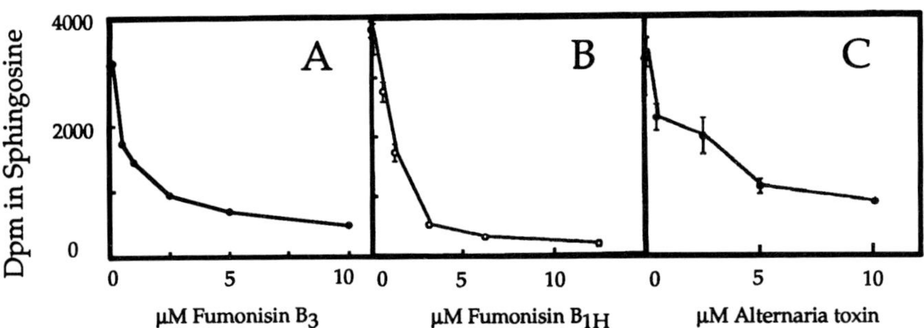

FIG. 9. Concentration dependence of inhibition of *de novo* sphingosine biosynthesis in hepatocytes exposed to fumonisin B_3 (A), fumonisin B_{1H} (B), or *Alternaria* toxin (C). Isolated hepatocytes were treated as described in the legend to Fig. 3. Fumonisin B_{1H} was prepared by mild base cleavage of the tricarballyic acids from fumonisin B_1; full cleavage of the residues was confirmed by NMR and mass spectroscopy.

of other fumonisin-like compounds, such as the host-specific phytotoxins produced by another fungus, *Alternaria alternata f. sp. lycopersici,* that produces the macroscopic symptoms of stem canker disease in tomatoes (Bottini *et al.*, 1981) (Fig. 10). Resistance is conferred by a single Mendelian gene which is inherited as a dominant for resistance to the pathogen but expresses incomplete dominance to the toxin (Clouse and Gilchrist, 1987). For example, the relative phytotoxicity of *Alternaria* toxin in a detached leaflet assay bioassay is 20 μM for plants with the dominant gene (Asc/Asc) and 20 nM for the sensitive isoline (asc/asc) (Gilchrist *et al.*, 1992). Removal of the ester-linked sidechain increases the concentration needed for necrosis by 400-fold (Gilchrist *et al.*, 1992); fumonisin B_1, but not B_{1H}, is also phytotoxic in this assay. Fumonisin B_1, *Alternaria* toxin, and (to a lesser extent) triacetylphytosphingosine are phytotoxic in a duckweed bioassay (Vesonder *et al.*, 1992).

These compounds have been shown to be cytotoxic by *in vitro* assays of various mammalian cells in culture, with IC_{50} ranging from 2- to 7-fold higher than fumonisin B_1 (Shier *et al.*, 1991). We find that *Alternaria*

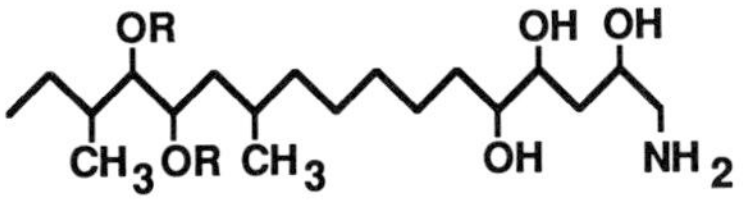

OR OH OH

CH_3 OR CH_3 OH NH_2

$R = COCH_2CH(COOH)CH_2COOH$

FIG. 10. Structure of *Alternaria* toxin.

toxin is also an inhibitor of sphingolipid biosynthesis with an IC_{50} of approximately 1 μM (Fig. 9C), which is about 10-fold higher than the IC_{50} for fumonisin B_1. (In preliminary studies we have also found that *Alternaria* toxin inhibits ceramide synthase *in vitro*.) Therefore, it appears that inhibition of sphingolipid metabolism is involved in the cytotoxicity of these mycotoxins. It is not known how common this structural motif is among other fungi, plants, and organisms. However, considering that it is encountered in organisms as different as the *Fusarium* and *Alternaria* molds, there is a reasonable likelihood that there are additional naturally occurring inhibitors of this pathway produced by other organisms.

VII. Implications for Diseases Caused by Fumonisins and Related Mycotoxins

Disruption of this pathway is an attractive mechanism for the pathological effects of fumonisins because sphingolipids are thought to be involved in diverse and complex ways in the regulation of cell behavior. These include the regulation of the properties of cell surface receptors, including direct interactions with the receptors, and the activation and inhibition of protein kinases and other intracellular signaling systems; the binding of cytoskeletal proteins and participation in cell–cell communication and cell–substratum interactions; and the regulation of these and other aspects of cell growth, differentiation, and transformation (Fishman and Brady, 1976; Sweeley, 1980, 1985; Hannun and Bell, 1989; Hakomori, 1990; Merrill, 1991). The various steps at which these might occur are shown in Fig. 11 and include:

A. Inhibition of ceramide synthase leads to accumulation of sphinganine (A in Fig. 11), which is a very bioactive molecule. Since the initial discovery that long-chain (sphingoid) bases inhibit protein kinase C (Hannun *et al.*, 1986), these compounds have been found also to inhibit phosphatidic acid phosphatase (Lavie and Liscovitch, 1990; Jamal *et al.*, 1991; Mullman *et al.*, 1991), Na/K ATPase (Oishi *et al.*, 1990), c-*src* and v-*src* kinases (Igarashi *et al.*, 1989), and CTP:phosphocholine cytidylyltransferase (Sohal and Cornell, 1990) and to activate the EGF receptor (Faucher *et al.*, 1988; Wedegaertner and Gill, 1989; Davis, this volume). Accumulation of sphinganine in cells exposed to fumonisins might be responsible for cell death since long-chain bases can be highly cytotoxic (Merrill, 1983; Stevens *et al.*, 1990). Long-chain bases have also been shown to be mitogenic for Swiss 3T3 cells at low concentrations (Zhang *et al.*, 1990; Spiegel *et al.*, this volume); therefore, this mitogenicity may account for the

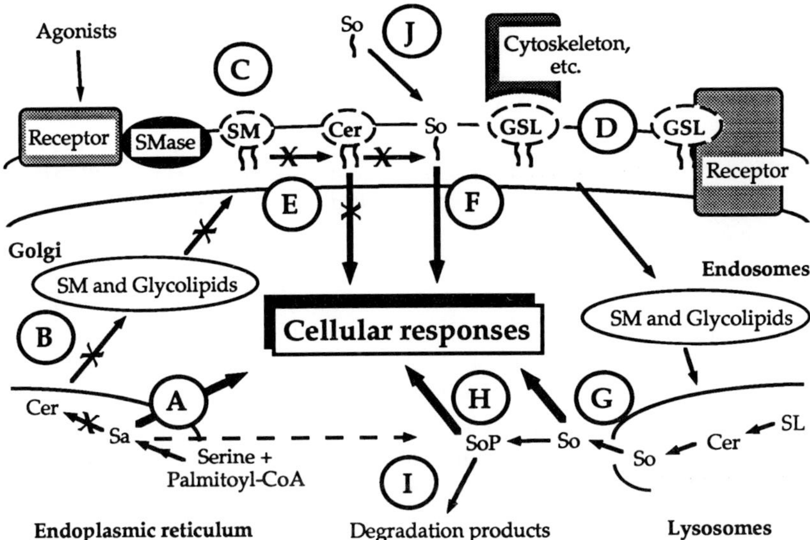

FIG. 11. Hypotheses for the mechanism(s) by which fumonisins are toxic and carcinogenic. By inhibition of ceramide synthase, sphinganine accumulates and may affect the various targets that are inhibited (e.g., protein kinase C, phosphatidic acid phosphatase, Na/K ATPase, EGF receptor) (A) and formation of more complex sphingolipids is blocked (B), resulting in reduced levels of sphingomyelin (C) and glycosphingolipids (D) which serve as membrane attachment sites for cytoskeletal proteins and cellular and extracellular receptors. Reduced levels of membrane sphingolipids may affect signal transduction systems that utilize ceramides (E) or, perhaps, sphingosine (F). Sphingosine levels may be increased (or reduced) based on the effects of fumonisins on sphingolipid turnover (G). Sphingosine (and sphinganine) can also be converted to the 1-phosphates (H), which have been reported to affect cytosolic calcium levels. The long-chain bases can also be cleaved (I) to ethanolamine phosphate and long-chain fatty aldehydes, which may affect other pathways. Animals consuming fumonisins exhibit elevated serum sphingosine and sphinganine, which will presumably be taken up by cells (J) and may invoke cellular responses.

carcinogenicity of the fumonisins. In support of this hypothesis, we have found (Schroeder *et al.*, 1993) that fumonisin B_1 causes sphinganine accumulation and mitogenesis in Swiss 3T3 cells.

B. Formation of more complex sphingolipids is blocked (B, in Fig. 11), resulting, perhaps, in accumulation of various precursors for sphingolipid synthesis (such as fatty acyl-CoAs, activated sugars, etc.). In addition, it has been suggested that the transport of proteins and sphingolipids through the secretory pathway may be coupled to sphingolipid synthesis (Rosenwald *et al.*, 1992). If so, fumonisins would be predicted to alter this process.

C. A reduction in the amount of sphingomyelin (C) will remove a membrane lipid with physical properties different from phosphoglycerolipids (Merrill and Jones, 1990), as well as a possible participant in a "sphingomyelin cycle" that is activated by 1α, 25-dihydroxyvitamin D_3, tumor necrosis factor, and γ-interferon (Okazaki *et al.*, 1989, 1990; Kim *et al.*, 1991).

D. A decrease in the amount of glycosphingolipids (D) may disrupt the interactions between the cell surface and cytoskeletal proteins, as well as remove modulators of cell surface receptors. For example, ganglioside G_{M3} appears to downregulate the EGF receptor (Bremer *et al.*, 1986; Song *et al.*, 1991; Davis, this volume); therefore, inhibition of G_{M3} synthesis might deplete the cell of G_{M3} and prove growth stimulatory.

E. A reduction in the amount of complex sphingolipids should be accompanied by less turnover of these compounds to bioactive "second messenger" ceramides (E) (Okazaki *et al.*, 1989, 1990; Kim *et al.*, 1991), which stimulate protein phosphatase (Dobrowsky and Hannun, 1992) and activate kinases (Goldkorn *et al.*, 1991; Mathias *et al.*, 1991), and sphingosine.

F. Little is known about factors that lead to the hydrolysis of complex sphingolipids to sphingosine (F). However, recent work has suggested that a substantial portion of the free sphingosine of cells is located in plasma membranes and may be formed there by sphingolipid hydrolases with neutral to alkaline pH optima (Slife *et al.*, 1989).

G. The cellular levels of free sphingosine (G) arising from the turnover of more complex sphingolipids may increase (as is seen when ponies are fed fumonisins) (Wang *et al.*, 1992) or decrease (as seen in cultured hepatocytes) (Wang *et al.*, 1991), perhaps depending on whether there is cell injury leading to membrane turnover and whether or not a significant amount is usually reacyclated.

H. Sphingosine can also be converted to sphingosine-1-phosphate (H), an intermediate long-chain base degradation (I) that has also been implicated as a mediator of the release of calcium from intracellular stores (Ghosh *et al.*, 1990; Zhang *et al.*, 1991; Spiegel, this volume) which appears to stimulate increases in intracellular calcium.

I. Sphingosine (and sphinganine)-1-phosphate may be cleaved to ethanolamine phosphate and a long-chain fatty aldehyde, affecting other lipid metabolic pathways, such as phosphatidylethanolamine and phosphatidylcholine, and the synthesis of plasmalogens (Stoffel *et al.*, 1968; Keenan, 1972; Van Veldhoven and Mannaerts, 1991).

J. Some of the effects of fumonisins may be due to the increase in serum long-chain bases (Wang *et al.*, 1992; Riley *et al.*, 1993), which will probably be taken up by cells and may have any of the effects that have been observed in the many *in vitro* studies where exogenous long-chain bases were added to cells.

In addition to these mechanistic considerations, the fact that the fumonisins inhibit sphingolipid biosynthesis raises several questions about the physiological significance of these compounds. Fumonisins have been found in corn meal and grits in the United States at up to 2.7 ppm (Sydenham *et al.,* 1991). These levels appear to be somewhat lower than the ones that cause diseases in agricultural animals; however, the relative sensitivity of humans is not known and epidemiological studies have implicated the consumption of *F. monoliforme* with an increased incidence of esophageal cancer (Li *et al.,* 1980; Yang, 1980; Marasas, 1982; Sydenham *et al.,* 1990a). The availability of a relatively simple and sensitive marker for fumonisin exposure (the sphinganine to sphingosine ratio in serum) should be valuable in conducting more conclusive epidemiological studies of the possible link between fumonisin consumption and human cancer. Relatively few persons in the United States consume large amounts of contaminated corn on a regular basis, but such analyses would help determine if these low levels of exposure are adequate to affect sphingolipid metabolism and, therefore, pose a potential health risk. Furthermore, one suspects that even episodic exposure to an inhibitor of sphingolipid biosynthesis could have deleterious consequences during periods of particularly active sphingolipid synthesis, such as during pregnancy and various stages of early life.

Acknowledgments

The authors are grateful to several valuable collaborators in various aspects of this work, namely, Drs. C. W. Bacon, V. Beasley, W. Haschek, R. LaRocque, D. C. Liotta, D. Menaldino, W. P. Norred, R. D. Plattner, P. F. Ross, J. J. Schroeder, E. R. Smith, J. L. Showker, G. Van Echten, K. A. Voss, T. Wilson, and H. Yoo. This work was supported by funds from the USDA and the NIH (grant GM33369).

References

Bezuidenhout, C. S., Gelderblom, W. C. A., Gorst-Allman, C. P., Horak, R. M., Marasas, W. F. O., Spiteller, G., and Vleggaar, R. (1988). *J. Chem. Soc., Chem. Commun.,* pp. 743–745.

Bottini, A. T., Bowen, J. R., and Gilchrist, D. G. (1981). *Tetrahedron Lett.* **22,** 2723–2726.

Bremer, E. G., Schlessinger, J., and Hakomori, S.-I. (1986). *J. Biol. Chem.* **261,** 2434–2440.

Cawood, M. E., Gelderblom, W. C. A., Vleggaar, R., Behrend, Y., Thiel, P. G., and Marasas, W. F. O. (1991). *J. Agric. Food Chem.* **39,** 1958–1962.

Clouse, S. D., and Gilchrist, D. G. (1987). *Phytopathology* **77,** 80–82.

Dobrowsky, R. T., and Hannun, Y. A. (1992). *J. Biol. Chem.* **267,** 5048–5051.

Faucher, M., Girones, N., Hannun, Y. A., Bell, R. M., and Davis, R. J. (1988). *J. Biol. Chem.* **263,** 5319–5327.

Fishman, P. H., and Brady, R. O. (1976). *Science* **194,** 906–915.

Gelderblom, W. C. A., Jaskiewicz, K., Marasas, W. F. O., Thiel, P. G., Horak, R. M., Vleggaar, R., and Kriek, N. P. J. (1988). *Carcinogenesis* (*London*) **9,** 1405–1409.

Gelderblom, W. C. A., Kriek, N. P. J., Marasas, W. F. O., and Thiel, P. G. (1991). *Carcinogenesis* (*London*) **12,** 1247–1251.

Gelderblom, W. C. A., Marasas, W. F. O., Vleggar, R., Thiel, P. G., and Cawood, M. E. (1992). *Mycopathologia* **117,** 11–16.

Ghosh, T. K., Bian, J., and Gill, D. L. (1990). *Science* **248,** 1653–1656.

Gilchrist, D. G., Ward, B., Moussato, V., and Mirocha, C. J. (1992). *Mycopathologia* **117,** 57–64.

Goldkorn, T., Dressler, K. A., Muindi, J., Radin, N. S., Mendelsohn, J., Menaldino, D., Liotta, D., and Kolesnick, R. N. (1991). *J. Biol. Chem.* **266,** 16092–16097.

Hakomori, S.-I. (1990). *J. Biol. Chem.* **265,** 18713–18716.

Hannun, Y. A., and Bell, R. M. (1989). *Science* **243,** 500–507.

Hannun, Y. A., Loomis, C. R., Merrill, A., and Bell, R. M. (1986). *J. Biol. Chem.* **261,** 12604–12609.

Harrison, L. R., Colvin, B., Greene, J. T., Newman, L. E., and Cole, J. R. (1990). *J. Vet. Diagn. Invest.* **2,** 217–221.

Hope, M. J., and Cullis, P. R. (1987). *J. Biol. Chem.* **262,** 4360–4366.

Igarashi, Y., Hakomori, S.-I., Toyokuni, T., Dean, B., Fujita, S., Sugimoto, M., Ogawa, T., El-Ghendy, K., and Racker, E. (1989). *Biochemistry* **28,** 6796–6800.

Jamal, H., Martin, A., Gomez-Munoz, A., and Brindley, D. N. (1991). *J. Biol. Chem.* **266,** 2988–2996.

Karlsson, K.-A. (1970). *Chem. Phys. Lipids* **5,** 6–43.

Keenan, R. W. (1972). *Biochim. Biophys. Acta* **270,** 383–386.

Kellerman, T. S., Marasas, W. F. O., Pienaar, J. G., and Naude, T. W. (1972). *Onderstepoort J. Vet. Res.* **39,** 205–208.

Kellerman, T. S., Marasas, W. F. O., and Thiel, P. G. (1990). *Onderstepoort J. Vet. Res.* **57,** 269–275.

Kim, M.-Y., Linardic, C., Obeid, L., and Hannun, Y. A. (1991). *J. Biol. Chem.* **266,** 484–489.

Kriek, N. P. J., Kellerman, T. S., and Marasas, W. F. O. (1981). *Onderstepoort J. Vet. Res.* **48,** 129–131.

Lavie, Y., and Liscovitch, M. (1990). *J. Biol. Chem.* **265,** 3868–3872.

Li, M., Lu, S., Ji, C., Wang, Y., Wang, M., Cheng, S., and Tian, G. (1980). *In* "Genetic and Environmental Factors in Experimental and Human Cancer" (H. V. Gelboin *et al.*, eds). pp. 139–148. Jpn. Sci. Soc. Press, Tokyo.

Marasas, W. F. O. (1982). *In* "Cancer of the Esophagus" (C. J. Pfeiffer, ed.), Vol. 1, pp. 29–40. CRC Press, Boca Raton, FL.

Marasas, W. F. O., Nelson, P. E., and Toussoun, T. A., (1984a). "Toxigenic Fusarium Species: Identity and Mycotoxicology." Pennsylvania State Univ. Press, University Park.

Marasas, W. F. O., Kriek, N. P. J., Fincham, J. E., and van Rensburg, S. J. (1984b). *Int. J. Cancer* **34,** 383–387.

Marasas, W. F. O., Kellerman, T. S., Gelderblom, W. C. A., Coetzer, J. A. W., Thiel, P. G., and van der Lugt, J. J. (1988). *Onderstepoort J. Vet. Res.* **55,** 197–203.

Mathias, S., Dressler, K. A., and Kolesnick, R. N. (1991). *Proc. Natl. Acad. Sci. U.S.A.* **88,** 10009–10013.

Merrill, A. H., Jr. (1983). *Biochim. Biophys. Acta* **754,** 284–291.

Merrill, A. H., Jr. (1991). *J. Bioenerg. Biomembr.* **23,** 83–104.

Merrill, A. H., Jr., and Jones, D. D. (1990). *Biochim. Biophys. Acta* **1044,** 1–12.

Merrill, A. H., Jr., and Wang, E. (1986). *J. Biol. Chem.* **261,** 3764–3769.

Merrill, A. H., Jr., Nixon, D. W., and Williams, R. D. (1985). *J. Lipid Res.* **26,** 617–622.

Merrill, A. H., Jr., Wang, E., Mullins, R. E., Jamison, W. C. L., Nimkar, S., and Liotta, D. C. (1988). *Anal. Biochem.* **17,** 373–381.

Merrill, A. H., Jr., Nimkar, S., Menaldino, D., Hannun, Y. S., Loomis, C., Bell, R. M., Tyagi, S. R., Lambeth, J. D., Stevens, V. L., Hunter, R., and Liotta, D. C. (1989). *Biochemistry* **28,** 4360–4366.

Messmer, T. O., Wang, E., Stevens, V. L., and Merrill, A. H., Jr. (1989). *J. Nutr.* **119,** 534–538.

Mullman, T. J., Siegel, M. I., Egan, R. W., and Billah, M. M. (1991). *J. Biol. Chem.* **266,** 2013–2016.

Norred, W. P., Wang, E., Yoo, H., Riley, R. T., and Merrill, A. H. (1992). *Mycopathologia* **117,** 73–78.

Oishi, K., Zheng, B., and Kuo, J. F. (1990). *J. Biol. Chem.* **265,** 70–75.

Okazaki, T., Bell, R. M., and Hannun, Y. A. (1989). *J. Biol. Chem.* **264,** 10976–19080.

Okazaki, T., Bielawska, A., Bell, R. M., and Hannun, Y. A. (1990). *J. Biol. Chem.* **265,** 15823–15831.

Ong, D. E., and Brady, R. N. (1973). *J. Biol. Chem.* **248,** 3884–3888.

Riley, R. T., An, N.-H., Showker, J. L., Yoo, H.-S., Norred, W. P., Chamberlain, W. J., Wang, E., Merrill, A. H., Motelin, G., Beasley, V. R., and Haschek, W. M. (1993). *Toxicol. Appl. Pharmacol.* **118,** 105–112.

Rosenwald, A. G., Machamer, C. E., and Pagano, R. E. (1992). *Biochemistry* **31,** 3581–3590.

Ross, P. F., Rice, L. G., Plattner, R. D., Osweiler, G. D., Wilson, T. M., Owens, D. L., Nelson, H. A., and Richard, J. L. (1991a). *Mycopathologia* **114,** 129–135.

Ross, P. F., Rice, L. G., Reagor, J. C., Osweiler, G. D., Wilson, T. M., Nelson, H. A., Owens, D. L., Plattner, R. D., Harlin, K. A., Richard, J. L., Colvin, B. M., and Banton, M. I. (1991b). *J. Vet. Diagn. Invest.* **3,** 238–241.

Schroeder, J. J., and Merrill, A. H., Jr. (1993). *FASEB J.* **7**(4), A216. [Abs 4139]

Shier, W. T., Abbas, H. K., and Mirocha, C. J. (1991). *Mycopathologia* **116,** 97–104.

Slife, C. W., Wang, E., Hunter, R., Wang, S., Burgess, C., Liotta, D. C., and Merrill, A. H., Jr. (1989). *J. Biol. Chem.* **264,** 10371–10377.

Sohal, P. S., and Cornell, R. B. (1990). *Biochem. Biophys. Res. Commun.* **170,** 162–168.

Song, W. X., Vacca, M. F., Welti, R., and Rintoul, D. A. (1991). *J. Biol. Chem.* **266,** 10174–10181.

Stevens, V. L., Nimkar, S., Jamison, W. C., Liotta, D. C., and Merrill, A. H., Jr. (1990). *Biochim. Biophys. Acta* **1051,** 37–45.

Stoffel, W., Sticht, G., and LeKim, D. (1968). *Hoppe-Seyler's Z. Physiol. Chem.* **349,** 1745–1748.

Sweeley, C. C., ed. (1980). "Cell Surface Glycolipids." Am. Chem. Soc., Washington, DC.

Sweeley, C. C. (1985). *In* "Biochemistry of Lipids and Membranes" (D. E. Vance and J. E. Vance, eds.), pp. 361–403. Benjamin/Cummings, Menlow Park, CA.

Sydenham, E. W., Gelderblom, W. C. A., Thiel, P. G., and Marasas, W. F. O. (1990a). *J. Agric. Food Chem.* **38,** 285–290.

Sydenham, E. W., Thiel, P. G., Marasas, W. F. O., Shephard, G. S., Van Schalkwyk, D. J., and Koch, K. R. (1990b). *J. Agric. Food Chem.* **38,** 1900–1903.

Sydenham, E. W., Shephard, G. S., Thiel, P. G., Marasas, W. F. O., and Stockenstrom, S. (1991). *J. Agric. Food Chem.* **39,** 2014–2018.

Thiel, P. G., Marasas, W. F. O., Sydenham, E. W., Shephard, G. S., and Gelderblom, W. C. A. (1992). *Mycopathologia* **117,** 3–10.

Van Veldhoven, P. P. V., and Mannaerts, G. P. (1991). *J. Biol. Chem.* **266,** 12502–12507.

Vesonder, R. F., Peterson, R. E., Labeda, D., and Abbas, H. K. (1992). *Arch. Environ. Contam. Toxicol.* **23,** 464–467.

Voss, K. A., Norred, W. P., Plattner, R. D., and Bacon, C. W. (1989). *Food Chem. Toxicol.* **27,** 89–96.

Voss, K. A., Plattner, R. D., Bacon, C. W., and Norred, W. P. (1990). *Mycopathologia* **112,** 81–92.

Wang, E., Norred, W. P., Bacon, C. W., Riley, R. T., and Merrill, A. H., Jr. (1991). *J. Biol. Chem.* **266,** 14486–14490.

Wang, E., Ross, P. F., Wilson, T. M., Riley, R. T., and Merrill, A. H., Jr. (1992). *J. Nutr.* **122,** 1706–1716.

Wedegaertner, R. B., and Gill, G. N. (1989). *J. Biol. Chem.* **264,** 11346–11353.

Wiebking, T. S., Ledoux, D. R., Bermudez, A. J., Turk, J. R., Rottinghaus, G. E., Wang, E., and Merrill, A. H., Jr. (1993). *Poult. Sci.* **72,** 456–466.

Wilson, T. M., Ross, P. F., Rice, L. G., Osweiler, G. D., Nelson, H. A., Owens, D. L., Plattner, R. D., Reffiardo, C., Noon, T. H., and Pickrell, J. W. (1990). *J. Vet. Diagn. Invest.* **2,** 213–216.

Wilson, T. M., Ross, P. F., Owens, D. L., Rice, L. G., Green, S. A., Jenkins, S. J., and Nelson, H. A. (1992). *Mycopathologia* **117,** 115–120.

Yang, C. S. (1980). *Cancer Res.* **40,** 2633–2644.

Yoo, H., Norred, W. P., Wang, E., Merrill, A. H., Jr., and Riley, R. T. (1992). *Toxicol. Appl. Pharmacol.* **114,** 9–15.

Zhang, H., Buckley, N. E., Gibson, K., and Spiegel, S. (1990). *J. Biol. Chem.* **265,** 76–81.

Zhang H., Desai, N. N., Olivera, A., Seki, T., Brooker, G., and Spiegel, S. (1991). *J. Cell Biol.* **114,** 155–167.

Extracellular Sialidases

CHARLES C. SWEELEY

Department of Biochemistry
Michigan State University
East Lansing, Michigan 48824

I. Introduction and Overview

Sialidase (neuraminidase, *N*-acylneuraminosyl glycohydrolase, EC 3.2.1.18) catalyzes the hydrolytic removal of sialic acid residues from oligosaccharides, gangliosides, and glycoproteins. The enzyme is widely distributed in microorganisms and vertebrates. Sialidase activity in bacteria was first discovered in cell-free culture filtrates of *Vibrio cholerae* (Burnet and Stone, 1947) and *Clostridium perfringens* (McCrea, 1947) on the basis of the effects these filtrates had on viral agglutinability and receptor properties of erythrocytes. The liberation of sialic acid from brain gangliosides (Rosenberg *et al.*, 1956) and ovomucin (Gottschalk and Lind, 1949) led to the recognition of sialidase activity in certain viruses. The existence of mammalian sialidase activity was first reported by Warren and Spearing (1960), and the relative distribution of sialidase activity was reported in various mammalian organs by Carubelli *et al.* (1962), using an assay consisting of sialyllactose as the substrate and supernatant fractions of tissue homogenates after centrifugation. The early literature on bacterial, viral, avian, and mammalian sialidases has been summarized in detail by Rosenberg and Schengrund (1976).

The reaction catalyzed by sialidase leads to the hydrolytic removal of α-glycosidically linked sialic acid residues from sialylated glycoconjugates at a somewhat acidic pH, as shown in Fig. 1. The glycosidic linkage to R′ in Fig. 1 is to C-2 or C-6 of a sugar unit of the oligosaccharide chain or to C-8 of another sialic acid residue. Although the first-formed product is the α-anomeric form of sialic acid, mutarotation leads to an anomeric mixture consisting predominantly of the β anomer. Details of stereochemistry, kinetic parameters, and substrate specificity of primarily bacterial sialidases have been reviewed by Corfield and Schauer (1982). Very briefly, the reaction in Fig. 1 has greater rates with α2–3 linked sialic acids than with α2–6 or α2–8 linked sialic acids. *N*-Acetylneuraminic acid residues are removed somewhat more readily than *N*-glycolylneuraminic acid residues. Substitution on the side chain or at C-4 by *O*-acetyl groups

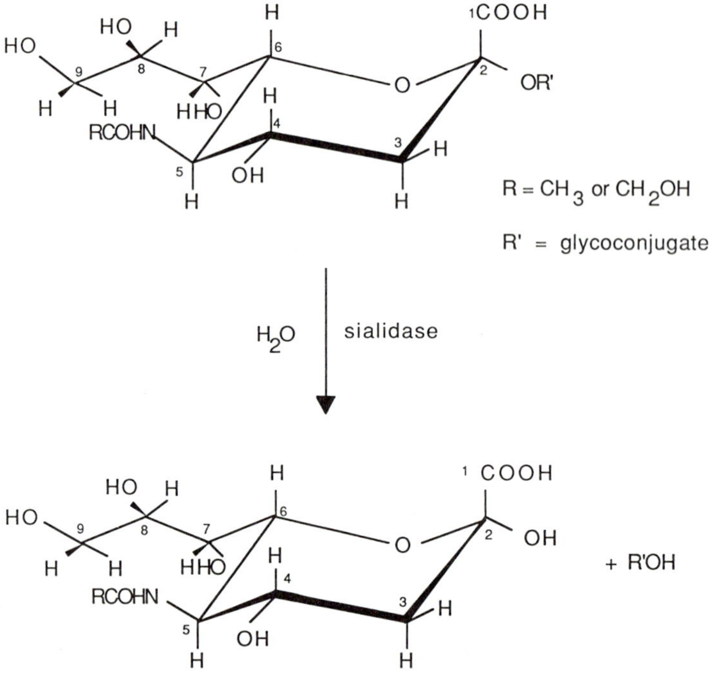

R = CH_3 or CH_2OH

R′ = glycoconjugate

H_2O sialidase

+ R′OH

FIG. 1. A sialic acid residue, containing an *N*-acetyl or *N*-glycolyl group at C-5 (RCO-), is hydrolyzed by sialidase at the α-ketosidic linkage (C-2) to liberate free sialic acid (α-anomeric form) and the desialylated glycoconjugate (R′OH). Sialic acid is shown in the ${}_5C^2$ conformation.

diminishes or completely blocks the sialidase-catalyzed hydrolysis. Detergents are sometimes required, especially with gangliosides, and some sialidases are activated by divalent cations. Some sialidases show relatively broad specificities, while others may have marked specificities for a particular kind of glycoconjugate (glycoprotein, ganglioside, or oligosaccharide).

The sialidase activity of influenza virus is an integral membrane tetrameric glycoprotein, which mediates the attachment of the virus to host cells via a sialoglycoconjugate, thus facilitating the process by which invasion and destruction of the tissue occur. "Antigen drift" resulting from point mutations of the viral genome has been reflected by the presence of two variant primary structures of the enzyme in human isolates, the three-dimensional structure of one of which (A/Tokyo/3/67) has been described by Varghese and Colman (1991) along with those of the wild-type and several variants of avian type A (N9 subtype) influenza virus (Tulip *et al.*, 1991). Several sialidases, isolated from clostridia and an actinomycete, have been cloned and sequenced (Henningsen *et al.*, 1991). From these reports of primary and tertiary structure, information is emerging about highly conserved sequences and active site residues. The nature of substrate binding and the mechanism of hydrolysis will certainly be forthcoming soon.

It has been known for many years that mammalian sialidase activity is localized in more than one subcellular fraction. Using gangliosides or sialyllactose as a substrate, sialidases have been found in lysosomes, cytosol, plasma membrane, and Golgi apparatus. Lysosomal activity has been further separated into intralysosomal and lysosomal membrane forms. The intralysosomal (soluble) form of sialidase isolated from human placenta exists as a multienzyme complex with β-galactosidase and a "protective protein" which expresses carboxypeptidase activity (Potier *et al.*, 1990a). The sialidase subunit of this complex is a 60-kDa protein that appears to result from partial proteolytic processing of lysosomal prosaposin (Potier *et al.*, 1990b).

Studies indicate that subconfluent cultures of human skin fibroblasts accumulate sialidase activity in the culture medium (Usuki *et al.*, 1988). This extracellular sialidase activity, as well as that on the plasma membrane of the fibroblasts, was measured using radiolabeled G_{M3} or an enzyme-coupled assay (Ogura *et al.*, 1992). More than one form of soluble sialidase was found in the conditioned medium of human fibroblasts, based both on pH optima and chromatographic resolution. Relatively nothing is known as yet about the cellular origin of the extracellular sialidases nor whether they have a biological role in the extracellular milieu.

The primary aims of this review are to compare the extracellular sialidase activities with other mammalian sialidases to the extent that it is possible at this time, and to discuss the potential biological roles that extracellular sialidases may have in the modulation of plasma membrane sialoglycoconjugates. Methods of analysis of sialidases and more detailed consideration of the various subcellular forms of sialidase activity are included, but only a few references are given to the literature prior to 1980. Additional detail and other considerations about sialidases can be obtained in reviews by Rosenberg and Schengrund (1976), Corfield and Schauer (1982), Tettamanti *et al.* (1981), Schauer (1982), and Gottschalk and Drzenick (1972).

II. Methods of Analysis

Methods for the determination of sialidase activity depend on the measurement of liberated sialic acid or the desialylated product of the reaction. Substrates have included glycoproteins, radiolabeled and unlabeled gangliosides or oligosaccharides such as sialyllactose, and artificial substrates containing chromogenic or fluorigenic aglycone moieties. Buffers are needed to maintain the incubation mixture at the desired pH, which will be optimal in the range from pH 4.0 to 6.5. Detergents are necessary in some cases to activate the sialidase.

A. Measurement of Liberated Sialic Acid

Colorimetric methods such as the thiobarbituric acid assay described by Warren (1959) and modified by Aminoff (1961) have been widely used to determine free sialic acid in the incubation mixtures of sialidases with glycoprotein, glycolipid, or oligosaccharide substrates. Samples after incubation with a sialidase are oxidized by periodate in strong acid, treated with arsenite, and developed with thiobarbituric acid, which yields a red chromophore that can be extracted into an organic solvent and measured spectrophotometrically at 549 nm. The optical density at 532 nm is used to correct for interference with other sugars. The results with this assay are good when carefully controlled, but the assay may lead to unreliable results with crude enzyme preparations unless elaborate chromatographic steps are taken to purify the chromophore.

Substrates labeled with ^{3}H or ^{14}C in sialic acid residue have also been used to monitor sialidase activity. Chromatographic separation of the labeled substrate and the liberated sialic acid is necessary and may be tedious and time-consuming. There are very few commercially available radiolabeled substrates and they must therefore be prepared and purified by the investigator. Biosynthesis of such substrates from a radiolabeled precursor such as [1-^{14}C]-*N*-acetylmannosamine yields glycoconjugates with nearly exclusive enrichment of the label in the sialic acid residue. The desired glycoconjugate must be separated from other labeled glycoconjugates, however, before it can be used as a substrate for quantitative assays of sialidase activity.

B. Radiolabeled Substrates

Substrates radiolabeled in the desialylated product have also been used. For example, the sphingoid base (sphingosine) in gangliosides can be selectively reduced with a palladium catalyst and tritium gas, generated *in situ* from sodium borotritide in a strictly anaerobic environment, as described by Schwarzmann (1978). This procedure introduces tritium atoms at the olefinic center of the sphingoid base and also at sites of unsaturation of the fatty acyl residues of the ceramide. Alternatively, the allylic hydroxyl group at C-3 of the sphingosine base can be selectively oxidized to a ketone and then reduced back to an OH group with sodium borotritide (Ghidoni *et al.*, 1981), recognizing that the substrate is converted to a racemic mixture by this procedure, a factor that will probably not be significant for sialidase assays. Oligosaccharides such as sialyllactose can be reduced to their corresponding ^{3}H-labeled alditols, as described by Bhavanandan *et al.* (1975).

C. Artificial Substrates

Colorimetric procedures using substrates such as 2-(*p*-nitrophenyl)-α-D-*N*-acetylneuraminic acid (pNP-NeuAc) have been employed to measure sialidase activity (Privalova and Khorlin, 1969; Tuppy and Palese, 1969). Hydrolysis of the substrate gives *p*-nitrophenol, which can be determined colorimetrically at an alkaline pH. Fluorometric assays with 2-(4-methylumbelliferyl)-α-D-*N*-acetylneuraminic acid (4MU-NeuAc) are more commonly used, however, since the liberated product, 4-methylumbelliferone, has an intense fluorescent emission at 450 nm at pH 10 or higher (Potier *et al.*, 1979). Assays carried out with this reagent must be corrected for nonenzymatic hydrolysis of the substrate, which can be appreciable at pH below 6.0. In addition, the commercial substrate may contain some free 4-methylumbelliferone and should be further purified before use, especially when attempting to measure low levels of sialidase activity.

D. Coupled Enzyme Assays

A novel approach was described for use in the assay of crude samples with low levels of sialidase activity (Ogura *et al.*, 1992). Ganglioside G_{D1a} was coated onto enzyme immunoassay plate wells and incubated with *Vibrio cholerae* sialidase and with various column fractions of sialidase activity derived from the medium of cultured human skin fibroblasts. The amount of product, G_{M1} ganglioside, was then measured by complex formation with cholera toxin B subunit conjugated with horseradish peroxidase, and the amount of bound peroxidase was determined by a colorimetric enzymatic assay. This procedure can be used with detergents, providing studies are carried out to determine the optimal level of detergent for activating the sialidase while keeping the solubilization of the G_{D1a} from the polystyrene surface as low as possible. The procedure has many advantages over others in that the reagents are commercially available, the assay is highly specific and free of interferences, and assays can be carried out at very high sensitivities (about 3 fmol of G_{M1} formed per minute could be detected).

In another approach, the product after sialidase treatment of a suitable substrate can be bound to a peroxidase-conjugated peanut agglutinin lectin specific for terminal galactose residues, followed by determination of bound peroxidase (Lambré *et al.*, 1991).

III. Subcellular Localization of Mammalian Sialidases

The localization of mammalian sialidase activity within the cell has been a subject of continuing interest for more than 20 years. Sometimes conflicting results and the use of incompletely separated organelles have confused the picture, but it is clear that there are several distinctly different sialidases in many cell types, with different substrate specificities, and that these activities reside primarily in lysosomes, plasma membrane, and the cytosolic fraction. In addition, one or more soluble sialidases are released into the medium by fibroblasts, KB cells, and perhaps other types of cells as well. Membrane-bound sialidase activities have been very difficult to purify because of poor results in attempts to solubilize them with various detergents and their lability in membrane-free environments.

A. Lysosomal Sialidases

Many tissues contain lysosomal sialidases, which generally have optimal activities between pH 4.0 and pH 4.5 with the 4MU-NeuAc substrate or sialyllactose. References to the literature prior to 1980 are cited in reviews by Rosenberg and Schengrund (1976) and Corfield and Schauer (1982). It was discovered in 1982 that a lysosomal sialidase could be purified as a complex with β-galactosidase and a protective protein (Verheijen *et al.*, 1982). This complex has been of considerable interest because of the combined deficiency of sialidase and β-galactosidase activities in patients with galactosialidosis (Wenger *et al.*, 1978; Miyatake *et al.*, 1979; Lowden and O'Brien, 1979), which results from a deficiency of the protective protein. This component of the complex has been shown to be a 32-kDa glycoprotein with carboxypeptidase activity (Verheijen *et al.*, 1985). The primary structures of the human (Galjart *et al.*, 1988) and mouse protective protein (Galjart *et al.*, 1990) have been determined by cDNA cloning.

The sialidase activity of human placenta was purified as the complex with β-galactosidase and protective protein, and was dissociated from the complex by potassium thiocyanate treatment and separated from the other subunits of the complex by affinity chromatography (Verheijen *et al.*, 1987). The dissociated sialidase was catalytically inactive by itself and showed two major polypeptide bands on sodium dodecyl sulfate polyacrylamide gel electrophoresis (SDS-PAGE) at apparent masses of 76 and 66 kDa. Antibodies against the dissociated inactive sialidase fraction were specific and immunoprecipitated sialidase activity in the undissociated complex. Immunoblotting suggested that the 76-kDa protein associates with the 32-kDa protective protein (carboxypeptidase) for the expression of sialidase activity. The sialidase of human placenta was also characterized by Hiraiwa *et al.* (1988), who reported 55,000-fold purification and determined substrate specificity using sialooligosaccharides, sialoglycoproteins, and gangliosides. Their preparation gave several bands on SDS-PAGE analysis, two of which (78 and 46 kDa) were suggested to be the essential components of the sialidase enzyme. The 46-kDa protein was subsequently identified as an α-*N*-acetylgalactosaminidase (Tsuji *et al.*, 1989). The multienzyme complex from human placental lysosomes was recently shown to consist of a core hexamer of sialidase (66 kDa) and β-galactosidase (63 kDa) protomers, surrounded by five carboxypeptidase heterodimers (32 and 20 kDa subunits) (Potier *et al.*, 1990a).

A lysosomal sialidase from rat liver, partially purified by Miyagi and Tsuiki (1984), was reported to be a monomer of 60 kDa that had no requirement for aggregation into a complex to be catalytically active. This activity, which was measured with sialyllactose and 4MU-NeuAc, was observed to be increased in lysosomes from rat hepatomas as compared with normal tissue (Miyagi *et al.*, 1984).

The catabolism of gangliosides by sialidase has been considered to occur primarily at the plasma membrane. It is claimed, however, that

G_{M3} ganglioside can also be hydrolyzed by a sodium cholate-activated lysosomal sialidase of human skin fibroblasts (Zeigler *et al.*, 1989).

B. Membrane-Associated Sialidases

Fractionation by discontinuous sucrose density gradient centrifugation gave evidence for the localization of ganglioside sialidase activity in the plasma membrane of rat liver (Schengrund *et al.*, 1972). Cultured hamster fibroblasts transformed with herpes simplex virus have approximately equal specific activities of ganglioside sialidase in the lysosomal and plasma membrane fractions (Schengrund *et al.*, 1976). The plasma membrane sialidase of rat liver hydrolyzed mixed brain gangliosides, α2–3 sialyllactose, and orosomucoid at relative rates of 100, 60, and 3, respectively, at pH 4.5 (Miyagi and Tsuiki, 1986). Of interest was the finding that G_{M1} and G_{M2} gangliosides were able to be hydrolyzed at significant rates to give asialo-G_{M1} and asialo-G_{M2} (Miyagi and Tsuiki, 1986), as was reported for G_{M2} ganglioside by Kolodny *et al.* (1971). However, most earlier work suggests that mammalian sialidases are unable to hydrolyze the internal sialic acid residue of ganglio-tri- and ganglio-tetra-type gangliosides [Gal-β1–3-GalNAc-β1–4(NeuAc-α2–3]-Gal-β1-4-Glc-R and GalNAc-β1–4 [NeuAc-α2–3)-Gal-β1–4-Glc-R]. The plasma membrane-associated sialidase activity was lower in rat hepatoma than in normal rat liver (Sagawa *et al.*, 1988).

Erythrocytes contain a membrane-associated ganglioside sialidase with very poor activity toward 4MU-NeuAc, oligosaccharides, and glycoproteins, whereas the sialidase of platelets and leukocytes appears to correspond to the sialidase of hepatic lysosomal matrix in its specificity (Sagawa *et al.*, 1990).

Brain tissue contains ganglioside sialidase activity in lysosomes and synaptosomal membranes. In bovine gray matter, the membrane-associated activity of synaptosomes requires detergent and acts on both endogenous and exogenous gangliosides (Schengrund and Rosenberg, 1970; Tettamanti *et al.*, 1972), whereas a neuronal perikarya fraction was practically devoid of sialidase activity (Schengrund and Nelson, 1976). Evidence for the presence of an intrinsic ganglioside sialidase in the myelin fraction of rat brain was reported by Yohe *et al.* (1986). The highest activity was found in myelin prepared from cerebral hemispheres and was higher in the first month of life than in adult rats. The relationship of the fluctuations in myelin-associated sialidase to developmental changes in the ganglioside profile of myelin has been discussed by Saito and Yu (1992). In comparison with primary neurons and astrocytes, several lines of neuroblastoma cells and an oligodendrocytoma line were found to have higher levels of ganglioside sialidase, whereas in glioma and Schwannoma cells the activity was lower (Moran *et al.*, 1986).

Isolated oligodendroglial cells of rat brain were found to contain sialidase activity with G_{M3} and sialyllactitol as substrates; assays with the two substrates exhibited somewhat different developmental profiles, suggesting that these cells may contain at least two different sialidases (Saito *et al.*, 1992). By a different approach, involving comparisons of substrate specificity, chromatographic resolution, and immunological precipitation, Miyagi *et al.* (1990a) found evidence for two rat brain sialidases, both of which were maximally active at pH 5.0 and exhibited $M_r = 70{,}000$ by gel filtration. Both sialidases hydrolyzed gangliosides but only one was active with oligosaccharides, glycoproteins, and 4MU-NeuAc. The two sialidases were immunologically distinct. Subcellular distribution suggested that the sialidase with narrow specificity for gangliosides was localized in the synaptosomal membrane, whereas the protein with broad specificity was in both synaptosomal and lysosomal membranes.

Incubation of cerebellar granule cells from 8-day-old rats with exogenous gangliosides in the presence and absence of chloroquine or at permissive and nonpermissive temperatures for receptor-mediated endocytosis (37 and 4°C) indicated that G_{D1a} and G_{D1b} are converted to G_{M1} by a sialidase presumed to be localized in the plasma membrane (Riboni *et al.*, 1991). Preliminary evidence suggests that the plasma membrane sialidase may be anchored in the membrane by a glycosylphosphatidylinositol (Chiarini *et al.*, 1990). Crude pellets as well as purified synaptosomal membranes from pig forebrain released sialidase upon treatment with a phosphatidylinositol-specific phospholipase C. The time course and yield of sialidase activity with different levels of added phospholipase C were similar to those obtained with 5′-nucleotidase, which is known to be anchored in the membrane by glycosylphosphatidylinositol (Low, 1987). After solubilization and ammonium sulfate precipitation, the solubilized sialidase exhibited two activities with optima at pH 4.2 and 6.6 with the 4MU-NeuAc substrate, suggesting the possibility that the plasma membrane of pig forebrain contains two glycosylphosphatidylinositol-anchored sialidases (Chiarini *et al.*, 1990).

C. Cytosolic Sialidase

The occurrence of soluble cytosolic sialidase activity, alluded to by Tulsiani and Carubelli (1970), was partially characterized in rat forebrain and its developmental profile was determined from birth to 150 days (Venerando *et al.*, 1982) using a sialyllactose assay at pH 5.8. This sialidase was subsequently purified to apparent homogeneity from rat liver cytosol by Miyagi and Tsuiki (1985). The final preparation, which represented nearly 83,000-fold purification, moved as a single protein band on SDS-PAGE with an apparent M_r = 43,000. The purified enzyme catalyzed the liberation of sialic acid from 4MU-NeuAc and all sialooligosaccharides, sialoglycoproteins, and gangliosides tested, except submaxillary mucins and G_{M1} and G_{M2}. The pH optimum was at pH 6.5 with sialyllactose and at pH 6.0 with orosomucoid and gangliosides. Greatest rates were observed with glycoconjugates containing α2–3 linked sialic acid. Bile acids or certain other detergents were required in assays with gangliosides. A sialidase isolated from the cytosolic fraction of rat skeletal muscle had properties similar to those of the rat liver cytosolic enzyme (Miyagi *et al.*, 1990b) and could be inhibited and immunoprecipitated with antiserum from mice immunized with the skeletal muscle enzyme. The rat liver cytosolic sialidase was similarly inhibited, but this antiserum failed to cross-react with rat liver lysosomal and rat brain synaptosomal sialidases. Further, antisera against rat brain synaptosomal and lysosomal sialidases were unable to immunoprecipitate the cytosolic sialidase activity, strengthening the conclusion that there are at least four distinctly different sialidases in rat tissues, which are probably the products of different genes.

IV. Extracellular Sialidases

A. Source of Activity

In 1988, we reported the occurrence of sialidase activity in the culture medium of human skin fibroblasts (Usuki *et al.*, 1988; Sweeley and Usuki, 1988; Usuki and Sweeley, 1988). The level of activity at pH 4.5 was very low compared to that of the plasma membrane sialidase, and assays with 4MU-NeuAc were inconsistent and unreliable. Using ^{3}H-labeled G_{M3} as a substrate, we found that the sialidase activities in conditioned culture medium, after 24 hours of incubation with the fibroblasts, were 4.1 and 39 pmol/hour/ml of medium in medium from sparse and confluent cultures, respectively (Usuki *et al.*, 1988). On a per-cell basis, the dense cultures contained only about 40% of the sialidase activity calculated from the per-cell levels of sparse cultures, suggesting that more sialidase was accumulating in the medium when the cells were actively dividing. Activity versus pH profiles of the sialidase in the conditioned medium were bimodal, with optimal ganglioside sialidase activity at pH 4.5 and pH 6.5 (a "neutral form"). The sialidase activity at pH 6.5 was approximately equal to that at pH 4.5 in sparse cultures but was virtually absent in the medium of contact-inhibited cells and could not be detected with ^{3}H-labeled sialyllactitol as substrate. Also, there was no separate sialidase activity at pH 6.5 in the plasma membrane. It was concluded from these studies that human fibroblasts release sialidase activity into the medium during growth, and that two or more sialidases are accumulated in the medium. A similar observation concerning the cell density-dependent level of sialidase activity was reported by Yogeeswaran and Hakomori (1975) but was attributed to the plasma membrane sialidase. The "neutral form" of sialidase seemed especially interesting, since its activity was proportional to cell number in actively growing cultures of fibroblasts, but could not be detected in cultures of contact-inhibited cells, which were presumably blocked in the G_1 or quiescent phase of the cell cycle.

A more sensitive ganglioside sialidase assay, developed by Ogura *et al.* (1992), was utilized for an evaluation of the chromatographic properties of the "neutral form" of sialidase activity in the conditioned medium of cultured human fibroblasts. The cells were maintained in Dulbecco's modified Eagle's medium supplemented with 5% fetal bovine serum and were incubated in a humidified 5% carbon dioxide atmosphere. The conditioned medium was collected from preconfluent cultures that had been incubated for 24 to 48 hours. After concentration of the medium on a PM-10 membrane, the concentrate was applied to a gel filtration column (1 × 65 cm) of Sephacryl S-200 HR and proteins were eluted with 20 m*M* phosphate buffer (pH 6.5) containing 0.2 *M* sodium chloride. Sialidase activity with G_{D1a} as the substrate was determined by the coupled enzyme procedure (Ogura *et al.*, 1992) and protein was estimated as absorbance at 280 nm. The results which we reported are reproduced in Fig. 2. The elution profile revealed two peaks of sialidase activity at pH 6.5, with molecular weights estimated at about 16 and 47–69 kDa. The sialidase activity of the higher molecular weight form was increased about 6-fold by the addition of Triton CF-54, whereas the 16-kDa form did not require detergent. Similar studies were not carried out at pH 4.5 and the relationship of these two sialidases to the previously described pH 4.5 form (Usuki *et al.*, 1988) is not known.

Soluble sialidases in the extracellular compartment may be derived by one of the transport pathways for the secretion of lysosomal enzymes. It has long been known that cultured fibroblasts secrete lysosomal enzymes and that there is a receptor-mediated endocytotic mechanism for their uptake, involving mannose-6-phosphate residues on their carbohydrate chains (see Fukuda, 1991 for review). It would not be surprising to detect a lysosomal sialidase activity in the conditioned medium of fibroblasts, derived from this pathway of secretion. The acidic form of ganglioside sialidase activity in conditioned medium (Usuki *et al.*, 1988) may therefore be of the lysosomal type. Further characterization of purified protein will be needed to clarify this possibility.

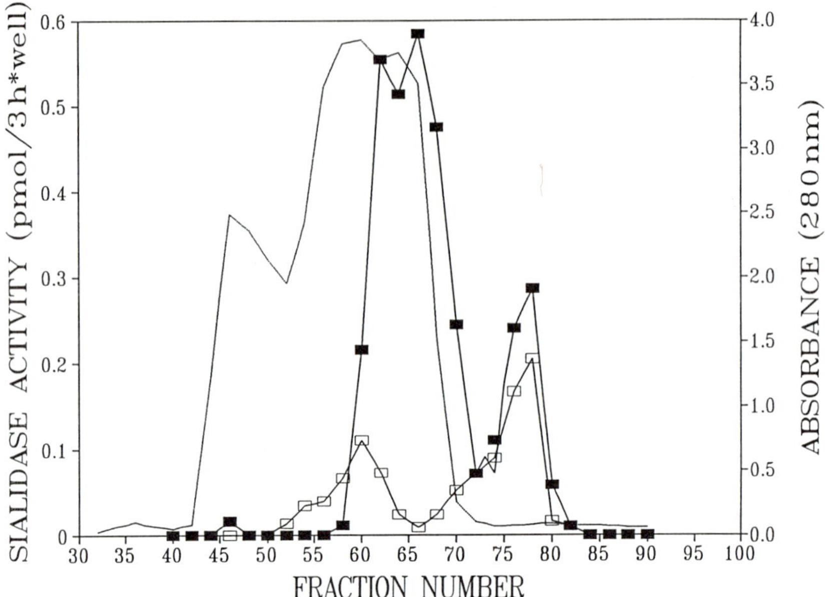

FIG. 2. Elution profile of protein and sialidase activity from conditioned medium of human skin fibroblasts on a Sephacryl S-200 HR column. Sialidase activity was monitored at pH 6.5 with G_{D1a} ganglioside using a coupled enzyme procedure (Ogura *et al.*, 1992). ——, A_{280}; ■, sialidase activity with 0.025% (w/v) Triton CF-54; □, sialidase activity at pH 6.5 with no added detergent. Reproduced from Ogura *et al.* (1992) with permission of the authors and publisher.

The accumulation of ganglioside sialidase activity in the conditioned medium of fibroblasts may also be accounted for by a proteolytic cleavage of the plasma membrane-bound sialidase, releasing part of the enzyme, including the active site, into the medium. The plasma membrane sialidase has previously been suggested to be involved in cell growth and transformation by Schengrund *et al.* (1976) and Yogeeswaran and Hakomori (1975). Its transfer to the medium via a proteolytic cleavage is especially attractive in view of the low molecular mass (16 kDa) of one of the ganglioside sialidases reported by Ogura *et al.* (1992). Comparison of the primary sequences of the 16-kDa sialidase in the medium and the plasma membrane sialidase will be necessary to evaluate this possible mechanism.

B. Functional Role of Sialidases in Transmembrane Signaling and Modulation of Growth

The stimulation of cell proliferation by growth factors involves a cascade of protein kinase-mediated intracellular events that are initiated at the cell surface by growth factor activation of a tyrosine kinase domain on the cytosolic side of epidermal growth factor (EGF) receptor and structurally related growth factors (Cadena and Gill, 1992). Laine and Hakomori (1973) observed that exogenous addition of G_{M3} to cultured cells inhibited their growth by extension of the G_1 phase of the cell cycle. In terms of a possible mechanism, gangliosides have been shown to inhibit the autophosphorylation of growth factor receptors in a structure-specific manner. Bremer *et al.* (1984) reported that exogenous G_{M3} added to cultures of KB and A431 cells binds to EGF receptor and downregulates growth by negatively modulating the tyrosine kinase autophosphorylation of the receptor. In a similar manner, platelet-derived growth factor (PDGF) receptor kinase activity could be inhibited, but the most effective ganglioside in this case was G_{M1} (Bremer *et al.*, 1986). Insulin growth factor receptor was also modulated by a ganglioside, and the effect seemed to be specific for sialoparagloboside (NeuAcα2–3Galβ1–4GlcNAcβ1–3Galβ1–4GlcCer; Nojiri *et al.*, 1991).

We have speculated that there must be a mechanism to relieve the inhibition of growth factor receptor tyrosine kinases by gangliosides if the modulation reported by Bremer *et al.* (1984, 1986) and Nojiri *et al.* (1991) is a normally operational mode of growth control. Since G_{M3} is a common constituent of cell membranes, it can be argued that a similar effect of endogenous G_{M3} may be involved in the regulation of the cell cycle. If so, then EGF receptor function would be permanently downregulated without a pathway by which the effect of G_{M3} can be overcome. Although it is possible that the ganglioside and receptor could be physically separated in some manner, perhaps after the complex is internalized, it seems more reasonable to expect that the G_{M3} may be metabolized at the cell surface by an extracellular sialidase. To test this possibility, turnover studies of G_{M3} were carried out with pulse doubly labeled G_{M3} that was synthesized in the fibroblasts from labeled serine and labeled *N*-acetylmannosamine, which are specific precursors of the sphingoid base and the sialic acid residues, respectively. The results showed that about 35% of the labeled sialic acid of G_{M3} was lost, with the appearance of labeled free sialic acid in the medium, during a 24-hour chase period without measureable loss of labeled sphingosine from the G_{M3} pool (Usuki *et al.*, 1988). A model, shown in Fig. 3, was proposed to account for the positive and negative modulation of EGF receptor tyrosine kinase activity by EGF and G_{M3}, respectively (Usuki and Sweeley, 1988). According to this model, the inhibitory effect of G_{M3} is abolished by a sialidase-catalyzed conversion of G_{M3} to lactosylceramide (LacCer), which might be further metabolized at the cell surface or internally. Recycling of LacCer via endosomes and Golgi to G_{M3} and return to the cell surface may also occur. A possible alternative mechanism has been proposed (Hanai *et al.*,

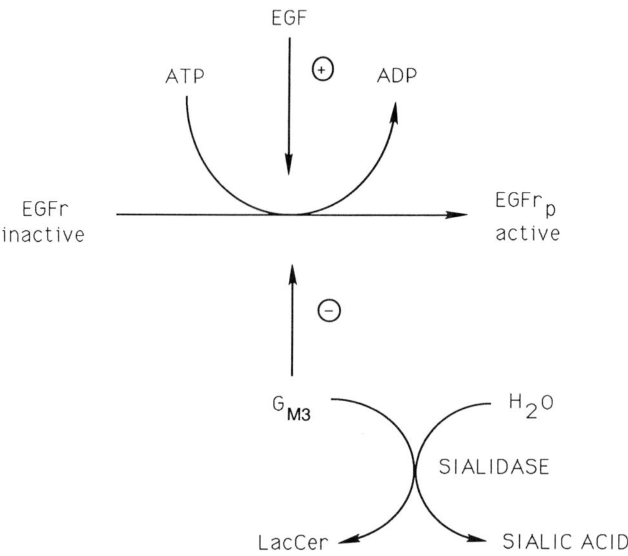

FIG. 3. A hypothetical model for the regulation of growth in fibroblast cultures, based on the negative modulation of epidermal growth factor receptor (EGF_r) autophosphorylation by G_{M3}, as described by Bremer *et al.* (1984), and relieved by metabolism of G_{M3} to LacCer by an extracellular sialidase in the medium or on the plasma membrane. This model was described previously by Usuki and Sweeley (1988).

1988) involving *N*-deacetylation of the sialic acid moiety of G_{M3} ganglioside.

Further evidence for a physiological role of an extracellular sialidase in the modulation of cell growth was suggested by the finding that exogenously added sialidase from *Clostridium perfringens* stimulated the growth of human skin fibroblasts (Ogura and Sweeley, 1992). Preconfluent cultures of fibroblasts were incubated for 24–48 hours in the presence or absence of several levels of the bacterial sialidase, after which cell density and DNA synthesis from [^{3}H]thymidine were monitored. At a level of 100 milliunits per milliliter, the sialidase increased DNA synthesis 1.9-fold and cell density 1.4-fold over controls at 24 and 48 hours, respectively. When the cells were pulse-labeled with [^{14}C]serine prior to sialidase treatment, there was a marked decrease in labeled G_{M3} and a corresponding increase in labeled LacCer following sialidase treatment of the cells for 24 hours. Vaheri *et al.* (1972) also reported that a bacterial sialidase stimulated DNA synthesis in contact-inhibited cultures of chick fibroblasts. These results are not conclusively in support of the model in Fig. 3, of course, since other cell surface sialoglycoconjugates are probably desialylated as well. It would be interesting to examine the cell surface ganglioside and other glycoconjugates in fibroblasts from patients with a genetic defect in the formation of mannose-6-phosphate residues. It is characterized by generalized oversecretion of lysosomal enzymes and an increased density at which contact inhibition of growth occurs.

C. Other Effects of Extracellular Sialidases

The accumulation of sialidases in the extracellular fluid could be expected to exert a biological effect wherever gangliosides have a functional role, such as in cell adhesion and neuritogenesis as well as in growth factor receptor modulation. Treatment of neuroblastoma cells in culture with a bacterial sialidase stimulated neurite outgrowth (Wu and Ledeen, 1991). The effect of sialidase depended on exogenous Ca^{2+} and could be blocked by cholera toxin B subunit, suggesting the possibility that neuritogenesis was the result of an increased level of G_{M1} ganglioside on the cell surface as a result of sialidase-catalyzed conversion of G_{D1a} and G_{D1b} to G_{M1}. A modest increase in the influx of Ca^{2+} was also noted by Wu and Ledeen (1991) in the treated cells. A similar finding was reported by Yee *et al.* (1989), who found that exogenously added bacterial sialidase removed about 20% of the total cellular content of sialic acid from guinea pig cardiac myocytes and concomitantly and selectively increased Ca current of a transient Ca channel.

Sialidase treatment causes lymphocytes to disappear from circulation (Fischer *et al.*, 1991), a phenomenon that had previously been shown with rat erythrocytes (Janick *et al.*, 1978). The sialidase-treated lymphocytes were found to bind to homologous hepatocytes, Kupffer cells, and peritoneal macrophages by a galactose-specific mechanism, and were subsequently released into the circulation. The authors (Fischer *et al.*, 1991) found that binding could be inhibited by 2-deoxy-2,3-dehydro-*N*-acetylneuraminic acid, an effective inhibitor of sialidase activity, showing a sialic acid involvement in the binding mechanism, and speculated that release into the circulation followed resynthesis of the sialoglycoconjugates on the lymphocyte surface, possibly by a membrane-bound ectosialyltransferase.

Treatment of isolated membranes or tissue homogenates of guinea pig lung with a bacterial sialidase reduced the number of muscarinic superhigh density-affinity binding sites (Haddad and Gies, 1992, and references therein), an effect that could account for virus-induced airway hyperresponsiveness. It can be expected that other consequences of viral infection can be explained by the action of the viral sialidases on the cell surface sialoglycoconjugates.

V. Conclusions and Future Directions

Much has been learned about the substrate specificity and subcellular distribution of multiple forms of sialidases. Their purification to apparent homogeneity has been achieved in some instances and antibodies have been prepared. Advances within the next few years will include the purification of other sialidases and will lead to cloning and sequence determination. The situation with respect to the soluble sialidases accumulating in the medium of cultured fibroblasts is much less clear. It isn't known how general the release of sialidases is among various cell types and the origin of these sialidases remains to be determined. It would seem reasonable to expect that they are at least partially derived by a vesicle-mediated transport system from Golgi apparatus and trans-Golgi network and that these forms are likely to be lysosomal in nature. It may be that some of the sialidases in the medium are products of proteolytic action at the extracellular surface, in which case they should have sequence homology to portions of a plasma membrane sialidase. Finally, it is plausible that there are completely novel secreted sialidases which are exported to carry out specific cell surface-mediated functional roles. More intensified research is needed to distinguish between these possibilities and also to determine further whether extracellular sialidases have a role in modulating transmembrane signaling events.

References

Aminoff, D. (1961). *Biochem. J.* **81,** 384–392.

Bhavanandan, V. P., Yeh, A. K., and Carubelli, R. (1975). *Anal. Biochem.* **69,** 385–394.

Bremer, E. G., Hakomori, S.-I., Bowen-Pope, D. F., Raines, E., and Ross, R. (1984). *J. Biol. Chem.* **259,** 6818–6825.

Bremer, E. G., Schelessinger, J., and Hakomori, S.-I. (1986). *J. Biol. Chem.* **261,** 2434–2440.

Burnet, F. M., and Stone, J. D. (1947). *J. Exp. Biol. Med. Sci.* **25,** 227–233.

Cadena, D. L., and Gill, G. N. (1992). *FASEB J.* **6,** 2332–2337.

Carubelli, R., Trucco, R. E., and Caputto, R. (1962). *Biochim. Biophys. Acta* **60,** 196–197.

Chiarini, A., Fiorilli, A., Siniscalco, C., Tettamanti, G., and Venerando, B. (1990). *J. Neurochem.* **55,** 1576–1584.

Corfield, A. P., and Schauer, R. (1982). *In* "Sialic Acids: Chemistry, Metabolism and Function" (R. Schauer, ed.), pp. 225–241. Springer-Verlag, Vienna.

Fischer, C., Kelm, S., Ruch, B., and Schauer, R. (1991). *Carbohydr. Res.* **213,** 263–273.

Fukuda, M. (1991). *J. Biol. Chem.* **266,** 21327–21330.

Galjart, N. J., Gillemans, N., Harris, A., van der Horst, G. T. J., Verheijen, F. W., Galjaard, H., and d'Azzo, A. (1988). *Cell (Cambridge, Mass.)* **54,** 755–764.

Galjart, N. J., Gillemans, N., Meijer, D., and d'Azzo, A. (1990). *J. Biol. Chem.* **265,** 4678–4684.

Ghidoni, R., Sonnino, S., Masserini, M., Orlando, P., and Tettamanti, G. (1981). *J. Lipid Res.* **22,** 1286–1295.

Gottschalk, A., and Drzenick, R. (1972). *In* "Glycoproteins: Their Composition, Structure and Function" (A. Gottschalk, ed.), 2nd ed., Part A, pp. 381–402. Elsevier, Amsterdam and New York.

Gottschalk, A., and Lind, P. E. (1949). *Nature (London)* **164,** 232–233.

Haddad, E. B., and Gies, J. P. (1992). *Eur. J. Pharmacol.* **211,** 273–276.

Hanai, N., Dohi, T., Nores, G. A., and Hakomori, S.-I. (1988). *J. Biol. Chem.* **263,** 6296–6301.

Henningsen, M., Roggentin, P., and Schauer, R. (1991). *Hoppe-Seyler's Z. Physiol. Chem.* **372,** 1065–1072.

Hiraiwa, M., Nishizawa, M., Uda, Y., Nakajima, T., and Miyatake, T. (1988). *J. Biochem. (Tokyo)* **103,** 86–90.

Jancik, J. M., Schauer, R., Andres, K. H., and von During, M. (1978). *Cell Tissue Res.* **186,** 209–226.

Kolodny, E. H., Kanfer, J., Quirk, J. M., and Brady, R. O. (1971). *J. Biol. Chem.* **246,** 1426–1431.

Laine, R. A., and Hakomori, S. I. (1973). *Biochem. Biophys. Res. Commun.* **54,** 1039–1045.

Lambré, C. R., Terzidis, H., Greffard, A., and Webster, R. G. (1991). *Clin. Chim. Acta* **198,** 183–194.

Low, M. G. (1987). *Biochem. J.* **244,** 1–13.

Lowden, J. A., and O'Brien, J. S. (1979). *Am. J. Hum. Genet.* **31,** 1–18.

McCrea, J. F. (1947). *Aust. J. Exp. Biol. Med. Sci.* **25,** 127–136.

Miyagi, T., and Tsuiki, S. (1984). *Eur. J. Biochem.* **141,** 75–81.

Miyagi, T., and Tsuiki, S. (1985). *J. Biol. Chem.* **260,** 6710–6716.

Miyagi, T., and Tsuiki, S. (1986). *FEBS Lett.* **206,** 223–228.

Miyagi, T., Goto, T., and Tsuiki, S. (1984). *Gann* **75,** 1076–1082.

Miyagi, T., Sagawa, J., Konno, K., Handa, S., and Tsuiki, S. (1990a). *J. Biochem. (Tokyo)* **107,** 787–793.

Miyagi, T., Sagawa, J., Konno, K., and Tsuiki, S. (1990b). *J. Biochem. (Tokyo)* **107,** 794–798.

Miyatake, T., Atsumi, T., Obayashi, T., Mizuno, Y., Ando, S., Ariga, T., Matsui-Nakamura, K., and Yamada, T. (1979). *Ann. Neurol.* **6,** 232–244.

Moran, N. M., Breen, K. C., and Regan, C. M. (1986). *J. Neurochem.* **47,** 18–22.

Nojiri, H., Stroud, M., and Hakomori, S.-I. (1991). *J. Biol. Chem.* **266,** 4531–4537.

Ogura, K., and Sweeley, C. C. (1992). *Exp. Cell Res.* **199,** 169–173.

Ogura, K., Ogura, M., Anderson, R. L., and Sweeley, C. C. (1992). *Anal. Biochem.* **200,** 52–57.

Potier, M., Mameli, L., Belisle, M., Dallaire, L., and Melancon, S. B. (1979). *Anal. Biochem.* **94,** 287–296.

Potier, M., Michaud, L., Tranchemontagne, J., and Thauvette, L. (1990a). *Biochem. J.* **267,** 197–202.

Potier, M., Lamontagne, S., Michaud, L., and Tranchemontagne, J. (1990b). *Biochem. Biophys. Res. Commun.* **173,** 449–456.

Privalova, I. M., and Khorlin, A. (1969). *Izv. Akad. Nauk. SSSR, Ser. Khim.*, p. 2727.

Riboni, L., Prinetti, A., Bassi, R., and Tettamanti, G. (1991). *FEBS Lett.* **287,** 42–46.

Rosenberg, A., and Schengrund, C.-L. (1976). *In* "Biological Roles of Sialic Acid" (A. Rosenberg and C.-L. Schengrund, eds.), pp. 295–359. Plenum, New York..

Rosenberg, A., Howe, C., and Chargaff, E. (1956). *Nature (London)* **177,** 234–235.

Sagawa, J., Miyagi, T., and Tsuiki, S. (1988). *Jpn. J. Cancer Res.* **79,** 69–73.

Sagawa, J., Miyagi, T., and Tsuiki, S. (1990). *J. Biochem. (Tokyo)* **107,** 452–456.

Saito, M., and Yu, R. K. (1992). *J. Neurochem.* **58,** 83–87.
Saito, M., Sato-Bigbee, C., and Yu, R. K. (1992). *J. Neurochem.* **58,** 78–82.
Schauer, R. (1982). *Adv. Carbohydr. Chem. Biochem.* **40,** 131–234.
Schengrund, C.-L., and Nelson, J. T. (1976). *Neurochem. Res.* **1,** 171–180.
Schengrund, C.-L., and Rosenberg, A. (1970). *J. Biol. Chem.* **245,** 6196–6200.
Schengrund, C.-L., Jensen, D. S., and Rosenberg, A. (1972). *J. Biol. Chem.* **247,** 2742–2746.
Schengrund, C.-L., Rosenberg, A., and Repman, M. A. (1976). *J. Cell Biol.* **70,** 555–561.
Schwarzmann, G. (1978). *Biochim. Biophys. Acta* **529,** 106–114.
Sweeley, C. C., and Usuki, S. (1988). *In* "New Trends in Ganglioside Research: Neurochemical and Neuroregenerative Aspects" (R. W. Ledeen, E. L. Hogan, G. Tettamant, A. J. Yates, and R. K. Yu, eds.), Vol. 14, pp. 308–315. Liviana Press, Springer-Verlag, Pa'dova, Berlin.
Tettamanti, G., Morgan, I. G., Gombos, G., Vincendon, G., and Mandel, P. (1972). *Brain Res.* **47,** 515–518.
Tettamanti, G., Durand, P., and DiDonato, S. (1981). "Sialidases and Sialidoses: Perspectives in Inherited Metabolic Diseases," Vol. 4. Edi Ermes, Milan.
Tsuji, S., Yamauchi, T., Hiraiwa, M., Isobe, T., Okiyama, T., Sakimura, K., Takahashi, Y., Nishiawa, M., Uda, Y., and Miyatake, T. (1989). *Biochem. Biophys. Res. Commun.* **163,** 1498–1504.
Tulip, W. R., Varghese, J. N., Baker, A. T., van Donkelaar, A., Laver, W. G., Webster, R. G., and Colman, P. M. (1991). *J. Mol. Biol.* **221,** 487–497.
Tulsiani, D. R. P., and Carubelli, R. (1970). *J. Biol. Chem.* **245,** 1821–1827.
Tuppy, H., and Palese, P. (1969). *FEBS Lett.* **3,** 72–75.
Usuki, S., and Sweeley, C. C. (1988). *Indian J. Biochem. Biophys.* **25,** 102–105.
Usuki, S., Lyu, S.-C., and Sweeley, C. C. (1988). *J. Biol. Chem.* **263,** 6847–6853.
Vaheri, A., Rouslahti, E., and Nordling, S. (1972). *Nature (London) New Biol.* **238,** 211–212.
Varghese, J. N., and Colman, P. M. (1991). *J. Mol. Biol.* **221,** 473–486.
Venerando, B., Goi, G. C., Preti, A., Fiorilli, A., Lombardo, A., and Tettamanti, G. (1982). *Neurochem. Int.* **4,** 313–320.
Verheijen, F. W., Brossmer, R., and Galjaard, H. (1982). *Biochem. Biophys. Res. Commun.* **108,** 868–875.
Verheijen, F. W., Palmeri, S., Hoogevan, A. T., and Galjaard, H. (1985). *Eur. J. Biochem.* **149,** 315–321.
Verheijen, F. W., Palmeri, S., and Galjaard, H. (1987). *Eur. J. Biochem.* **162,** 62–67.
Warren, L. (1959). *J. Biol. Chem.* **234,** 1971–1975.
Warren, L., and Spearing, C. W. (1960). *Biochem. Biophys. Res. Commun.* **3,** 489–492.
Wenger, D. A., Tarby, T. J., and Wharton, C. (1978). *Biochem. Biophys. Res. Commun.* **82,** 589–595.
Wu, G. W., and Ledeen, R. W. (1991). *J. Neurochem.* **56,** 95–104.
Yee, H. F., Jr., Weiss, J. N., and Langer, G. A. (1989). *Am. J. Physiol.* **256,** C1267–C1272.
Yogeeswaran, G., and Hakomori, S.-I. (1975). *Biochemistry* **14,** 2151–2156.
Yohe, H. C., Saito, M., Ledeen, R. W., Kunishita, T., Sclafini, J. R., and Yu, R. K. (1986). *J. Neurochem.* **46,** 623–629.
Zeigler, M., Sury, V., and Bach, G. (1989). *Eur. J. Biochem.* **183,** 455–458.

ADVANCES IN LIPID RESEARCH, VOL. 26

Sphingolipids with Inositolphosphate-Containing Head Groups

ROBERT L. LESTER AND ROBERT C. DICKSON

Department of Biochemistry
University of Kentucky College of Medicine
Lexington, Kentucky 40536

I. Introduction

Since recent studies with yeast have shown the essentiality and possible function of one or more inositolphosphorylceramides, now appears to be an appropriate time for a review of this group of sphingolipids. The generalized structure of these lipids is ceramide-P-myoinositol-X, with X referring to polar substituents.

These compounds had been referred to as "phytoglycolipids" and "mycoglycolipids," indicating their origins. Since it is now clear that the distribution of these compounds is not limited to plants or fungi, we propose the neutral, generic designation, inositolphosphorylceramides, abbreviated here as InsPCers. Inositolphosphorylceramide itself will be abbreviated as IPC. We will refer to myoinositol as simply inositol.

II. Plant InsPCers

Studies on InsPCers derived from plants commenced over 50 years ago in a number of laboratories with the recognition of incompletely characterized inositol-containing lipid fractions from seed oils that yielded various sugars such as galactose and arabinose on acid hydrolysis. This material has been reviewed by Celmer and Carter (1952), Hawthorne (1960), and Allen and Good (1965). Much of this early work was hampered by inadequate purification methods applied to exceedingly complex mixtures of InsPCers.

H. Carter and associates at the University of Illinois carried out pioneering studies in the 1950s and 1960s on the so-named "phytoglycolipide" purified from seeds of corn, flax, and soybeans (Carter *et al.*, 1958b) and from bean leaves (Carter and Koob, 1969). Strong alkaline hydrolysis of these lipid mixtures yielded a series of (phospho)oligosaccharides of different sizes containing glucosaminido-glucuronido-inositol with variable amounts of galactose, mannose, arabinose, and fucose (Carter *et al.*, 1964) as well as (phospho)ceramide (Carter *et al.*, 1958b). Ceramides were mainly composed of phytosphingosine (D-*ribo*-1,3,4-trihydroxy-2-amino-octadecane) or dehydrophytosphingosine (D-*ribo*-1,3,4-trihydroxy-2-amino-8-*trans*-octadecene) (Carter and Hendrickson, 1963) and mainly a hydroxy C_{24} fatty acid (Carter and Koob, 1969). Detailed structural analysis was carried out only on the tetrasaccharide component derived from alkaline hydrolysis of corn seed InsPCers, resulting in the proposed structure (Carter *et al.*, 1969):

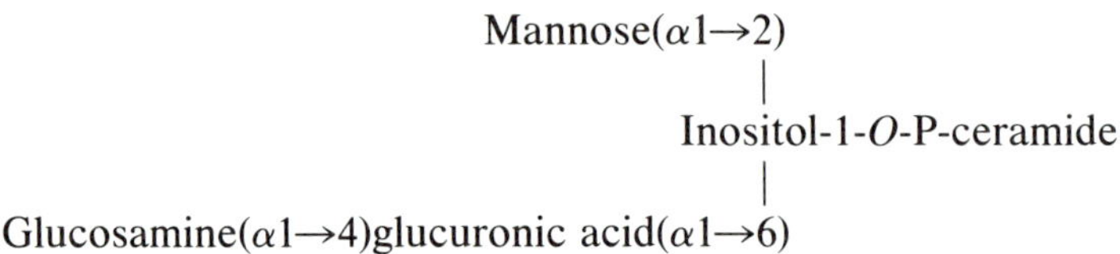

Wagner *et al.* (1969) purified a mixture of InsPCers from a commercial preparation of peanut lecithin which resembled the lipid and carbohydrate composition of the earlier described InsPCers from soy beans (Carter and Hendrickson, 1963).

Kaul and Lester (1975) devised a relatively mild extraction procedure that resulted in an InsPCer concentrate yielding about 100 μmol of P/kg fresh weight of tobacco leaves. This procedure avoided the potentially harsh treatments employed by Carter *et al.* (1958a), involving treatment with alkali to destroy acyl esters or refluxing with HCl-ethanol (Carter and Koob, 1969). From the tobacco InsPCer concentrate, pure lipids were obtained which fell into two classes, the PSL-I group containing *N*-acetylglucosamine and the PSL-II group containing unacetylated glucosamine (Kaul and Lester, 1975, 1978). The structure of PSL-I was determined to be (Hsieh *et al.*, 1978):

$$\begin{array}{r} \text{Inositol-1-}O\text{-P-ceramide} \\ | \quad\quad\quad\quad\quad\quad \\ N\text{-Acetylglucosamine}(\alpha 1\rightarrow 4)\text{glucuronic acid}(\alpha 1\rightarrow 2) \end{array}$$

PSL-II is therefore:

$$\begin{array}{r} \text{Inositol-1-}O\text{-P-ceramide} \\ | \quad\quad\quad\quad\quad\quad \\ \text{Glucosamine}(\alpha 1\rightarrow 4)\text{glucuronic acid}(\alpha 1\rightarrow 2) \end{array}$$

The ceramides from both PSL-I and PSL-II comprised roughly equal amounts of phytosphingosine and dehydrophytosphingosine with OH-fatty acids ranging from C_{22} to C_{26}, with the OH-C_{24} homolog predominating. Thus, the hydrophobic components of the tobacco leaf InsPCers resemble those previously observed by Carter *et al.* from other plant sources. The other purified lipids (PSL-I/II-ABC) all have the trisaccharide components of PSL-I and PSL-II and are presumed to be derivatives of PSL-I and PSL-II; these are characterized as to composition but not as to sugar linkages and anomeric configuration (Kaul and Lester, 1978):

Compound	Structure
PSL-IA	PSL-I-[Arabinose$_2$Galactose$_2$]
PSL-IB	PSL-I-[Arabinose$_3$Galactose$_2$]
PSL-IC	PSL-I-[Arabinose$_4$Galactose$_2$]
PSL-IIA	PSL-II-[Arabinose$_3$Galactose]
PSL-IIB	PSL-II-[Arabinose$_{2 \text{ or } 3}$Galactose$_2$]
PSL-IIC	PSL-II-[Arabinose$_2$Galactose$_2$Mannose]

The InsPCers purified from tobacco leaves accounted for 50–60% of the InsPCer concentrate; many more additional components more complex than ceramide-P-nonasaccharides were not resolvable at that time (Kaul and Lester, 1978). Modern HPLC supports could probably resolve more of these components as intact lipids.

The complexity of tobacco leaf InsPCers was further probed by the work of Hsieh *et al.* (1981). These workers reduced the glucuronic acid carboxyl group of the InsPCer concentrate, and liberated the oligosaccharides with strong alkali, followed by peracetylation and separation by reversed-phase high-performance liquid chromatography; at least 24 components were evident in the chromatographic profile. Alkaline hydrolysis and peracetylation completely eliminated the distinction between the PSL-I and the PSL-II series so that the complexity may be even greater than that observed. A major tetrasaccharide was identified:

Galactose(β1→4)*N*-acetylglucosamine(α1→4)glucuronic acid (α1→2)inositol

A minor tetrasaccharide was also partially characterized:

N-Acetylglucosamine(α1→4)glucuronic acid (α1→?)inositol(?←1)mannose

From black gram sprouts (*Vigna mungo*), Kondo and Nakano (1987) obtained a material from the upper phase of a Bligh–Dyer (1969) lipid extract that was further purified by gel filtration and ion-exchange chromatography, both with aqueous buffers. Analysis indicated the presence of the core plant InsPCer components, ceramide, P, inositol, glucuronic acid, and hexosamine as well as 5.3% protein. In addition, large amounts of galactose and arabinose were present, which their analysis suggests may consist of a β1,6-galactan substituted at some of the three positions with arabinose. No mannose was found. It is difficult to judge the purity of this material since the components, InsPCers and proteins, would be expected to exist together as mixed micelles under the conditions of isolation. No direct evidence for a covalent linkage between protein and lipid was given.

Thus far, only crop plants have been examined for the presence of InsPCers; however, they are probably widespread if not ubiquitous in plants. The complexity of the InsPCers in plants may be comparable to that of the glycosphingolipids in animals. The plant InsPCers await further studies of their structures, organ and intracellular distribution, biosynthesis, and function(s).

III. InsPCers of Fungi and Yeast

A. Analytical Methods

1. *Extraction of InsPCers*

A systematic and quantitative study of the extraction of InsPCers from *Saccharomyces cerevisiae* and from the mycelial phase of *Neurospora crassa* was carried out by Hanson and Lester (1980). Cells uniformly labeled with [^{3}H]inositol were used to gauge the extraction efficiency of a variety of procedures, some previously employed by others to extract phospholipids. The best method involved treating the cells with 5% trichloroacetic acid at 0°C to destroy phospholipases, known to be activated by organic solvents during lipid extraction (Letters, 1968), followed by extraction with a slightly alkaline, warm, water-rich mixture of ethanol–diethylether–water–pyridine (Method IIIB, Hanson and Lester, 1980). This method approached 100% extraction of the labeled inositol; it is possible that inositol-containing lipid–protein anchors contribute to the minor amounts not extracted. Some water-poor solvents completely failed to extract the InsPCers. The adopted procedure has been used for InsPCer extraction from other fungi such as *Histoplasma capsulatum* (Barr and Lester, 1984) and from fresh tobacco leaves (Kaul and Lester, 1975) and has been used for the extraction of lipophosphoglycan from *Leishmania donovani* (Orlandi and Turco, 1987). In the absence of labeling of InsPCer components, efficacy of InsPCer extraction could be judged by monitoring total long-chain base and/or very-long-chain fatty acids (Dickson *et al.*, 1990).

2. *Purification of InsPCers*

Since the purification of InsPCers offers challenges no different from any very polar, acidic lipid, only details specific for InsPCers will be discussed. First, InsPCers precipitate almost quantitatively at low temperature after adjusting to pH 5 the initial lipid extracts from *S. cerevisiae* (Smith and Lester, 1974), *H. capsulatum* (Barr and Lester, 1984), *N. crassa* (Lester *et al.*, 1974), and tobacco leaves (Kaul and Lester, 1975). Second, mild alkaline methanolysis is a useful step to destroy the ester-containing lipids in crude or semipurified extracts (Barr and Lester, 1984); however, stronger alkaline conditions should be avoided owing to the lability of unsubstituted inositol attached by a phosphodiester bond (Smith and Lester, 1974). Finally, liquid chromatography on silica gel columns is a valuable technique for isolating InsPCers (Barr and Lester, 1984); low levels of salt in the eluting solvents are required to chromatograph macroscopic quantities of the very polar InsPCers. A solvent polar enough to dissolve a practical amount of the very polar InsPCers yields insufficient retention. Inclusion of salt increases the retention dramatically, presumably by providing a charged surface for the negatively charged InsPCers. In contrast, the salt had little influence on the retention of neutral lipids (R. L. Lester and G. B. Wells, unpublished).

B. Structures of InsPCers of Yeast and Fungi

1. Saccharomyces cerevisiae

The first report of an InsPCer in yeast was that of Wagner and Zofcsik (1966a,b,c), who characterized a so-called "mycoglycolipid" in *S. cerevisiae* and *Candida utilis* as

Ceramide-P-inositol-mannose

The ceramide was composed of C_{18} and C_{20} phytosphingosines, C_{18} dihydrosphingosine, and principally OH-C_{24}, OH-C_{26} (*C. utilis*), and C_{26} and OH-C_{24} (*S. cerevisiae*) fatty acids. Subsequent work with *S. cerevisiae* (Steiner *et al.*, 1969; Smith and Lester, 1974) indicated that the mycoglycolipid of Wagner and Zofcsik largely arose by alkaline hydrolysis (1 *N* KOH, 37°C, 24 hours) of a major sphingolipid with the composition mannose(inositol-P)$_2$ceramide, hereafter abbreviated as M(IP)$_2$C. These alkaline conditions were developed by Schmidt *et al.* (1946) and used widely in procedures for sphingomyelin isolation and analysis. The procedure of Schmidt *et al.* was based on the observation of Thudichum (1901) that sphingomyelin is base-stable relative to other phospholipids. In fact, the other major InsPCer in *S. cerevisiae,* IPC, was completely destroyed by strong alkaline treatment (Smith and Lester, 1974). The facile formation of a cyclic inositol phosphate accounts for the breakdown of these InsPCers; the stability of the mannose-inositol-P-ceramide (abbreviated MIPC) suggests that the mannose is vicinal to the phosphate and is thus unable to form a cyclic inositol-P:

$$\mathrm{M(IP)_2C} + \mathrm{H_2O} \xrightarrow{\mathrm{OH^-}} \text{cyc-1,2-Inositol-P} + \mathrm{MIPC}$$

$$\mathrm{IPC} + \mathrm{H_2O} \xrightarrow{\mathrm{OH^-}} \text{cyc-1,2-Inositol-P} + \text{ceramide}$$

$$\mathrm{MIPC} + \mathrm{H_2O} \xrightarrow{\mathrm{OH^-}} \text{No reaction}$$

Small amounts of MIPC were found in *S. cerevisiae* untreated with strong alkali; however, M(IP)$_2$C and IPC were the major InsPCers. Collectively, the InsPCers comprised about 40% of the inositol in the lipid extract and about 15% of the total yeast phospholipid P (Smith and Lester, 1974).

Molecular species of IPC and MIPC with different levels of hydroxylation of the ceramide components were separated with difficulty on low-performance silica gel chromatographic supports by Smith and Lester (1974). Subsequent work with high-performance chromatographic supports (Wells and Lester, 1983; G. B. Wells and R. L. Lester, unpublished) showed that four major variously hydroxylated species of each of the InsPCers can be detected. In the order of elution from the column, these are as follows.

Species	Long-chain bases	Fatty acids
I	$C_{18/20}$ *erythro*-dihydrosphingosines	26 : 0
IIa	$C_{18/20}$ *erythro*-dihydrosphingosines	OH-26 : 0
IIb	$C_{18/20}$ phytosphingosines	26 : 0
III	$C_{18/20}$ phytosphingosines	OH-26 : 0
IV	$C_{18/20}$ phytosphingosines	$(OH)_2$-26 : 0

Molecular species IIa and IIb have the same total level of hydroxylation and are inseparable with standard methods; however, ordinarily they do not coexist. Species I is found only under anaerobic culture conditions (Wells and Lester, 1983). We propose this system of nomenclature to be superior to that used earlier by Smith and Lester (1974).

Preliminary nuclear magnetic resonance (NMR) analysis of the polar head group of $M(IP)_2C$ as well as other data (Shelling *et al.*, unpublished; R. L. Lester *et al.*, unpublished) suggest the following structure for $M(IP)_2C$.

Inositol-1-P-(6)mannose(α1,2)inositol-1-P-(1)ceramide

2. *Aspergillus niger*

Hackett and Brennan (1977) offered some evidence for a substance from *A. niger* that labels with [^{3}H]inositol and that has some of the chromatographic and chemical properties of inositol-P-ceramide. A brief communication by Brennan and Roe (1975) described the isolation of a glycosphingolipid that contained, in unspecified molar ratios, P, ceramide, inositol, galactose, mannose, and small amounts of glucosamine. This compound could be related to the InsPCers from *H. capsulatum* (see below).

3. *Histoplasma capsulatum*

From the yeast phase of the dimorphic, pathogenic fungus *H. capsulatum*, IPC as well as three other novel InsPCers were identified (Barr and Lester, 1984; Barr *et al.*, 1984):

Compound V $\quad$ $Man_p(\alpha 1 \rightarrow 3)$
$\qquad\qquad\qquad\qquad$ |
$\qquad\qquad\qquad Man_p(\alpha 1 \rightarrow 2$ or $6)$inositol-P-ceramide

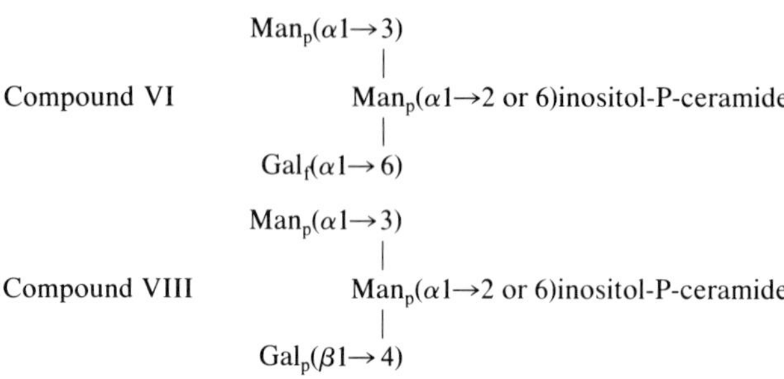

The ceramides were composed principally of C_{18} phytosphingosine and OH-C_{24} fatty acid. Compound VIII was the major InsPCer in both the yeast and the mycelial phase. Compounds V and VI were absent from the mycelial phase and the major *S. cerevisiae* InsPCer, $M(IP)_2C$, was absent from both phases. Antibodies reacting with these lipids were detected in sera from patients with histoplasmosis.

4. Neurospora crassa

Several strains of *N. crassa* exhibited an InsPCer with the composition

$$(\text{Inositol-P})_2\text{-ceramide}$$

that accounted for 30–60% of the lipid-extractable inositol (Lester *et al.*, 1974). The compound was labile to alkali (1 *N* KOH, 37°C, 15 hours), yielding inositol monophosphates; the logical inference from the limited data was that the compound could be formulated as

Inositol-P-inositol-P-ceramide

with the details of the phosphate linkages undetermined. The ceramide was composed principally of C_{18} phytosphingosine and OH-C_{24} fatty acid.

5. Phytophtora capsici

A substance that elicits a defense reaction in peppers against *P. capsici* infection was isolated from this pathogenic fungus and was identified as an IPC (Lhomme *et al.*, 1990). The ceramide was unusual for fungal InsPCers: *N*(4-hydroxy-2-docosenoyl)C_{16}-sphingosine. The nature of the biological specificity of this sphingolipid is unclear, since these workers (Pivot *et al.*, 1991) have also elicited a defense reaction in peppers and wheat against *P. capsici* infection with a quite different lipid isolated from this fungus: dihexadecanoyl phosphatidylcholine.

IV. InsPCers in Protozoa

A. *Leishmania* sp.

Kaneshiro *et al.* (1986) showed that about 40% of the lipid-extractable inositol of *Leishmania donovani* promastigotes exhibited chromatographic and chemical properties of an IPC. The principal constituents of the ceramide were stearic acid and C_{16} and C_{18} sphingosines. *L. donovani* is the causative agent of the human disease visceral leishmaniasis.

Mass spectral evidence was offered in support of the occurrence in promastigotes of *Leishmania mexicana mexicana* of an InsPCer with a ceramide composed mainly of stearic acid and a C_{16} sphingosine (Singh *et al.*, 1988).

B. *Tritrichomonas foetus*

Singh *et al.* (1991) have studied the lipids resistant to mild alkali in *T. foetus,* a flagellated protozoan responsible for spontaneous abortion in cattle. IPC species as well as novel InsPCers were observed that both had ceramides consisting of palmitic acid and either C_{18} sphinganine or C_{18} sphingosine. The novel InsPCer was partially characterized as

```
Fucose-inositol-P-ceramide
                |
                Phosphoethanolamine (± N-acetyl)
```

C. *Acanthamoeba castellanii*

The soil amoeba *A. castellanii* was shown to contain complex related InsPCers, termed lipophosphoglycans, which constituted 29% of the plasma membrane mass. These were partially characterized (Dearborn *et al.*, 1976):

```
             Galactosamine
                |
Ceramide-P-inositol-[2-aminoethylphosphonate or 1-OH-2-aminoethylphosphonate]
                |
             [Glucose2mannose5xylose2]
                    and
             [Glucose12mannose6galactose2]
```

The major ceramide components were 2-OH normal fatty acids of chain lengths from 22 to 28 carbons and 2-OH branched fatty acids with chain lengths of 21 to 25 carbons; the major long-chain bases were C_{24} and C_{25} phytosphingosines.

D. *Trypanosoma cruzi*

Beginning with the work of Lederkremer *et al.* (1976), a complex InsPCer in *T. cruzi* termed "lipopeptidophosphoglycan" has been the subject of numerous investigations; pertinent references can be found in Previato *et al.* (1990). *T. cruzi* is the causative agent of Chagas' disease and the surface glycoconjugates are actively studied because of their possible role in pathogenesis. The structure is thought to be (Previato *et al.*, 1990):

```
                                                              P(6←1)2-Aminoethylphosphonate
                                                              |
Galf(β1→3)Manp(α1→2)Manp(α1→2)Manp (α1→6)Manp(α1→4)GlcNH2
                         | ?   or    | ?                      |
                         Galf(β1→3)                           Inositol-P-ceramide
```

The ceramide is composed mainly of C_{18} sphinganine and C_{18} sphingosine with palmitic and lignoceric acids. The structure, -(Man)$_n$-GlcNH$_2$-inositol-P-, is found in phosphatidylinositol-linked membrane protein anchors (Ferguson and Williams, 1988) and in the surface glycoconjugate termed lipophosphoglycan from *L. donovani* (Turco *et al.*, 1989).

V. Metabolism of InsPCers

A. Synthesis of Long-Chain Bases

This review is limited to the organisms in which InsPCers have been identified; information is available only for *Hansenula ciferrii* and *S. cerevisiae*. For a general review of long-chain base synthesis, see Merrill and Jones (1990).

1. *Hansenula ciferrii*

At least three steps are required for the synthesis of phytosphingosine, the predominant long-chain base in InsPCers:

i. Serine Palmitoyltransferase (SPT):

```
CH3(CH2)14-C-SCoA + HOOC-CH-CH2OH ———→ CH3(CH2)14-C-CH-CH2OH + CO2 + CoA
           ||              |                        || |
           O               NH2                      O NH2

  Palmitoyl-CoA         L-Serine                D-3-Ketosphinganine
```

ii. 3-Ketosphinganine (3-KDS) Reductase:

$$\underset{\substack{\| \ \ | \\ \mathrm{O\ NH_2}}}{\mathrm{CH_3(CH_2)_{14}\text{-}C\text{-}CH\text{-}CH_2OH}} + \mathrm{NADPH} + \mathrm{H^+} \longrightarrow \underset{\substack{| \quad | \\ \mathrm{HO\ \ NH_2}}}{\mathrm{CH_3(CH_2)_{14}\text{-}CH\text{-}CH\text{-}CH_2OH}} + \mathrm{NADP^+}$$

iii. Phytosphingosine Synthase:

$$\underset{\substack{\| \quad | \\ \mathrm{HO\ \ NH_2}}}{\mathrm{CH_3(CH_2)_{14}\text{-}CH\text{-}CH\text{-}CH_2OH}} + \mathrm{O_2} + ?? \longrightarrow \underset{\substack{| \quad | \quad | \\ \mathrm{HO\ HO\ \ NH_2}}}{\mathrm{CH_3(CH_2)_{13}\text{-}CH\text{-}CH\text{-}CH\text{-}CH_2OH}} + ??$$

D-4-Hydroxysphinganine
(Phytosphingosine)

Steps i and ii were first defined in the 1960s by E. E. Snell and coworkers with crude membranes of the fungus *H. ciferrii* (for summary, see Snell *et al.*, 1970). Subsequent work with other biological systems has confirmed these findings (reviewed in Merrill and Jones, 1990). The enzymes have never been purified.

Phytosphingosine is not limited to plant, fungal, and protozoan sphingolipids but has a very widespread distribution, including higher animals (Nishimura, 1987), fish (Li *et al.*, 1984), amphibians (Nohara-Uchida and Ohashi, 1987; Hidari *et al.*, 1991), and invertebrates such as sea urchins (Kochetkov *et al.*, 1976; Kubo *et al.*, 1990) and starfish (Sugita, 1979). Nevertheless, phytosphingosine synthesis has not yet been demonstrated *in vitro*. *In vivo* studies with *H. ciferrii* (Stoffel *et al.*, 1968a) and in the rat (Crossman and Hirschberg, 1977) showed conversion of labeled *erythro*-dihydrosphingosine to phytosphingosine; molecular oxygen appears to be the source of the 4-oxygen of phytosphingosine (Kulmacz and Schroepfer, 1978). One can speculate that a mixed-function oxidase system might catalyze step iii.

2. *Saccharomyces cerevisiae*

a. Serine Palmitoyltransferase. SPT activity was demonstrated in crude membranes from *S. cerevisiae* (Pinto *et al.*, 1992b). The conversion of labeled serine into 3-keto-dihydrosphingosine was completely dependent on a fatty acyl-CoA; C_{14}, C_{16}, and C_{18} fatty acyl-CoAs gave rise to C_{16}, C_{18}, and C_{20} products, respectively. The abundant C_{20} homolog of phytosphingosine found in yeast thus originates from the SPT-catalyzed condensation of stearyl-CoA with serine.

Two potent antifungal agents have been shown to inhibit SPT activity in membranes of *S. cerevisiae* at nanomolar levels (Zweerink *et al.*, 1992):

sphingofungin B (2S-amino-3R,4R,5S,14-tetrahydroxyeicos-6-enoic acid) and sphingofungin C (2S-amino-5S-acetoxy-3R,4R,14-trihydroxyeicos-6-enoic acid). These agents inhibited inositol incorporation into sphingolipids but not into phosphatidylinositol; this specificity suggests these compounds may prove to be useful for the study of sphingolipid metabolism/function.

Free sphingolipid long-chain bases did not inhibit SPT activity *in vitro,* nor did they repress enzyme activity levels when added to growing cultures (Pinto *et al.,* 1992b); this contrasts with the downregulation of SPT activity observed by Mandon *et al.* (1991) in cultured neurons with the addition of sphingolipid bases.

b. 3-Ketodihydrosphingosine Reductase. Crude membranes of *S. cerevisiae* were shown to catalyze the stereospecific reduction of 3-ketodihydrosphingosine with NADPH, forming the *erythro* isomer of dihydrosphingosine (Pinto *et al.,* 1992b), a reaction analogous to that shown earlier with animal microsomes (Stoffel *et al.,* 1968b). Further confirmation of the pathway for phytosphingosine synthesis was the observation that mutants defective in SPT could grow and make phytosphingosine-containing sphingolipids when supplemented with either 3-ketodihydrosphingosine or dihydrosphingosine (Pinto *et al.,* 1992a). The fact that both the *erythro* and the unnatural *threo* isomers of dihydrosphingosine could support growth of the SPT mutant (Pinto *et al.,* 1992a) suggests that these isomers are interconvertible.

c. Genes Responsible for SPT Activity. Molecular cloning of the *S. cerevisiae* gene(s) encoding SPT activity was greatly facilitated by the isolation of a mutant strain that required a sphingolipid *l*ong-*c*hain *b*ase (*lcb*) (dihydrosphingosine or phytosphingosine) for growth (Wells and Lester, 1983). Isolation of an *lcb* mutant strain established that sphingolipids are essential for growth of *S. cerevisiae* which, in turn, suggested that one or more specific cellular functions required sphingolipids. The *lcb* strain also proved to be important because its behavior provided a simple procedure for enrichment of mutants defective in long-chain base biosynthesis. A buoyant density increase was noted when *lcb* mutant cells were starved for long-chain base. Consequently, mutant cells could be separated from normal cells by buoyant density centrifugation (Pinto *et al.,* 1992a). Using this enrichment procedure, numerous Lcb$^-$ mutants were isolated and placed into complementation groups, of which there turned out to be only two, *LCB1* and *LCB2* (Pinto *et al.,* 1992a). Both *lcb1* and *lcb2* mutant strains lack SPT activity (Pinto *et al.,* 1992b).

i. Characterization of the LCB1 gene. The *LCB1* gene was isolated from a *S. cerevisiae* recombinant genomic DNA library by complementation of the *Lcb*⁻ phenotype of an *lcb1* mutant strain (Buede *et al.*, 1991). The cloned gene restored SPT activity as expected if the gene encoded SPT or a subunit of the enzyme. Southern blot analysis demonstrated that there was but a single copy of *LCB1* in the *S. cerevisiae* haploid genome. The gene is predicted to encode a protein of 558 amino acid residues. Since SPT activity is found in the membrane fractions of *S. cerevisiae* (Pinto *et al.*, 1992b) and mammalian cells (Merrill and Jones, 1990), we anticipated that the predicted LCB1 protein would have at least one membrane-spanning domain. The algorithms of Kyte and Doolittle (1982) and Eisenberg *et al.* (1984) predicted two membrane-associated helices spanning residues 12 to 32 (IPIPAFIVTTSSYLWYYFNLV) and residues 344 to 373 (ATAIDITVGSMATALGSTGGFVLG).

Three results suggest that *LCB1* encodes SPT or a subunit of the enzyme (Buede *et al.*, 1991). First, the cloned *LCB1* gene complements the Lcb⁻ phenotype of an *lcb1* mutant and restores growth in media lacking long-chain base. Second, the cloned gene restores SPT activity in an *lcb1* deletion mutant. Finally, the predicted *LCB1* amino acid sequence shows amino acid similarity to the enzymes 5-aminolevulinic acid synthase (ALA synthase) and 2-amino-3-ketobutyrate-CoA ligase, enzymes that catalyze chemical reactions very similar to the one catalyzed by SPT. Moreover, all three enzymes use the cofactor pyridoxal phosphate.

ii. Characterization of the LCB2 gene. The *LCB2* gene of *S. cerevisiae* was isolated from a recombinant genomic DNA library by complementation of the Lcb⁻ phenotype of an *lcb2*-defective strain (R. C. Dickson *et al.*, unpublished results). *LCB2,* like *LCB1,* is present as a single-copy gene in the yeast haploid genome. The predicted protein product of *LCB2* has 561 amino acids and is expected to contain two membrane-associated helices spanning residues 57 to 77 (PYYISLLTYLNYLILIILGHV) and residues 443 to 463 (LGFIVYGVADSPVIPLLLYCP). Comparison of the predicted LCB1 and LCB2 amino acid sequences showed 22.4% amino acid identity and 15.2% similarity over the entire sequences, which suggests that the proteins are related. The predicted LCB2 amino acid sequence shows about the same level of amino acid similarity to ALA synthase, as does the predicted LCB1 protein sequence. Taken together, the available data suggest that the LCB1 and LCB2 proteins are subunits of SPT. This possibility awaits biochemical confirmation. With the availability of both the *LCB* genes, it should be possible to overexpress the genes in yeast or bacteria, which ought to facilitate purification and characterization of SPT.

B. Synthesis of Very Long Chain (Hydroxy) Fatty Acids

Information is incomplete on the synthesis of the very long chain, often hydroxylated, fatty acids found in the InsPCers. In animals the 16:0 product of the fatty acid synthase can be elongated in mitochondria by a system that shares some of the β-oxidation pathway enzymes; elongation in the endoplasmic reticulum proceeds as acyl-CoA derivatives by condensation with malonyl-CoA and reduction with reduced pyridine nucleotides (Seubert and Podack, 1973; Bernert and Sprecher, 1977). In no case have individual enzymes been highly purified or the system reconstituted with solubilized enzymes.

Information on fatty acid elongation is even scantier in those organisms with InsPCers. Evidence for a malonyl-CoA-primed elongation system in plants has been presented (Harwood, 1979; Walker and Harwood, 1986). In yeast, evidence for fatty acid elongation comes from observing the conversion of labeled fatty acids to longer chain acids *in vivo* with wild-type *C. utilis* and *S. cerevisiae* (Fulco, 1967) and with a fatty acid synthetase mutant of *S. cerevisiae* (Orme *et al.*, 1972). *In vitro* evidence for fatty acid elongation in *S. cerevisiae* comes from the work of Blanchardie *et al.* (1977) showing that purified membrane fractions could convert labeled stearoyl-CoA in the presence of malonyl-CoA and NADPH to fatty acids with 20–30 carbons. However, Schweizer (1984) questions the relevance of malonyl-CoA-dependent chain elongation in yeast.

Fatty acid α-hydroxylation has been studied in animal systems; however, molecular mechanisms remain incompletely understood, although molecular oxygen and NADPH are required (Kishimoto, 1983; Shigematsu *et al.*, 1990). The nature of the substrate is unclear, i.e., whether it is a fatty acyl-CoA, a ceramide, or a more complex sphingolipid. The electron transport chain of events is also unclear.

In *Tetrahymena pyriformis*, Kaya *et al.* (1984) provided *in vivo* evidence for the direct hydroxylation of ceramide 2-aminoethylphosphonate, converting the bound palmitic acid to α-hydroxypalmitic acid. On raising the growth temperature from 15 to 39°C, a rapid increase in hydroxylated sphingolipid was observed. Hydroxylation of sphingolipid *in vitro* was not observed. Indirect evidence for direct sphingolipid hydroxylation comes from the observations of Dickson *et al.* (1990) with a mutant of *S. cerevisiae* that makes no detectable sphingolipid; nevertheless, this mutant contains normal levels of C_{26} fatty acid which is strikingly underhydroxylated. The origin of the 2-oxygen in hydroxy fatty acids in the InsPCers of *S. cerevisiae* appears to be molecular oxygen, since anaerobically cultured cells have no hydroxy fatty acids in the InsPCers (R. L. Lester *et al.*, unpublished).

C. Synthesis of Ceramide

Ceramide synthesis has been studied in higher animals; it occurs by the reaction of a fatty acyl-CoA with a free sphingolipid long-chain base (Kishimoto, 1983). No information is available concerning ceramide synthesis in the organisms with InsPCers.

D. Synthesis of InsPCers

In spite of the widespread distribution of these major membrane constituents, little is known about their assembly and metabolism; the only available information comes from studies in yeast. A pathway has been proposed as a working hypothesis (Becker and Lester, 1980) whereby phosphatidylinositol is the precursor of the inositol-P groups, reacting with ceramide and MIPC, and with GDP-mannose as a mannose donor:

i. Phosphatidylinositol + ceramide → Inositol-P-ceramide + diacylglycerol (DAG)
ii. Inositol-P-ceramide + GDP-mannose → Mannose-inositol-P-ceramide (MIPC) + GDP
iii. MIPC + phosphatidylinositol → $M(IP)_2C$ + DAG

In vivo evidence supporting this scheme was obtained by Angus and Lester (1972). Synthesis of $M(IP)_2C$ in *S. cerevisiae* was studied by uniformly labeling cells with $^{32}P_i$ and [^{3}H]inositol, chasing with unlabeled isotopes for several cell generations, and monitoring the changes in all the major phospholipids. Half of both the ^{32}P and [^{3}H]inositol that disappeared from the phosphatidylinositol pool accumulated in the $M(IP)_2C$ pool with the identical label ratio; the other half appeared in the culture medium as glycerophosphoinositol. These data are wholly compatible with phosphatidylinositol providing both inositol-P groups intact to form $M(IP)_2C$. Little turnover of $M(IP)_2C$ was observed; however, the IPC + MIPC fraction did exhibit turnover compatible with roles as precursors of $M(IP)_2C$. Previous work by Tanner (1968, 1969) is consistent with these conclusions. The synthesis of the InsPCers may occur in the Golgi apparatus because Puoti *et al.* (1991) have shown that synthesis of InsPCers from labeled inositol ceases in temperature-sensitive mutants that cannot carry out vesicular transport from the endoplasmic reticulum to the Golgi at the restrictive temperature.

In vitro evidence supporting reactions (i) and (ii) was obtained using crude membranes from *S. cerevisiae* which catalyze the incorporation of phosphoinositol into IPC and $M(IP)_2C$ from added labeled phosphatidylinositol (Becker and Lester, 1980). Presumably, endogenous ceramide and MIPC served as substrates. Work from this laboratory has shown that exogenous labeled ceramide can serve as a precursor of IPC (W. J. Pinto *et al.*, unpublished) and that mannose-labeled GDP-mannose can serve as a mannose donor (G. W. Becker *et al.*, unpublished) in this *in vitro* system. The possibility that mannosylphosphoryldolichol serves as the ultimate mannose donor cannot be ruled out. We are undertaking purification of the synthetic enzymes away from endogenous lipids in order to establish definitively the stoichiometry of reactions i–iii. Evidence for reaction (i) in *A. niger* was obtained by Hackett and Brennan (1977). The postulated phosphoinositol transferase reaction for IPC synthesis is analogous to the phosphocholine transferase in the formation of sphingomyelin (Merrill and Jones, 1990).

VI. Functions of Sphingolipids in *Saccharomyces cerevisiae*

All eukaryotic cells, including animals (Hakomori, 1983), higher plants (Laine *et al.*, 1980), and fungi (Brennan and Losel, 1978), contain sphingolipids, particularly in their plasma membrane. In animals, sphingolipids are thought to play roles as modulators of membrane signal transducers, resulting in the regulation of cell growth and differentiation, and as mediators of cell-to-cell or cell-to-substratum recognition (Hannun and Bell, 1989; Hakomori, 1990). Additional biological roles for sphingolipid intermediates and breakdown products including ceramide and sphingosine have been reported by several workers (reviewed in Hannun and Bell, 1989; Merrill, 1991). Further effort is needed, however, to establish the physiological function of most sphingolipids. A particularly nagging problem in deciphering the function of a specific sphingolipid is the presence of dozens, perhaps hundreds, of sphingolipids in most animal cells (Hakomori, 1983).

To try to overcome the problems posed by the presence of many sphingolipids in a cell, we have begun to examine the function of sphingolipids in *S. cerevisiae,* since there is only one major sphingolipid ($M(IP)_2C$) and two minor sphingolipids (IPC and MIPC). The discovery of sphingolipid long-chain base auxotrophs of *S. cerevisiae* unable to make sphingolipids owing to defective SPT activity (Wells and Lester, 1983; Pinto *et al.*, 1992a,b) opened the possibility of using molecular genetic techniques to study sphingolipid functions. The inability to grow and the loss of viability of such mutant strains if not supplemented with a sphingolipid long-chain base strongly suggests that there are one or more vital functions for yeast sphingolipid. Besides contributing to general membrane properties such as charge and fluidity, sphingolipids may contribute unique functions such as specific binding to a protein and membrane anchors for proteins (see below). Recent *in vitro* studies have demonstrated that the phospholipid requirement of the plasma membrane H^+-ATPase is effectively satisfied by certain InsPCers (Patton and Lester, 1992), suggesting that the physiological activity of the H^+-ATPase is dependent on InsPCers for normal function.

A. Sphingolipid Compensation Strains

One strategy for determining the physiological function(s) of sphingolipids was to compare the phenotype of a strain that has been manipulated so as to either contain or lack sphingolipids. Phenotypes that required sphingolipids, particularly ones dependent on the plasma membrane, could point the way to more detailed experiments which would reveal the function of a sphingolipid at the molecular level. To apply this strategy to *S. cerevisiae,* we isolated strains that can grow without making sphingolipids (Dickson *et al.,* 1990). Such strains have two essential mutations. First, the *lcb1* gene is deleted so that the strain cannot make the long-chain base component of sphingolipids. Second, the strains carry a semidominant mutation termed SLC (*s*phingo*l*ipid *c*ompensation) that enables the cell to suppress or bypass the *lcb1* defect and to grow without exogenous phytosphingosine. When exogenous phytosphingosine is supplied to the culture medium, the strains make a normal complement of sphingolipids.

Conditions that required sphingolipids for growth were searched for by comparing the behavior of SLC strains either containing or lacking sphingolipids. Sphingolipids were not required for growth under normal nonstressing culture conditions (Dickson *et al.,* 1990), but were required when the pH of the medium was reduced to 3.5, when the temperature of incubation was increased to 37°C, when the medium contained 0.75 *M* NaCl or KCl, and when the medium contained sodium acetate buffered at pH 6 (Patton *et al.,* 1992). These growth-inhibiting conditions seemed unrelated, but, when compared to previously published data (McCusker *et al.,* 1987), they showed a striking similarity to results obtained for some mutant strains defective in *pma1,* the gene coding for the plasma membrane H^+-ATPase, a major constituent of the plasma membrane of *S. cerevisiae*. This enzyme pumps protons out of the cell to create an electrochemical gradient that is utilized for transporting a variety of nutrients into the cell (for reviews, see Serrano *et al.,* 1989; Sigler and Hoefer, 1991). Furthermore, the InsPCers of *S. cerevisiae* are highly localized in the plasma membrane and represent a large fraction of its phospholipid (Patton and Lester, 1991). One interpretation of the behavior of SLC cells lacking sphingolipids is that the InsPCers are essential for H^+-ATPase activity when cells are subjected to environmental extremes.

This possibility was examined in more detail using whole cells to follow net extrusion of protons into the suspension medium (Patton *et al.*, 1992). The proton extrusion assay was initiated by adding glucose to the suspension medium. SLC cells containing sphingolipids behaved like wild-type cells and showed net proton extrusion even when the suspension medium was buffered at pH 4. In contrast, SLC cells lacking sphingolipids did not show net proton extrusion at pH 4 and, in addition, the suspension medium became alkaline. Alkalinization seemed to be due to increased permeability of sphingolipid-deficient cells to protons. Increased permeability was not due merely to formation of large holes or pores in the plasma membrane because cells did not leak proteins into the suspension medium and lactose did not freely enter cells. The effects of the low pH treatment occurred rapidly (within 1 minute) and were irreversible.

One explanation for these observations is that the SLC suppressor mutation is in *PMA1*. This is not the case, however, because the cloned *SLC1* suppressor gene is not *PMA1* (see below). At this time, it is not clear how the lack of sphingolipids affects membrane permeability and/or the H^+-ATPase under environmental extremes; there may or may not be one unifying explanation.

The predicted protein product of the *SLC1* suppressor gene (M. M. Nagiec, G. B. Wells, R. L. Lester, and R. C. Dickson, unpublished data) shows homology to the *PlsC* gene of *Escherichia coli* which encodes 1-acyl-*sn*-glycerol-3-phosphate acyltransferase (Coleman, 1992). Such homology suggests that the wild-type allele of *SLC1* may have a similar function in *S. cerevisiae*. It is pertinent to note that SLC strains accumulate significant quantities of novel lipids when cultured without a long-chain base: mono- and di-fattyacyl versions of phosphatidylinositol (PI), mannosyl-PI, and inositol-P-(mannosyl-PI), each containing one mole of C_{26} fatty acid, ordinarily found in yeast sphingolipids but not abundantly found in glycerophospholipids (Lester, *et al.*, 1993). Thus the polar head groups and hydrophobic portions of these novel lipids are strikingly similar to *S. cerevisiae* sphingolipids found in wild-type cells, and it was speculated that the novel lipids that structurally mimic sphingolipids partially compensate for some sphingolipid function(s) necessary for growth. A reasonable hypothesis is that the suppressor *SLC1* allele could be responsible for a fattyacyl transferase that inserts a C_{26} fatty acid into a precursor of the novel lipids.

B. InsPCers May Act as Anchors for Proteins

It is now well established in several organisms that some proteins are anchored to a cellular membrane by covalent attachment to a glycophosphatidylinositol (GPI) (reviewed in Ferguson and Williams, 1988). The lipid moiety in these anchors is normally either a diacylglycerol or an alkylacylglycerol. Two reports indicate that some proteins may be anchored by attachment to a ceramide-containing glycolipid. The contact site A glycoprotein of *Dictyostelium discoideum* is an adhesion molecule expressed at the aggregation stage of development. The protein is anchored to the plasma membrane. The fatty acid in the anchor lipid is not released by mild alkaline hydrolysis but is released by strong alkaline hydrolysis, indicating amide linkage to the lipid. Incubation of the anchor with sphingomyelinase releases a lipid with the chromatographic behavior of a ceramide. Thus, the available evidence suggests a ceramide-containing lipid anchor (Stadler *et al.*, 1989).

Several proteins in *S. cerevisiae* appear to be anchored to the plasma membrane by two different types of phosphoinositol-containing lipid anchors (Conzelmann *et al.*, 1992). GPI is one type of anchor and the other type contains ceramide in place of diacylglycerol. Evidence supports the idea that the GPI anchor is attached first to proteins and is then remodeled by replacement of the diacylglycerol with ceramide. The route of synthesis of these novel lipid anchors remains to be determined as do their precise structures and function and the function of the proteins they anchor. How widely distributed in nature are the InsPCer protein anchors also remains to be determined.

VII. Summary

InsPCers have been characterized in many plants, fungi, and protozoans but not in animals. There are no well-documented reports of the absence of InsPCers in organisms of these categories and one might possibly consider these lipids to be ubiquitous in plants, fungi, and protozoans. The polar headgroups of these lipids display quite heterogeneous structures depending on the source, including attachment to proteins as possible membrane anchors. The ceramides are with some exceptions composed of phytosphingosine and a very long-chain, usually hydroxylated, fatty acid. The vital nature of such sphingolipids in the plasma membrane is indicated in *S. cerevisiae*. Clearly, much remains to be discovered about the structure, metabolism, and function of the InsPCers.

Acknowledgments

Some of the work reported here was supported by USPHS Grants AI20600 and GM41302 and by NSF Grant EHR-9108764.

References

Allen, C. F., and Good, P. (1965). *J. Am. Oil Chem. Soc.* **42,** 610–614.
Angus, W. W., and Lester, R. L. (1972). *Arch. Biochem. Biophys.* **151,** 483–495.
Barr, K., and Lester, R. L. (1984). *Biochemistry* **23,** 5581–5588.
Barr, K., Laine, R. A., and Lester, R. L. (1984). *Biochemistry* **23,** 5589–5596.
Becker, G. W., and Lester, R. L. (1980). *J. Bacteriol.* **142,** 747–754.
Bernert, T. T., and Sprecher, H. (1977). *J. Biol. Chem.* **252,** 6736–6744.
Blanchardie, P., Carde, J.-P., and Cassagne, C. (1977). *Biol. Cell.* **30,** 127–136.
Bligh, E. G., and Dyer, W. J. (1959). *Can. J. Biochem. Physiol.* **31,** 911–917.
Brennan, P. J., and Losel, D. M. (1978). *Microb. Physiol.* **17,** 47–179.
Brennan, P. J., and Roe, J. (1975). *Biochem. J.* **147,** 179–180.
Buede, R., Rinker-Schaeffer, C., Pinto, W. J., Lester, R. L., and Dickson, R. C. (1991). *J. Bacteriol.* **173,** 4325–4332.
Carter, H. E., and Hendrickson, H. S. (1963). *Biochemistry* **2,** 389–393.
Carter, H. E., and Koob, J. L. (1969). *J. Lipid Res.* **10,** 363–369.
Carter, H. E., Celmer, W. D., Galanos, D. S., Gigg, R. H., Lands, W. E. M., Law, J. H., Mueller, K. L., Nakayama, T., Tomizawa, H. H., and Weber, E. (1958a). *J. Am. Oil Chem. Soc.* **35,** 335–343.
Carter, H. E., Gigg, R. H., Law, J. H., Nakayama, T., and Weber, E. (1958b). *J. Biol. Chem.* **233,** 1309–1314.
Carter, H. E., Betts, B. E., and Strobach, D. R. (1964). *Biochemistry* **3,** 1103–1107.
Carter, H. E., Strobach, D. R., and Hawthorne, J. N. (1969). *Biochemistry* **8,** 383–388.
Celmer, W. D., and Carter, H. E. (1952). *Physiol. Rev.* **32,** 167–196.
Coleman, J. (1992). *Mol. Gen. Genet.* **232,** 295–303.
Conzelmann, A., Puoti, A., Lester, R. L., and Desponds, C. (1992). *EMBO J.* **11,** 457–466.
Crossman, M. W., and Hirschberg, C. B. (1977). *J. Biol. Chem.* **252,** 5815–5819.
Dearborn, D. G., Smith, S., and Korn, E. D. (1976). *J. Biol. Chem.* **251,** 2976–2982.
de Lederkremer, R. M., Alves, M. J. M., Fonseca, G. C., and Colli, W. (1976). *Biochim. Biophys. Acta* **444,** 85–96.
Dickson, R. C., Wells, G. B., Schmidt, A., and Lester, R. L. (1990). *Mol. Cell. Biol.* **10,** 2176–2181.
Eisenberg, H., Schwartz, E., Komaromy, M., and Wall, R. (1984). *J. Mol. Biol.* **179,** 125–142.
Ferguson, M. A., and Williams, A. F. (1988). *Annu. Rev. Biochem.* **57,** 285–320.
Fulco, A. (1967). *J. Biol. Chem.* **242,** 3608–3613.
Hackett, J. A., and Brennan, P. J. (1977). *FEBS Lett.* **74,** 259–263.
Hakomori, S.-I. (1983). *In* "Sphingolipid Biochemistry" (J. N. Kanfer and S.-I. Hakomori, eds.), pp. 1–50. Plenum, New York.
Hakomori, S.-I. (1990). *J. Biol. Chem.* **265,** 18713–18716.
Hannun, Y. A., and Bell, R. M. (1989). *Science* **243,** 500–507.
Hanson, B. A., and Lester, R. L. (1980). *J. Lipid Res.* **21,** 309–315.
Harwood, J. L. (1979). *Prog. Lipid Res.* **18,** 55–86.
Hawthorne, J. N. (1960). *J. Lipid Res.* **1,** 255–280.
Hidari, K., Itonori, S., Sanai, Y., Ohashi, M., Kasama, T., and Hagai, Y. (1991). *J. Biochem. (Tokyo)* **110,** 412–416.
Hsieh, T. C.-Y., Kaul, K., Laine, R. A., and Lester, R. L. (1978). *Biochemistry* **17,** 3575–3581.
Hsieh, T. C.-Y., Laine, R. A., and Lester, R. L. (1981). *J. Biol. Chem.* **256,** 7747–7755.
Kaneshiro, E. S., Jayasimhulu, K., and Lester, R. L. (1986). *J. Lipid Res.* **27,** 1294–1303.
Kaul, K., and Lester, R. L. (1975). *Plant Physiol.* **55,** 120–129.

Kaul, K., and Lester, R. L. (1978). *Biochemistry* **17,** 3569–3575.
Kaya, K., Ramesha, C. S., and Thompson, G. A., Jr. (1984). *J. Biol. Chem.* **259,** 3548–3553.
Kishimoto, Y. (1983). *In* "The Enzymes" (P. D. Boyer, ed.), Vol. 16, pp. 357–407. Academic Press, New York.
Kochetkov, N. K., Smirnova, G. P., and Chekareva, N. V. (1976). *Biochim. Biophys. Acta* **424,** 274–283.
Kondo, Y., and Nakano, M. (1987). *Biochim. Biophys. Acta* **919,** 156–163.
Kubo, H., Irie, A., Inagaki, F., and Hoshi, M. (1990). *J. Biochem.* (*Tokyo*) **108,** 185–192.
Kulmacz, R. J., and Schroepfer, G. J. (1978). *J. Am. Chem. Soc.* **100,** 3963–3964.
Kyte, J., and Doolittle, R. F. (1982). *J. Mol. Biol.* **157,** 105–132.
Laine, R. A., Hsieh, T. C.-Y., and Lester, R. L. (1980). *ACS Symp. Ser.* **128,** 65–68.
Lester, R. L., Smith, S. W., Wells, G. B., Rees, D. C., and Angus, W. W. (1974). *J. Biol. Chem.* **249,** 3388–3394.
Lester, R. L., Wells, G. B., Oxford, G., and Dickson, R. C. (1993). *J. Biol. Chem.* **268,** 845–856.
Letters, R. (1968). *Bull. Soc. Chim. Biol.* **50,** 1385–1393.
Lhomme, O., Bruneteau, M., Costello, C. E., Mas, P., Molot, P., Dell, A., Tiller, P. R., and Michel, G. (1990). *Eur. J. Biochem.* **191,** 203–209.
Li, Y.-T., Hirabayashi, Y., DeGasperi, R., Yu, R. K., Ariga, T., Koerner, T. A. W., and Li, S.-C. (1984). *J. Biol. Chem.* **259,** 8980–8985.
Mandon, E. C., van Echten, G., Birk, R., Schmidt, R. R., and Sandhoff, K. (1991). *Eur. J. Biochem.* **198,** 667–674.
McCusker, J. H., Perlin, D. S., and Haber, J. E. (1987). *Mol. Cell. Biol.* **7,** 4082–4088.
Merrill, A. H., Jr. (1991). *J. Bioenerg. Biomembr.* **23,** 83–104.
Merrill, A. H., Jr., and Jones, D. D. (1990). *Biochim. Biophys. Acta* **1044,** 1–12.
Nishimura, K. (1987). *Comp. Biochem. Physiol. B* **86B,** 149–154.
Nohara-Uchida, K., and Ohashi, M. (1987). *J. Biochem.* (*Tokyo*) **102,** 923–932.
Orlandi, P. A., and Turco, S. J. (1987). *J. Biol. Chem.* **262,** 10384–10391.
Orme, T. W., McIntyre, J., Lynen, F., Kuhn, L., and Schweizer, E. (1972). *Eur. J. Biochem.* **24,** 407–415.
Patton, J. L., and Lester, R. L. (1991). *J. Bacteriol.* **173,** 3101–3108.
Patton, J. L., and Lester, R. L. (1992). *Arch. Biochem. Biophys.* **292,** 70–76.
Patton, J. L., Srinivasan, B., Dickson, R. C., and Lester, R. L. (1992). *J. Bacteriol.* **174,** 7180–7184.
Pinto, W. J., Srinivasan, B., Shepherd, S., Schmidt, A., Dickson, R. C., and Lester, R. L. (1992a). *J. Bacteriol.* **174,** 2565–2574.
Pinto, W. J., Wells, G. W., and Lester, R. L. (1992b). *J. Bacteriol.* **174,** 2575–2581.
Pivot, V., Bruneteau, M., Mas, P., Molot, P., and Michel, G. (1991). *C. R. Seances Acad. Sci., Ser. 3* **313,** 259–264.
Previato, J. O., Gorin, P. A. J., Mazurek, M., Xavier, M. T., Fournet, B., Wieruszesk, J. M., and Mendonça-Previato, L. (1990). *J. Biol. Chem.* **265,** 2518–2526.
Puoti, A., Desponds, C., and Conzelmann, A. (1991). *J. Cell Biol.* **113,** 515–525.
Schmidt, G., Benotti, J., Hershman, B., and Thannhauser, S. J. (1946). *J. Biol. Chem.* **166,** 505–511.
Schweizer, E. (1984). *New Compr. Biochem.* **7,** 59–83.
Serrano, R., Cid, A., Pardo, J. M., Portillo, F., and Vallejo, C. J. (1989). *In* "Plant Membrane Transport: The Current Position" (J. Dainty, M. I. De Michelis, E. Marre, and F. Rasi-Caldogno, eds.), pp. 439–448. Elsevier, Amsterdam.
Seubert, W., and Podack, E. R. (1973). *Mol. Cell. Biochem.* **1,** 29–40.
Shigematsu, H., Hisanari, Y., and Kishimoto, Y. (1990). *Int. J. Biochem.* **22,** 1427–1432.

Sigler, K., and Hoefer, M. (1991). *Biochim. Biophys. Acta* **1071,** 375–391.

Singh, B. N., Costello, C. E., Beach, D. H., and Holz, G. G. (1988). *Biochem. Biophys. Res. Commun.* **157,** 1239–1246.

Singh, B. N., Costello, C. E., and Beach, D. H. (1991). *Arch. Biochem. Biophys.* **286,** 409–418.

Smith, S. W., and Lester, R. L. (1974). *J. Biol. Chem.* **249,** 3395–3405.

Snell, E. E., Dimari, S. J., and Brady, R. N. (1970). *Chem. Phys. Lipids* **5,** 116–138.

Stadler, J., Keenan, T. W., Bauer, G., and Gerisch, G. (1989). *EMBO J.* **8,** 371–377.

Steiner, S., Smith, S. W., Waechter, C. J., and Lester, R. L. (1969). *Proc. Natl. Acad. Sci. U.S.A.* **64,** 1042–1048.

Stoffel, W., Sticht, G., and LeKim, D. (1968a). *Hoppe-Seyler's Z. Physiol. Chem.* **349,** 1149–1156.

Stoffel, W., Sticht, G., and LeKim, D. (1968b). *Hoppe-Seyler's Z. Physiol. Chem.* **349,** 1637–1644.

Sugita, M. (1979). *J. Biochem.* (*Tokyo*) **86,** 765–772.

Tanner, W. (1968). *Arch. Mikrobiol.* **64,** 158–172.

Tanner, W. (1969). *Ann. N. Y. Acad. Sci.* **165,** 726–742.

Thudichum, J. L. W. (1901). "Die chemische Konstitution des Gehirns des Menschen und der Tiere," p. 170. Franz Pietzcker, Tübingen.

Turco, S. J., Orlandi, P. A., Jr., Homans, S. W., Ferguson, M. A., Dwek, R. A., and Rademacher, T. W. (1989). *J. Biol. Chem.* **264,** 6711–6715.

Wagner, H., and Zofscik, W. (1966a). *Biochem. Z.* **344,** 314–316.

Wagner, H., and Zofscik, W. (1966b). *Biochem. Z.* **346,** 333–342.

Wagner, H., and Zofscik, W. (1966c). *Biochem. Z.* **346,** 343–350.

Wagner, H., Zofscik, W., and Heng, I. (1969). *Z. Naturforsch B: Anorg. Chem., Org. Chem., Biochem., Biophys., Biol.* **24B,** 922–927.

Walker, K. A., and Harwood, J. L. (1986). *Biochem. J.* **237,** 41–46.

Wells, G. B., and Lester, R. L. (1983). *J. Biol. Chem.* **258,** 10200–10203.

Zweerink, M. M., Edison, A. M., Wells, G. B., Pinto, W., and Lester, R. L. (1992). *J. Biol. Chem.* **267,** 25032–25038.

ADVANCES IN LIPID RESEARCH, VOL. 26

Sphingomyelinase D: A Pathogenic Agent Produced by Bacteria and Arthropods

A. P. TRUETT III AND L. E. KING, JR.

Division of Dermatology
Vanderbilt University Medical Center
Nashville, Tennessee 37232

I. Introduction

Increased interest in venom research over the past 20 years has led to the identification of venoms possessing sphingomyelinase activity. In bacteria, sphingomyelinase D (SMD) activity has been identified as a component of the exotoxin from *Corynebacterium pseudotuberculosis* and *Vibrio damsela*. SMD has also been identified as the toxic component of the venom from an arthropod, *Loxosceles reclusa* (brown recluse spider). *C. pseudotuberculosis* is a widely distributed pathogen among animals and commonly infects sheep. *V. damsela* is an aquatic bacterium increasingly reported to cause severe, occasionally life-threatening, wound infections in humans. The brown recluse spider, endemic to the southern United States, is well known in that region for a bite that induces severe and occasionally devastating sequelae. The brown recluse's bite has become a considerable cause of morbidity as this region's population has increased and its natural habitat is changing.

Although arising from divergent organisms, purified SMD have similar clinical and biochemical effects. Injection of very small amounts of the purified SMD into rabbits produces a severe and prolonged inflammatory response. Although the signaling mechanisms for the effects of these venoms have not been identified, evidence would suggest a marked disruption of the normal inflammatory process.

II. Bacterial Venoms

A. *Corynebacterium pseudotuberculosis* AND *C. ulcerans*

1. *Background*

Corynebacterium pseudotuberculosis (*ovis*), a pleomorphic gram-positive organism, is a well-recognized pathogen of domestic animals. Skin abscesses and caseous lymphadenitis predominate in sheep, the most common host, whereas both ulcerative lymphangitis and pectoral, inguinal, and abdominal abscesses are common features in horses (Doty *et al.*, 1964; Hedden *et al.*, 1986; Miers and Ley, 1980; Nairn and Robertson, 1974; Williamson and Nairn, 1980). Infection in animals is thought to occur when soil containing *C. pseudotuberculosis* contaminates a superficial wound. Sheep characteristically develop a caseous lymphadenitis (''cheesy gland'') in the area draining the infected site, but bacterial dissemination may occur (Knight, 1975; Smith, 1966). *C. pseudotuberculosis* has only rarely infected humans and almost all cases have involved prolonged exposure to sheep or infected horses. The majority of cases in both animals and humans have occurred in Australia. Of the reported cases in humans, granulomatous lymphadenitis has been almost exclusively the presenting diagnosis (Goldberger *et al.*, 1981). Because of frequent relapses, prolonged therapy with erythromycin or penicillin combined with surgical drainage or excision has been required for cure (Goldberger *et al.*, 1981; Peloux *et al.*, 1980). A single case of eosinophilic pneumonia has been reported (Keslin *et al.*, 1979).

Closely related to *Corynebacterium diphtheriae, C. ulcerans* and *C. pseudotuberculosis* are true diphtheroids. All three of these organisms have some biochemical characteristics in common and all may produce the diphtheria toxin (Wong and Groman, 1984). These three organisms can be distinguished from other corynebacteria by routine culture; however, more definitive identification requires an experienced reference laboratory (Coleman *et al.*, 1992). Isolates of *C. pseudotuberculosis* from domestic animals are phenotypically heterogeneous, but two biovars can be distinguished (Biberstein *et al.*, 1971; Songer *et al.*, 1988). Most isolates from horses reduce nitrate and are streptomycin susceptible (biovar *equi*), while those from sheep and goats fail to reduce nitrate and are streptomycin resistant (biovar *ovis*). *C. ulcerans* has been shown to be phenotypically heterogeneous and many strains produce the *C. pseudotuberculosis* exotoxin (Barksdale *et al.*, 1981; Carne and Onon, 1982). In a study of 10 strains of *C. ulcerans* (Carne and Onon, 1982), 5 produced diphtheria toxin and *C. pseudotuberculosis* toxin. Two strains produced only diphtheria toxin and 2 only *C. pseudotuberculosis* toxin.

Considerable evidence has accumulated suggesting that SMD has a primary role in the pathogenesis of caseous lymphadenitis (Lovell and Zaki, 1966; Soucek *et al.,* 1971). Purified SMD in small quantities from these exotoxins is lethal for animals, producing extensive lesions and rapid death (Hsu *et al.,* 1985; Lovell and Zaki, 1966). Other corynebacteria do not produce SMD (Barksdale *et al.,* 1981).

2. *Purification*

Two biological activities of *C. pseudotuberculosis*-culture supernatant facilitated purification of the exotoxin: dermonecrosis and inhibition of staphylococcal β-toxin (sphingomyelinase C) hemolysis of sheep erythrocytes. A protein was found to be released into the culture supernatant of *C. pseudotuberculosis* (Soucek *et al.,* 1971) which hydrolyzed sphingomyelin with the release of choline and *N*-acylsphingosylphosphate (ceramide phosphate) and hydrolyzed lysophosphatidylcholine to lysophosphatidic acid and choline. No hydrolysis of phosphatidylcholine occurred. Using sphingomyelin isolated from sheep erythrocytes and lysophosphatidylcholine prepared from egg, these investigators found a K_m of 6.2 m*M* for sphingomyelin and 8.3 m*M* for lysophosphatidylcholine. The primary sphingosine base in sphingomyelin and ceramide phosphate was 4-sphingenine, with trace amounts of sphinganine. Fractionation of toxin preparations subsequently demonstrated that the toxic and enzymatic activities reside in the same protein (Goel and Singh, 1972; Soucek and Soucková, 1974a). The primary fatty acids were $C_{24:1}$ and $C_{16:0}$ acids, ~47 and 24%, respectively. In subsequent investigations, p*I* values of 9.6 (Soucek and Soucková, 1974a), 9.8 (Linder and Bernheimer, 1978), and 9.1 (Onon, 1979) have been found. The estimated molecular weight has varied widely. Initial studies suggested a molecular weight of 16,000 to 18,000 by thin-layer gel filtration (Souková and Soucek, 1974) and 14,500 by sedimentation coefficients (Onon, 1979). Purification to near-homogeneity generated a molecular weight of 31,000, based on sodium dodecyl sulfate (SDS) polyacrylamide gel electrophoresis (Linder and Bernheimer, 1978). Cloning and expressing of the SMD gene has generated a 31-kDa protein (Hodgson *et al.,* 1990; Songer *et al.,* 1990). The enzyme is basic, with estimated p*I* values from 9.1 to 9.7 (Hsu *et al.,* 1985; Linder and Bernheimer, 1978; Onon, 1979; Soucek and Soucková, 1974b). Although cofactor requirements have not been studied in detail, the enzyme requires calcium and magnesium ions for activity (Hsu *et al.,* 1985).

The SMD was subsequently shown to be responsible for erythrocyte-protecting activity against staphylococcal β-toxin (sphingomyelinase C) hemolysis (Goel and Singh, 1972; Soucek and Soucková, 1974a; Soucková and Soucek, 1972).

3. *Toxic Effect in Laboratory Rodents and Domestic Animals*

Intradermal injection of SMD toxin into rabbits produces an extensive edematous swelling with ill-defined margins followed by a central area of necrosis and hemorrhage. Subcutaneous injection of lethal doses initially produces extensive edema, necrosis, and hemorrhage (Carne and Onon, 1982).

The interaction of SMD from *C. pseudotuberculosis* with erythrocytes is complex. As stated above, coincubation of the purified enzyme with erythrocytes protects against hemolysis by staphylococcal β-toxin, a sphingomyelinase C (Linder and Bernheimer, 1978). Additionally, the toxin protects against hemolysis of erythrocytes by helianthin: a nonenzymatic but sphingomyelin-binding cytolysin produced by the sea anemone *Stoichactis helianthus* (Linder and Bernheimer, 1978). In contrast, hemolysis of sheep blood agar develops when *C. pseudotuberculosis* (*ovis*) and *Rhodococcus equi* grow in close proximity. The toxin from *R. equi* was subsequently purified and found to be a phospholipase C with ceramide phosphate, phosphatidic acid, phosphatidylcholine, phosphatidylethanolamine, phosphatidylserine, and sphingomyelin as substrates (Bernheimer *et al.,* 1980). The *R. equi* phospholipase C had no demonstrable effect on intact erythrocytes. These investigators demonstrated evidence consistent with a model in which sphingomyelin is hydrolyzed by SMD to yield ceramide phosphate, which is subsequently hydrolyzed by phospholipase C to ceramide. In addition to the two enzymes and magnesium ions, a fourth condition, a physical disruption to the membrane such as chilling, which is capable of causing dislocation of *in situ* ceramide, was needed on the erythrocytes in order for maximum hemolysis to occur.

Acute infection of sheep with *C. pseudotuberculosis* is characterized by lesions that reflect a suspected effect of the SMD on erythrocytes. These *in vivo* features include severe hemolytic anemia, with a marked fall in erythrocyte numbers, hemoglobin, and packed cell volume. In lambs, purified exotoxin induces hemorrhage, hemolytic anemia, hemoglobinuria, dark red fluid in body cavities, lung edema, and icterus (Hsu *et al.,* 1985). Experimentally infected sheep have been shown to have alterations in morphology and phospholipid composition of erythrocytes (Brogden and Engen, 1990). Brogden *et al.* (1990) speculated that these changes were due to the unusually low phosphatidylcholine : sphingomyelin ratio (1 : 12) in sheep in comparison to erythrocytes of other mammalian species (e.g., 3 : 2 in humans) which may result in the increased susceptibility of the sheep erythrocytes to sphingomyelin hydrolysis (Zwall *et al.*, 1973). Incubation of sheep erythrocytes with broth culture filtrates or purified exotoxin resulted in progressive alteration of erythrocyte morphology. After 30 minutes, pits were apparent and, by 60 minutes, these had enlarged, leading to the appearance of an extensively scalloped membrane surface. Over this period, there was an extensive loss of membrane sphingomyelin, with $\leq 1\%$ remaining at 60 minutes. Despite the extensive shape changes, no erythrocyte hemolysis occurred *in vitro* despite the hemolysis that occurs when the purified exotoxin is injected *in vivo*. These SMD induced changes are similar to that observed after incubating erythrocytes with phospholipase C from *Clostridium perfringens*.

Sphingomyelin from endothelial cells was hydrolyzed in a similar manner to erythrocytes in sheep and rabbit aorta exposed to SMD purified from *C. pseudotuberculosis*, suggesting that the toxic effect (e.g., increased vascular permeability) may be due to alterations of endothelial cells (Carne and Onon, 1978).

The inflammatory process during experimentally induced *C. pseudotuberculosis* pyogranulomas in lambs has recently been studied (Guilloteau *et al.*, 1990). Autologous blood polymorphonuclear leukocytes (PMNs) labeled with 99*m*-technetium or 111-indium were reinjected intravenously into infected lambs. Foci of labeled PMNs were observed to be localized to the site of bacterial inoculation as well as in the draining lymph nodes during the 4 days following inoculation. Histopathological examination performed 32 hours after *C. pseudotuberculosis* inoculation confirmed the intense accumulation of PMNs at these sites. Foci of labeled PMNs were not observed during the later phases after inoculation despite the continued presence of pyogranulomas. Histopathological examination of these lesions revealed layers of fibroblasts, lymphocytes, and macrophages surrounding a necrotic center with few PMNs.

The pulmonary alveolar macrophage (PAM) has been implicated as the primary target cell of *C. pseudotuberculosis* from cases of chronic pneumonia in sheep with caseous lymphadenitis (Ellis, 1987). Further research demonstrated marked tumor necrosis factor α (TNF-α) production in PAM cultures infected with *C. pseudotuberculosis* (Ellis *et al.*, 1991). The production of TNF-α was similar in magnitude to and occurred over a similar time course as lipopolysaccharide (LPS), a potent stimulus of the immune system. TNF-α production appeared independent of phagocytosis and did not depend on the presence of live organisms.

4. *Cloning*

The SMD gene from *C. pseudotuberculosis* has independently been cloned and expressed in *Escherichia coli* by two groups of investigators

(Hodgson *et al.*, 1990; Songer *et al.*, 1990). The first group identified a recombinant bacteria by synergistic hemolysis with *R. equi* factor (phospholipase C) of sheep erythrocytes. Subcloning yielded a 1.8-kb insert. The recombinant protein had a molecular weight of 31,000, possessed SMD activity, reacted with antibodies in serum from a sheep naturally infected with *C. pseudotuberculosis,* and reacted with serum that contained phospholipase D-neutralizing antibodies. These investigators probed Southern blots of biovars *equi* and *ovis*. In biovar *ovis* isolates, a single band of about 4.8 kb was found; however, in biovar *equi* isolates, a 1.9-kb band was found, suggesting the genome organization of SMD is different for isolates from the two biovars. No homology to chromosomal DNA of *Staphylococcus aureus* or *Arcanobacterium hyamolyticum* (which produces a phospholipase D) was demonstrated.

The second group identified a 31.4-kDa protein that reacted with SMD-specific antibodies. Northern and Southern blot analyses suggested that the gene is transcribed from mRNA approximately 1.1-kb in length and that it is present in a single copy within the *C. pseudotuberculosis* genome. Sequence analysis revealed a region of similarity with phospholipase A_2 (Table I): A conserved region that contains both the calcium-binding and the active site.

B. *Vibrio damsela*

1. *Background*

Vibrio damsela, a pleomorphic, gram-negative rod, is a recently characterized, halophilic bacterium that causes wound infections and fatal disease in temperate-water damselfish (Love *et al.*, 1981), neritic sharks (Grimes *et al.*, 1984), and humans (Clarridge and Zighelboim-Daum, 1985; Coffey *et al.*, 1986; Fernandez and Pankey, 1975; Love *et al.*, 1981; Morris *et al.*, 1982). All cases of human infection have been wound contamination

Table I

AMINO ACID SEQUENCE HOMOLOGY BETWEEN *C. pseudotuberculosis* SMD AND *Laticauda laticaudata* PHOSPHOLIPASE A_2[a]

	111				
Sphingomyelinase D	ARKYLEPAGV	RVLYGFYKTV	GGPAWKTITA	DLRDGEAVAL	SGPAQ
	* * *	** *	** |	*	*
Phospholipase A_2	ADKGKRPRWH	YMDYGCYCGP	GGSGTPVDEL	DRCCKTHDQC	YAQAE
	12				

[a] Underlined residues in phospholipase A_2 are involved in calcium binding. Table adapted from (Hodgson *et al.*, 1990).

due to marine water or fish contact. *V. damsela* was first identified in 1981 as a cause of skin ulcers on a damselfish (Love *et al.*, 1981). Initial reports described a self-limited localized infection (Morris *et al.*, 1982); however, later reports describe devastating infections with severe necrotizing disease requiring limb amputation and/or extensive surgical debridement (Coffey *et al.*, 1986; Fernandez and Pankey, 1975). Several deaths have been reported (Coffey *et al.*, 1986; Fernandez and Pankey, 1975).

Histology of infected bite sites showed diffuse, necrotizing inflammation of the lower dermis, subcutaneous fat, and superficial skeletal muscle. The inflammatory cells were composed almost entirely of PMNs in association with the areas of necrosis. Within the areas of intense inflammation and necrosis, there was medial necrosis of medium-sized blood vessels, with an intramural PMN infiltrate and intraluminal thrombi (Coffey *et al.*, 1986). Tissue damage in at least one patient was postulated to be extensively toxin-mediated as tissue gram stains and cultures were negative (Coffey *et al.*, 1986). *V. damsela* has been found to be sensitive *in vitro* to synthetic penicillin, cephalosporins, tetracycline, gentamycin, and chloramphenicol.

The organism is grown easily in culture on most routine bacterial culture agars and is strongly β-hemolytic positive on rabbit blood agar, but weakly hemolytic when grown on sheep blood agar (Clarridge and Zighelboim-Daum, 1985). Found in brackish or salt water, the major distinguishing features of *V. damsela* are its sensitivity to the vibriostatic agent 0129 and a requirement of 1–6% NaCl for growth (Tison and Kelly, 1984).

Virulent strains of *V. damsela* produce large amounts of an extracellular, heat-labile cytolysin (damselysin) that is active against rabbit erythrocytes (Kothary and Kreger, 1985; Kreger, 1984). The presence of damselysin produced by 19 isolates correlated with virulence *in vitro;* however, there was no correlation between the severity of the disease in the original cases and the hemolytic activity of the isolates (Kreger, 1984). Despite increased growth of the organism, $\geq$0.8% Na^+ in the culture medium decreased the yield of damselysin (Kreger, 1984).

Kreger *et al.* (1987) investigated the inhibition of hemolysis produced by staphylococcal β-toxin (phospholipase C) and helianthin. Exposure of sheep erythrocytes to subnanogram amounts of damselysin markedly protected them from lysis by staphylococcal β-toxin and helianthin. Exposure to damselysin did not protect sheep erythrocytes from lysis by sphingomyelin-independent cytolysins (staphylococcal α- and Δ-toxin). Sheep erythrocytes and erythrocyte ghosts treated with sublytic amounts of damselysin exhibited a dose-dependent depletion of sphingomyelin. The only lipid product of this incubation was ceramide phosphate.

2. *Purification*

The ability of damselysin to lyse washed mouse erythrocytes facilitated assaying its activity and subsequent purification (Kreger *et al.,* 1988).

Damselysin is very heat labile (37°C, 30 minutes) but crystalline bovine albumin (1 mg/ml) prevents its inactivation. The toxin is stable between pH 7 and 9 (4°C, 24 hours) and is inactivated by pronase and trypsin. Its activity is not affected by phospholipids, dithiothreitol, chelating agents (EDTA or EGTA, 1 m*M*), or divalent cations (Ca^{2+}, Mg^{2+}, and Zn^{2+}; 1 m*M*).

The p*I* of damselysin is ~5.6 and it has a molecular weight of ~69,000 as determined by SDS-PAGE and 52,000 determined by gel filtration with Sephadex G-100.

Damselysin has phospholipase D substrate activity against phosphatidylcholine and phosphatidylethanolamine (prepared from Madin–Darby canine kidney cells), sphingomyelin, platelet-activating factor, and lysoplatelet-activating factor (Daniel *et al.,* 1988; Kreger *et al.,* 1987; Wilcox *et al.,* 1987).

3. *Toxic Effect*

The lethal activity of the purified damselysin preparation was heat labile (56°C, 30 minutes) and was observed after intraperitoneal, intravenous, and subcutaneous infection in mice. The LD_{50} by the intraperitoneal, intravenous, and subcutaneous routes were 1, 2, and 18 μg/kg per mouse, respectively (Kreger *et al.,* 1987). Mice injected with 1 LD_{50} of damselysin are lethargic and have ruffled fur, encrustations around their eyelids, and severe local edema before their death, which is similar to that observed in mice infected with *V. damsela* (Kreger, 1984).

The toxin from *V. damsela* has been shown to cause cytolysis of erythrocytes from 4 of 19 animal species (Kreger *et al.,* 1988). Erythrocytes from mouse and rat were most sensitive; whereas sheep erythrocytes were markedly resistant. The purified toxin is also cytolytic for Chinese hamster ovary cells in tissue culture. Cytolysis occurs as a multihit, at least two-step process consisting of a temperature-independent, toxin-binding step followed by a temperature-dependent, membrane-perturbation step (Kothary and Kreger, 1985). The rate of erythrocyte lysis is optimal between 37 and 47°C and pH 7 to 9. Damselysin is unique among these three SMD-containing toxins in that it is the only one that causes hemolysis of erythrocytes *in vitro* in the absence of serum or plasma cofactors.

4. *Cloning*

Chromosomal DNA from a markedly hemolytic strain of *V. damsela* was digested and inserted into *E. coli* (Cutter and Kreger, 1990). Subcloning yielded a 5.9-kb insert and mapping localized the damselysin gene to a ~1.5-kb region of the insert. The insert did not hybridize with species other than *V. damsela*. Colony hybridization of various strains of *V. damsela* demonstrated that only highly hemolytic strains hybridized with a gene-specific probe. Western blot analysis demonstrated that the *E. coli*-expressed protein migrated with purified damselysin.

III. Arthropod Venom

Brown Recluse Spider

1. Background

Spiders from the family Loxocelidae are distributed widely throughout North and South America and are responsible for virtually all documented cases of spider bites leading to significant skin necrosis. *Loxosceles reclusa* is one of six species identified within the United States and is present in greatest concentration in the south-central area of the United States. Nocturnal hunters by habit, *L. reclusa* are generally not aggressive toward man and attack only when trapped or crushed.

Spiders of the genus *Loxosceles* are identified most readily by the dark-brown violin-shaped marking on the cephalothorax. This marking leads to the common names fiddleback spider and violin spider.

Bites from *L. reclusa* in mammals range in severity from negligible irritation to full-thickness skin necrosis to severe systemic symptoms leading to death. In humans the causes for the wide variation in outcomes remain unknown; however, bites in fatty areas such as the abdomen, buttocks, and thighs develop necrosis with greater frequency than bites elsewhere (Auer and Hershey, 1974). Other potential factors include host susceptibility and quantity of venom injected.

In the 10% of patients who are destined to have a more significant wound, symptoms usually begin within 6–12 hours. Local wound signs and symptoms may include pruritus, pain, and swelling, induration, erythema, and blister or pustule formation. If these changes have not occurred by 48–96 hours, patients usually will not suffer any significant amount of necrosis (King and Rees, 1986). More severe bites also develop at the exact bite site, a deep blue to purple mottled central area ("incipient necrosis" due to thrombosis) that is surrounded by a blanched halo (vasoconstriction). Local signs and symptoms may be associated with a variety of systemic symptoms. These symptoms may include headache, fever, malaise (41%), anorexia, arthralgias, and a fleeting generalized maculopapular rash (29%) (Rees *et al.*, 1983, 1987; Wasserman and Anderson, 1983).

Systemic loxoscelism is the name given to the much more uncommon presentation of *L. reclusa* bites with marked systemic features and little or no skin involvement. Systemic loxoscelism usually occurs in children and most commonly presents as massive hemolysis (Wasserman and Anderson, 1983). Other systemic findings that may be detected include acute renal failure secondary to massive hemoglobinuria, leukocytosis, leukopenia, thrombocytopenia, disseminated intravascular coagulation, convulsions, coma, and, very rarely, death (Rees *et al.*, 1983; Taylor and Denny, 1966; Vorse *et al.*, 1972). Systemic loxoscelism is more often associated with an unapparent or minor-appearing local bite wound which may not become apparent for 2 to 3 days after the bite, whereas the bite site in the dermonecrotic form is noticed within 12 to 24 hours.

A very common cutaneous complication of a severe bite is the prolonged time the wound takes to heal. In more severe bites, it may take 2 to 3 months for a well-delineated eschar to develop and then even more time for the skin to heal completely. In some patients, skin grafts are repeatedly rejected and pyoderma gangrenosum develops as a long-term sequela or complication (Rees *et al.*, 1985b).

The histological changes in the bite site are dermal edema and thickening of the endothelium of blood vessels, collection of inflammatory cells, vasodilatation, intravascular clotting, degeneration of blood vessel walls, and hemorrhage into the dermis. The accumulation of PMNs is especially marked, and within 3 to 5 days dermal liquefaction and abscess formation are noted (Atkins *et al.*, 1958; Pizzi *et al.*, 1957).

Early reports advocated early excision of the bite site; however, subsequent work suggests that this is contraindicated (Rees *et al.*, 1985a). Initial therapy is focused on preventing spread of the venom effects by immobilizing the bite site and applying ice compresses, which theoretically hinders enzyme function (King, 1985). Patients who apply ice regularly generally have better outcomes (Campbell *et al.*, 1987). Dapsone (4,4′-diaminodiphenylsulfone) has been recommended as treatment for necrotic arachnidism (King and Rees, 1983). Dapsone is a poorly understood inhibitor of PMN function (Lang, 1979); however, studies have shown that depletion of PMNs or inhibition of their function limits the severity of necrosis in experimental bite wounds (Smith and Micks, 1970).

2. *Venom*

The average adult spider contains approximately 4 μl of venom that has a protein content in the range of 65–100 μg (Forrester *et al.*, 1978).

The enzymes alkaline phosphatase, 5′-ribonucleotide phosphohydrolase, and hyaluronidase, have been detected in this venom; however, none of the enzymes has been demonstrated to cause typical skin lesions.

Hyaluronidase appears to facilitate the spread of the toxin through the tissue (Wright *et al.*, 1973).

3. *Purification*

Purification of sphingomyelinase D from venom has been reported by a number of investigators (Kurpiewski *et al.*, 1981b; Rees *et al.*, 1984). Using L-α-[palmitoyl-1-^{14}C]lysophosphatidylcholine as a substrate, the first group identified a protein of molecular weight 32,000. Isoelectric focusing generated four active forms of the sphingomyelinase with p*I* values of 8.7, 8.4, 8.2, and 7.8. All four forms hydrolyzed sphingomyelin to form ceramide phosphate and choline and were immunologically cross-reactive. Intradermal injection of 1 μg of purified SMD caused typical lesions in rabbits after 8 days.

Using [^{14}C]sphingomyelin as a substrate, the second group purified a 34-kDa protein that caused dermonecrosis typical of *L. reclusa* venom. An IgG antibody was generated against this protein that inhibited the dermonecrotic activity of the purified toxin and venom. Moreover, incubation of venom with this antibody abolished sphingomyelinase activity as well. By electron microscopy, the purified protein was demonstrated to bind to human erythrocyte membranes.

Injection with 10 μg of purified toxin caused erythema, edema, and local warmth by 12 hours (Rees *et al.*, 1984). Characteristically, a papule formed which was dusky purple and evolved into local necrosis. Histological evaluation demonstrated vascular thrombosis, platelet thrombi, and intense PMN infiltration throughout the dermis and subcutaneous tissue.

4. *Mechanisms of Action*

a. PMNs. The typical histological findings suggested that the recruitment of PMNs may be essential to the pathophysiology of the brown recluse bites. Depletion of circulating PMNs in rabbits by a single intravenous injection of nitrogen mustard inhibited the typical early leukocyte infiltration and hemorrhage seen after intradermal injection of venom (Smith and Micks, 1970). Both edema and necrosis were greatly reduced. Involvement of circulating complement components was suggested by depletion of complement in guinea pigs by the parenteral administration of zymosan or aggregated human γ-globulin which resulted in inhibition of leukocyte infiltration and hemorrhage, and in marked reduction of the edema (Smith and Micks, 1970).

The incubation of *L. reclusa* venom with complement generated a chemotactic activity for guinea pig PMNs which had been harvested from peritoneal exudate cells (Gebel *et al.*, 1979a). Venom–guinea pig complement mixtures were 42% more chemotactic than complement alone, but venom alone was not chemotactic. Both C4- and properdin-deficient sera were chemotactic after exposure to venom. These investigators have also shown that incubation of as little as 0.02 μg protein/ml of partially purified SMD can totally inactivate one CH50 of guinea pig complement *in vitro* (Gebel *et al.*, 1979b). Further research has suggested that venom activates the alternate complement pathway (Kurpiewski *et al.*, 1981a).

In contrast to these effects on guinea pig PMNs, Majeski *et al.* (1977) demonstrated a concentration-dependent inhibition of human PMN chemotaxis *in vitro*. No effect on PMN viability or bactericidal activity was observed. More recent observations with the purified SMD have similarly shown inhibition of PMN chemotaxis (Babcock *et al.*, 1986). No effect of the purified toxin was observed on the microbicidal ability of PMNs, the release of enzymes, or the adherence to glass beads.

b. Erythrocytes. Early investigators demonstrated that *L. reclusa* venom had a direct hemolytic effect on human erythrocytes in the presence or absence of plasma (Denny *et al.*, 1964; Kniker *et al.*, 1969); however, contamination of venom preparation was subsequently suggested (Geren *et al.*, 1973, 1976). More purified preparations of venom later confirmed a hemolytic effect of venom on human erythrocytes; however, sheep erythrocytes were markedly more sensitive (Forrester *et al.*, 1978). The optimal pH was 7.1 and calcium concentration was 6 to 10 m*M* for lysis. Phospholipid analysis demonstrated only the breakdown of sphingomyelin with the formation of ceramide phosphate and choline.

In separate work (Morgan *et al.*, 1978), venom produced hemolysis at 37°C but not at 4°C in the presence of human sera. When sera were heat-inactivated, no hemolysis occurred. Addition of EDTA to sera inhibited hemolysis. Subsequently, it was shown that when serum was incubated with agarose gel beads to which specific rabbit antiserum to human C-reactive protein was coupled, hemolysis was greatly inhibited (Hufford and Morgan, 1981). Moreover, human cord serum that is deficient in C-reactive protein regained the ability to stimulate hemolysis following addition of purified C-reactive protein. This hemolysis was concentration-dependent on C-reactive protein. The concentration of C-reactive protein required for hemolysis was within the human physiological range. However, these investigators speculated that variations in hemolysis occurring with *L. reclusa* bites may be the result of wide variation in the concentration of normal human C-reactive protein.

A single patient with massive hemolysis due to a *L. reclusa* bite has been investigated (Hardman *et al.*, 1983). Her erythrocytes were found to be reactive with anti-human IgG. After pretreating normal human erythrocytes with *L. reclusa* venom and human sera *in vitro*, these investigators found anti-C-reactive protein, anti-C3, and anti-IgG on the erythrocyte membrane's surface by indirect antiglobulin tests.

c. Platelets. The purified SMD was shown to aggregate platelets similarly to the physiological agonist ADP (Kurpiewski *et al.*, 1981b). No platelet lysis was observed. Concurrent with aggregation, serotonin was released following treatment with SMD or ADP. Aggregation required the presence of Ca^{2+} and was inhibited by EDTA.

Using a refined technique, Rees *et al.* (1988) demonstrated that SMD induced platelet aggregation and secretion of serotonin and the prostaglandin thromboxane B_2 in human plasma but not in buffer or in human neonate plasma. The aggregation was inhibited by apyrase, an ADP-degrading enzyme. Aggregation and formation of thromboxane B_2 and serotonin were inhibited by addition of the nonsteroidal anti-inflammatory drug indomethacin; however, the PAF receptor antagonist BN 52021 had no effect. Moreover, the aggregation curve was different for purified SMD than with PAF, which suggested that production of PAF is not responsible for aggregation. The results did suggest the secretion of ADP as well as activation of the prostaglandin cascade participate in aggregation in response to SMD from *L. reclusa*.

On the basis of earlier work implicating C-reactive protein as cofactor for SMD-induced hemolysis Rees *et al.* (1988) investigated whether C-reactive protein was a necessary component for SMD to cause platelet aggregation. In initial studies, partially purified C-reactive protein induced platelet aggregation in a dose-, time-, and concentration-dependent manner. Subsequent studies using more purified C-reactive protein that was shown to be free of a common trace contaminant, serum amyloid protein (SAP), did not induce platelet aggregation (Gates and Rees, 1990). SAP, however, caused concentration-dependent aggregation of platelets in the presence of purified SMD. The concentration curve was similar to that observed for adult plasma and maximal levels of aggregation and serotonin secretion were obtained with less than physiological concentrations (5 μg/ml) of SAP.

IV. Conclusions

Characterization of SMD from these three venoms has revealed considerable similarity (Table II). Most notable is the markedly similar clinical effect of intradermal injection of the three venoms. On the basis of size and preferred substrates, SMD from *C. pseudotuberculosis* and *L. reclusa* would appear to be similar. However, using IgG fractions from sheep

Table II
CHARACTERISTICS OF PURIFIED SPHINGOMYELINASE D FROM VENOMS

	C. pseudotuberculosis	*V. damsela*	*L. reclusa*
Molecular weight	30,000	69,000	34,000
p*I*	9.1–9.7	5.6	7.8–8.7
Substrates	Sphingomyelin, lysophosphatidylcholine	Sphingomyelin, phosphatidylcholine, PAF, Lyso-PAF	Sphingomyelin, lysophosphatidylcholine

antiserum to *C. pseudotuberculosis,* these enzymes were shown to be immunologically distinct (Bernheimer *et al.*, 1985).

With the cloning of two of the three SMDs (and soon the third), information about relative similarity of these enzymes should be available. Additionally, sufficient quantities of SMD should be produced to expand knowledge of enzyme function, substrate, and cofactor requirements rapidly. Because of the intense inflammation, tissue damage, and disruption of wound healing generated by these three enzymes without evidence for a purely cytotoxic effect, an effect on and several disruption of normal cell-signaling is suggested. Study of these enzymes may therefore provide valuable information concerning sphingolipid metabolism in normal intracellular signaling.

References

Atkins, J. A., Wingo, C. W., Sodeman, W. A., and Flynn, J. E. (1958). *Am. J. Trop. Med. Hyg.* **7,** 165–184.

Auer, A. I., and Hershey, F. B. (1974). *Arch. Surg.* (*Chicago*) **108,** 612–618.

Babcock, J. L., Marmer, D. J., and Steele, R. W. (1986). *Toxicon* **24,** 783–790.

Barksdale, L., Linder, R., Sulea, I. T., and Pollice, M. (1981). *J. Clin. Microbiol.* **13,** 335–343.

Bernheimer, A. W., Linder, R., and Avigad, L. S. (1980). *Infect. Immun.* **29,** 123–131.

Bernheimer, A. W., Campbell, B. J., and Forrester, L. J. (1985). *Science* **228,** 590–591.

Biberstein, E. L., Knight, H. D., and Jang, S. (1971). *Vet. Rec.* **89,** 691–692.

Brogden, K. A., and Engen, R. L. (1990). *Am. J. Vet. Res.* **51,** 874–877.

Brogden, K. A., Engen, R. L., Songer, J. G., and Gallagher, J. (1990). *Microb. Pathog.* **8,** 157–162.

Campbell, D. S., Rees, R. S., and King, L. E. (1987). *Cutis* **39,** 113–114.

Carne, H. R., and Onon, E. O. (1978). *Nature* (*London*) **271,** 246–248.

Carne, H. R., and Onon, E. O. (1982). *J. Hyg.* **88,** 173–191.

Clarridge, J. E., and Zighelboim-Daum, S. (1985). *J. Clin. Microbiol.* **21,** 302–306.

Coffey, J. A. J., Harris, R. L., Rutledge, M. L., Bradshaw, M. W., and Williams, T. W. J. (1986). *J. Infect. Dis.* **153,** 800–802.

Coleman, G., Weaver, E., and Efstratiou, A. (1992). *J. Clin. Pathol.* **45,** 46–48.

Cutter, D. L., and Kreger, A. S. (1990). *Infect. Immun.* **58,** 266–268.

Daniel, L. W., King, L., and Kennedy, M. (1988). *In* "Microbial Toxins: Tools in Enzymology" (S. Harshman, ed.), pp. 298–301. Academic Press, Orlando, FL.

Denny, W. F., Dillaha, C. J., and Morgan, P. N. (1964). *J. Lab. Clin. Med.* **64,** 291–298.

Doty, R. B., Dunne, H. W., Hokanson, J. F., and Reid, J. J. (1964). *Am. J. Vet. Res.* **25,** 1679–1684.

Ellis, J. A. (1987). *Vet. Pathol.* **25,** 362–368.

Ellis, J. A., Lairmore, M. D., O'Toole, D. T., and Campos, M. (1991). *Infect. Immun.* **59,** 3254–3260.

Fernandez, C. R., and Pankey, G. A. (1975). *JAMA, J. Am. Med. Assoc.* **233,** 1173–1176.

Forrester, L. J., Barrett, J. T., and Campbell, B. J. (1978). *Arch. Biochem. Biophys.* **187,** 355–365.

Gates, C. A., and Rees, R. S. (1990). *Toxicon* **28,** 1303–1315.

Gebel, H. M., Campbell, B. J., and Barrett, J. T. (1979a). *Toxicon* **17,** 55–60.

Gebel, H. M., Finke, J. H., Elgert, K. D., Campbell, B. J., and Barrett, J. T. (1979b). *Am. J. Trop. Med. Hyg.* **28,** 756–762.

Geren, C. R., Chan, T. K., Ward, B. C., Howell, D. E., Pinkston, K., and O'Dell, G. V. (1973). *Toxicon* **11,** 471–479.

Geren, C. R., Chaw, T. K., Nowell, D. E., and O'Dell, G. V. (1976). *Arch. Biochem. Biophys.* **174,** 90–99.

Goel, M. C., and Singh, I. P. (1972). *J. Comp. Pathol.* **82,** 345–353.

Goldberger, A. C., Lipsky, B. A., and Plorde, J. J. (1981). *Am. J. Clin. Pathol.* **76,** 486–490.

Grimes, D. J., Stemmler, J., Hada, H., May, E. B., Maneval, D., Hetrick, F. M., Jones, R. T., Stoskopf, M., and Colwell, R. R. (1984). *Microb. Ecol.* **10,** 271–282.

Guilloteau, L., Pepin, M., Pardon, P., and Le, P. A. (1990). *J. Leukocyte Biol.* **48,** 343–352.

Hardman, J. T., Beck, M. L., Hardman, P. K., and Stout, L. C. (1983). *Transfusion (Philadelphia)* **23,** 233–236.

Hedden, J. A., Thomas, C. M., Songer, J. G., and Olson, G. B. (1986). *Am. J. Vet. Res.* **47,** 1265–1267.

Hodgson, A. L., Bird, P., and Nisbet, I. T. (1990). *J. Bacteriol.* **172,** 1256–1261.

Hsu, T. Y., Renshaw, H. W., Livingston, C. W., Jr., Augustine, J. L., Zink, D. L., and Gauer, B. B. (1985). *Am. J. Vet. Res.* **46,** 1206–1211.

Hufford, D. C. and Morgan, P. N. (1981). *Proc. Soc. Exp. Biol. Med.* **167,** 493–497.

Keslin, M. H., McCoy, E. L., McCusker, J. J., and Lutch, J. S. (1979). *Am. J. Med.* **67,** 228–231.

King, L. E., Jr. (1985). *JAMA, J. Am. Med. Assoc.* **254,** 2895–2896.

King, L. E., Jr., and Rees, R. S. (1983). *JAMA, J. Am. Med. Assoc.* **250,** 648.

King, L. E., Jr., and Rees, R. S. (1986). *J. Am. Acad. Dermatol.* **14,** 691–692.

Knight, H. D. (1975). *In* "Diseases Transmitted from Animals to Man" (W. T. Hubbert, W. F. McCulloch, and P. R. Schnurrenberger, eds.), pp. 263–270. Thomas, Springfield, IL.

Kniker, W. T., Morgan, P. N., Flanigan, W. J., Reagan, P. W., and Dillaha, C. J. (1969). *Proc. Soc. Exp. Biol. Med.* **131,** 1432–1434.

Kothary, M. H., and Kreger, A. S. (1985). *Infect. Immun.* **49,** 25–31.

Kreger, A. S. (1984). *Infect. Immun.* **44,** 326–331.

Kreger, A. S., Bernheimer, A. W., Etkin, L. A., and Daniel, L. W. (1987). *Infect. Immun.* **55,** 3209–3212.

Kreger, A. S., Kothary, M. H., and Gray, L. D. (1988). *In* "Microbial Toxins: Tools in Enzymology" (S. Harshman, ed.), pp. 176–189. Academic Press, Orlando, FL.

Kurpiewski, G., Campbell, B. J., Forrester, L. J., and Barrett, J. T. (1981a). *Int. J. Tissue React.* **3,** 39–45.

Kurpiewski, G., Forrester, L. J., Barrett, J. T., and Campbell, B. J. (1981b). *Biochim. Biophys. Acta* **678,** 467–476.

Lang, P. G., Jr. (1979). *J. Am. Acad. Dermatol.* **1,** 479–492.

Linder, R., and Bernheimer, A. W. (1978). *Biochim. Biophys. Acta* **530,** 236–246.

Love, M., Teebken-Fisher, D., Hose, J. E., Farmer, J. J., III, Hickman, F. W., and Fanning, G. R. (1981). *Science* **214,** 1139–1140.

Lovell, R., and Zaki, M. M. (1966). *Res. Vet. Sci.* **7,** 302–366.

Majeski, J. A., Stinnett, J. D., Alexander, J. W., and Durst, G. G. S. (1977). *Toxicon* **15,** 423–427.

Miers, K. C., and Ley, W. B. (1980). *J. Am. Vet. Med. Assoc.* **177,** 250–253.

Morgan, B. B., Morgan, P. N., and Bowling, R. E. (1978). *Toxicon* **16,** 85–88.

Morris, J. G. J., Miller, H. G., Wilson, R., Tacket, C. O., Hollis, D. G., Hickman, F. W., Weaver, R. E., and Blake, P. A. (1982). *Lancet* **1,** 1294–1297.

Nairn, M. E., and Robertson, J. P. (1974). *Aust. Vet. J.* **50,** 537–542.

Onon, E. O. (1979). *Biochem. J.* **177,** 181–186.

Peloux, Y., Maresca, C., and Oddou, J.-H. (1980). *Méditerr. Méd.* **234,** 7–12.

Pizzi, T., Zacria, J., and Schenome, H. (1957). *Biológica* **23,** 33–39.

Rees, R. S., O'Leary, J. P., and King, L. E. J. (1983). *J. Surg. Res.* **35,** 1–10.

Rees, R. S., Nanney, L. B., Yates, R. A., and King, L. J. (1984). *J. Invest. Dermatol.* **83,** 270–275.

Rees, R. S., Altenbern, D. P., Lynch, J. B., and King, L. E., Jr. (1985a). *Ann. Surg.* **202,** 659–663.

Rees, R. S., Fields, J. P., and King, L. E., Jr. (1985b). *South Med. J.* **78,** 283–287.

Rees, R. S., Campbell, D., Rieger, E., and King, L. E., Jr. (1987). *Ann. Emerg. Med.* **16,** 945–949.

Rees, R. S., Gates, C., Timmons, S., Des Pres, R. M., and King, L. E., Jr. (1988). *Toxicon* **26,** 1035–1045.

Smith, C. W., and Micks, D. W. (1970). *Lab. Invest.* **22,** 90–93.

Smith, J. E. (1966). *J. Appl. Bacteriol.* **29,** 119–130.

Songer, J. G., Beckenbach, K., Marshall, M. M., Olson, G. B., and Kelley, L. (1988). *Am. J. Vet. Res.* **49,** 223–226.

Songer, J. G., Libby, S. J., Iandolo, J. J., and Cuevas, W. A. (1990). *Infect. Immun.* **58,** 131–136.

Soucek, A., and Soucková, A. (1974a). *J. Hyg., Epidemiol., Microbiol., Immunol.* **18,** 327–335.

Soucek, A., and Soucková, A. (1974b). *Cesk. Epidemiol., Mikrobiol., Immunol.* **23,** 294–304.

Soucek, A., Michaleč, C., and Soucková, A. (1971). *Biochim. Biophys. Acta* **227,** 116–128.

Soucková, A., and Soucek, A. (1972). *Toxicon* **10,** 501–509.

Souková, A., and Soucek, A. (1974). *J. Hyg., Epidemiol., Microbiol., Immunol.* **18,** 336–341.

Taylor, E. H., and Denny, W. F. (1966). *South. Med. J.* **59,** 1209–1211.

Tison, D. L., and Kelly, M. T. (1984). *Diagn. Microbiol. Infect. Dis.* **2,** 263–276.

Vorse, H., Seccareccio, P., Woodruff, K., and Humphrey, G. B. (1972). *J. Pediatr.* **80,** 1035–1037.

Wasserman, G. S., and Anderson, P. C. (1983). *J. Toxicol. Clin. Toxicol.* **21,** 451–472.
Wilcox, R. W., Wykle, R. L., Schmitt, J. D., and Daniel, L. W. (1987). *Lipids* **22,** 800–807.
Williamson, P., and Nairn, M. E. (1980). *Aust. Vet. J.* **56,** 496–498.
Wong, T. P., and Groman, N. (1984). *Infect. Immun.* **43,** 1114–1116.
Wright, R. P., Elgert, K. D., Campbell, B. J., and Barrett, J. T. (1973). *Arch. Biochem. Biophys.* **159,** 415–426.
Zwall, R. F. A., Roelosfsew, B., and Colley, C. M. (1973). *Biochim. Biophys. Acta* **300,** 159–182.

ADVANCES IN LIPID RESEARCH, VOL. 26

Sphingolipid-like Molecule Linked to CD45, a Protein Tyrosine Phosphatase

AKIKO TAKEDA

Department of Pathology
Roger Williams Medical Center—Brown University
Providence, Rhode Island 02908

I. Introduction

The transmembrane glycoprotein CD45 is present in all hematopoietic cells except for erythrocytes, platelets, and their immediate progenitors. The cytoplasmic portion possesses protein tyrosine phosphatase (PTP) activity (Tonks *et al.*, 1988), and involvement of CD45 in signal transduction has been amply documented (Ledbetter *et al.*, 1988, 1991; Kiener and Mittler, 1989; Pingell and Thomas, 1989; Koretzky *et al.*, 1990, 1991; Samelson *et al.*, 1990a). CD45 is therefore thought to play an important role in leukocyte activation and growth regulation.

The present review describes an unusual sphingolipid-like structure covalently linked to the CD45 peptide (Takeda and Maizel, 1990) and discusses the potential roles of such lipid linkage. For reviews on the general structural and functional aspects of CD45, see the articles by Thomas (1989) and Trowbridge (1991).

A. General Background

The past decade has witnessed critical advances in understanding the mechanism of signal transduction in eukaryotes (Hunter, 1987; Berridge, 1989; Gilman, 1989; Krebs, 1989; Nishizuka, 1989). In general, extracellular signals, such as antigens, growth factors, hormones, and neurotransmitters, are recognized by specific receptors on the plasma membrane, and the receptors in turn trigger cascades of intracellular reactions involving protein phosphorylation. In spite of the recent advances, however, there are still numerous "missing links" within the various cascades of signal transduction.

One of the well-known signal transduction pathways involves tyrosine phosphorylation of cellular proteins mediated by a receptor which may be a protein tyrosine kinase (PTK) itself or may be associated with one (Yarden and Ullrich, 1988). This is the pathway utilized by the epidermal growth factor (EGF) receptor, which is a transmembrane protein composed of an extracellular ligand-binding portion, a single membrane-spanning segment, and an intracellular portion with PTK activity (Ullrich *et al.,* 1984; Schlessinger, 1986). On ligand-binding, the receptor elicits increased phosphorylation of phospholipase C (PLC)-γ1 (Nishibe *et al.,* 1990) and a redistribution of the lipase from the cytosol to the plasma membrane (Todderud *et al.,* 1990). The activated PLC-γ1 then hydrolyzes phosphatidylinositol 4,5-biphosphate (PIP_2) to 1,2-diacylglycerol (DAG) and inositol 1,4,5-triphosphate (IP_3). DAG activates protein kinase C (PKC), a serine/threonine kinase, while IP_3 triggers the release of an intracellular pool of calcium which, among other things, also activates PKC. Thus, these "second messengers" elicit phosphorylation of serine/threonine residues, causing activation of various key enzymes, and eventually leading to transcriptional activation and cell division. The initial signal could be amplified a multitude of times through the cascade.

Although there must be a vast number of variations in the ways different signals are transduced, significant parallels can often be drawn between different systems. Thus, the T cell antigen receptor (TCR) activation pathway (Klausner and Samelson, 1991) is thought to follow a cascade comparable to that of the EGF receptor (Yarden and Ullrich, 1988). The TCR complex, which does not possess PTK activity itself, however, is thought to be coupled to a PTK (or PTKs) in order to transduce signals. Two likely candidates are the PTK $p59^{fyn}$, which appears to be in association with the TCR complex (Samelson *et al.,* 1990b), and $p56^{lck}$, which is associated with the cytoplasmic domain of either the CD4 or CD8 molecule (Rudd *et al.,* 1988; Veillette *et al.,* 1988).

PTK phosphorylation reactions under normal conditions are believed to be highly restricted, i.e., a given PTK under a given set of conditions will phosphorylate specific proteins at specific sites (Hunter, 1987; Yarden and Ullrich, 1988). Furthermore, the proteins remain phosphorylated only transiently after stimulation. Mutation or overexpression of PTKs is often associated with neoplastic transformation. Therefore, strict regulation of tyrosine phosphorylation appears important for maintenance of normal growth control. In theory, PTPs possess counterbalancing powers to PTKs and might be critical for normal cell growth regulation, since PTPs dephosphorylate tyrosine residues (Fischer *et al.*, 1991). In fact, the possibility that PTP genes act as "tumor suppressor genes" was raised recently, because the gene for PTP-γ was frequently found to be deleted in renal cell carcinomas and lung carcinomas (LaForgia *et al.*, 1991). A family of molecules which contain homologous PTP domains, but otherwise are quite diverse in their structures, has been recognized recently (Fischer *et al.*, 1991). It includes small intracellular enzymes such as placental PTP-1B and larger transmembrane receptorlike molecules such as CD45.

B. Description of CD45

CD45 [also designated T200, B220, Ly5, and leukocyte common antigen (LCA)] is a transmembrane glycoprotein ubiquitously found in hematopoietic cells of the granulocyte, monocyte, and lymphoid series (Thomas, 1989; Trowbridge, 1991) (Fig. 1). The molecule has been studied extensively in human, rat, and mouse, and bears close similarity among different species. It is one of the major high-molecular-mass molecules of leukocytes, comprising up to 10% of their total membrane protein. Several isotypes of CD45 with a molecular mass range of 180–220 kDa are variably expressed in a cell type-specific manner. The extracellular domain consists of about 400–550 amino acids and exhibits this lineage- and developmental stage-specific structural variation, while the intracellular domain is common among the different isotypes. The isotypic variation of CD45 is a result of alternative mRNA splicing of exons which code for the amino-terminal portion of the extracellular domain (Saga *et al.*, 1986; Thomas *et al.*, 1987). The isotypic variations are often described by letters which represent the amino-terminal exons employed for the particular isotype, following the letter "R" (for restriction). For example, the isotype with no additional variant exons is "CD45 RO," while the isotype containing exon A is expressed as "CD45 RA." N-Linked glycosylation sites are abundant in the extracellular domains of all isotypes, while O-linked glycosylation sites are thought to occur predominantly in the higher molecular mass isotype (B220). The isotypically invariant carboxyl-terminal portion of the extracellular domain is rich in cysteine residues.

CD45 spans the plasma membrane only once by a short hydrophobic transmembrane segment separating a large extracellular domain from a large intracellular portion. The intracellular portion of CD45 consists of about 700 amino acids and contains two homologous tandem repeats of about 240 amino acids each. Although each of the repeating domains shares homology with the PTP family (Charbonneau *et al.*, 1988), only the domain

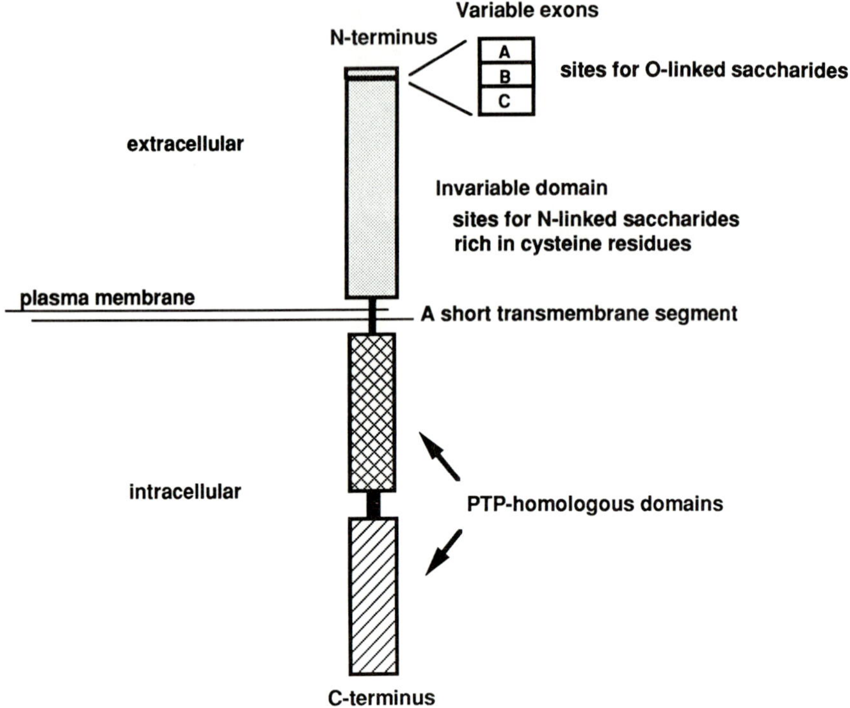

FIG. 1. Schematic representation of the CD45 structure. Alternative splicing of the three variable exons can potentially produce as many as eight isoforms of CD45.

closer to the amino-terminal appears to have PTP activity (Streuli *et al.*, 1989). However, the presence of the other domain closer to the carboxyl-terminal has been shown to be required for the PTP activity. Two cysteine residues which are present in each of these repeating domains are conserved among the family of PTPs. At least one of them is considered essential for PTP activity (Streuli *et al.*, 1989), presumably by providing a site for a covalent thiol-phosphate bond which is formed between the enzyme and its substrates (Guan and Dixon, 1991). Some serine residues in the intracellular portion are phosphorylated (Omary and Trowbridge, 1980). A selective decrease in serine phosphorylation at specific sites appears to accompany a decrease in the CD45 PTP activity in T cells (Ostergaard and Trowbridge, 1991). However, phosphorylation of CD45 carried out *in vitro* with various serine/threonine kinases does not seem to affect its PTP activity (Tonks *et al.*, 1990). In addition, transient and stimulation-dependent phosphorylation of tyrosine residues on CD45 has been reported (Stover *et al.*, 1991), indicating its involvement in an early stage of the signal transduction cascade. It is still not known whether tyrosine phosphorylation affects CD45 PTP activity.

Documentation of CD45 involvement in signal transduction preceded the recognition of its PTP activity. CD45 appeared to exert either a positive or a negative effect on stimulation, depending on the circumstances. For example, the stimulatory signals triggered by antibodies against CD3, TCR, CD2, or CD28 are abolished if those molecules were brought together with CD45 by antibody-mediated cross-linking (Ledbetter *et al.*, 1988, 1991; Kiener and Mittler, 1989; Samelson *et al.*, 1990a). In contrast, CD45 appears to enhance the stimulatory signal when coupled with CD4. Involvement of CD45 in signal transduction was established further when CD45-deficient T cells stimulated by anti-TCR or anti-CD2 antibodies failed to produce elevations in IP_3 and intracellular calcium (Pingell and Thomas, 1989; Koretzky *et al.*, 1990, 1991). Therefore, it has been proposed that CD45 participates in signal transduction by dephosphorylating key molecules of the cascade (Clark and Ledbetter, 1989; Fischer *et al.*, 1991; Klausner and Samelson, 1991). In fact, CD45 has already been shown to enhance the PTK activity of $p56^{lck}$ by dephosphorylating a tyrosine residue (Tyr^{505}) *in vitro* (Mustelin *et al.*, 1989; Ostergaard *et al.*, 1989; Ostergaard and Trowbridge, 1990). In summary, CD45 may play two seemingly contradictory roles in signal transduction, namely, (1) stimulate PTK by dephosphorylating its key tyrosine residue, thus participating in one of the early stages of the stimulatory pathway, and (2) abrogate the PTK effect by dephosphorylating its effectors such as PLC, thus providing "anti-oncogenic" regulation for normal cell growth.

Physical associations of CD45 with some of the membrane proteins involved in Tcell activation, such as CD2 (Schraven *et al.*, 1990), Thy-1, TCR (Volarevič *et al.*, 1990), and CD26 (Torimoto *et al.*, 1991) have been documented under certain circumstances. In addition, CD45 has been reported to interact with a cytoskeletal protein, fodrin (Bourguignon *et al.*, 1985). It is not known whether such associations affect the PTP activity. The structural and functional characteristics of CD45 raise the possibility that it functions as a receptor with the extracellular portion acting as a ligand-binding site. It has been recently proposed that CD22β, the B lineage-specific cell surface glycoprotein, interacts specifically with the CD45 of a subpopulation of $CD4^+$ T cells and serves as its ligand (Stamenkovič *et al.*, 1991).

The topographical arrangement of CD45 in the plasma membrane (Saga *et al.*, 1986; Thomas *et al.*, 1987) shares a similarity with that of the EGF receptor (Ullrich *et al.*, 1984), that is, both of them consist of a large, extracellular, cysteine-rich "receptor" portion, a short transmembrane stretch which spans the membrane only once, and a large intracellular portion with enzymatic activity. The PTK activity of the EGF receptor is allosterically regulated through interconversion between monomeric and dimeric forms (Schlessinger, 1986). The dimeric state has a higher affinity for the ligand as well as elevated levels of PTK activity and autophosphorylated residues. A comparable regulatory mechanism might be in effect for CD45 PTP activity, since oligomeric forms of CD45 have recently been detected. Monomeric and dimeric forms of CD45 in association with an additional 30-kDa protein have been reported in mouse lymphocytes (Takeda *et al.*, 1992). Furthermore, CD45, $p56^{lck}$, and a protein termed pp32 are reported to form a complex in human T lymphocytes (Schraven *et al.*, 1991). The pp32 protein coimmunoprecipitated with CD45, and could be phosphorylated *in vitro* by $p56^{lck}$ as well as dephosphorylated by CD45. On the other hand, a 32-kDa GTP-binding protein which belongs to the family of α-chains of G proteins has been found to be associated with the CD4-$p56^{lck}$ and CD8-$p56^{lck}$ of the TCR complexes (Telfer and Rudd, 1991) in human lymphocytes. It is possible that the "30-kDa" protein associated with CD45, the "pp32" protein, and the "32-kDa" GTP-binding protein are identical, since CD45 is involved in signal transduction mediated by TCR (Fischer *et al.*, 1991).

II. Sphingolipid-like Molecule Linked to CD45

A. Introduction

Four forms of lipid-linkage are known to occur among proteins of eukaryotic origin (Fig. 2); (i) myristoylation on amino-terminal glycine through amide linkage in such proteins as $p60^{src}$ (Sefton *et al.*, 1982), cAMP-dependent kinase (Carr *et al.*, 1982), and calcineurin B (Aitken *et al.*, 1982); (ii) fatty acylation of cysteine (in some cases, serine and threonine) side chains forming a thioester linkage as in the transferrin receptor (Omary and Trowbridge, 1981), HLA-B and DR heavy chains (Kaufman *et al.*, 1984), and p21Ras (Buss *et al.*, 1989); (iii) phosphatidylinositol (PI) as a part of a PI–glycan membrane anchor attached to the carboxyl-terminus of numerous membrane proteins such as Thy-1 (Tse *et al.*, 1985) and acetylcholine esterase (Roberts and Rosenberry, 1985); and (iv) polyisoprenoid attached to carboxyl-terminal cysteine forming a thioether linkage as reported for p21Ras (Hancock *et al.*, 1989) and unidentified proteins of Chinese hamster ovary cells and HeLa cells (Farnsworth *et al.*, 1990; Rilling *et al.*, 1990).

Enzymes involved in the formation of these lipid linkages display dis-

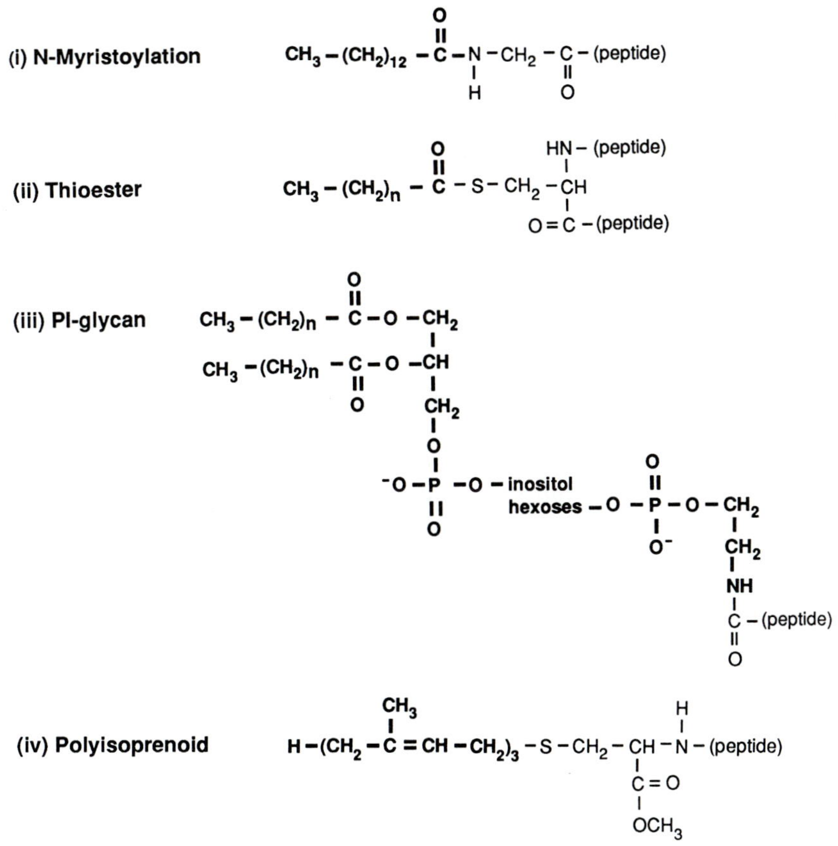

(i) **N-Myristoylation** $\mathbf{CH_3-(CH_2)_{12}-C(=O)}-N(H)-CH_2-C(=O)-$(peptide)

(ii) **Thioester** $\mathbf{CH_3-(CH_2)_n-C(=O)}-S-CH_2-CH(-HN-\text{(peptide)})(-C(=O)-\text{(peptide)})$

(iii) **PI-glycan**

$\mathbf{CH_3-(CH_2)_n-C(=O)-O-CH_2}$
$\mathbf{CH_3-(CH_2)_n-C(=O)-O-CH}$
$\mathbf{CH_2-O-P(=O)(-O^-)-O}$ **– inositol**
hexoses $\mathbf{-O-P(=O)(-O^-)-O-CH_2-CH_2-NH}-C(=O)-$(peptide)

(iv) **Polyisoprenoid** $\mathbf{H-(CH_2-C(CH_3)=CH-CH_2)_3}-S-CH_2-CH(-C(=O)-OCH_3)-N(H)-$(peptide)

FIG. 2. Typical lipid linkages to eukaryotic proteins. Lipid moieties are depicted in bold letters. Typically, $n = 12$–16 for type ii, and $n = 12$–20 for type iii.

tinctly different characteristics. The enzyme responsible for *N*-myristoylation (type i), *N*-myristoyltransferase, has a strong preference for 14-carbon acyl-CoA as a fatty acid donor, requires highly specific amino acid sequences with amino-terminal glycine as an acceptor for the fatty acid, and acts cotranslationally. On the other hand, acyltransferases responsible for fatty acyl ester formation (type ii) allow a broad fatty acid specificity, do not appear to require a particular amino acid sequence for the acyl acceptors, and act posttranslationally (Towler *et al.*, 1988). Carboxyl-terminal addition of a prefabricated PI-glycan moiety to proteins (type iii) occurs in the endoplasmic reticulum shortly (1–5 minutes) after the completion of peptide chain biosynthesis, replacing a short hydrophobic carboxyl-terminal peptide segment which is cleaved (Ferguson and Williams, 1988). The amino acid sequences at the site of cleavage and Pl-glycan attachment show only a weak pattern, no more specific than that of the amino-terminal "leader" sequences which mediate translocation of proteins through the endoplasmic reticulum membranes. Recent studies on p21Ras farnesylation disclosed that the polyisoprenoid attachment (type iv) is also a part of multiple steps of posttranslational modification involving the carboxyl-terminal (Gibbs, 1991). Namely, a farnesyl group is transferred to a cysteine residue in the carboxyl-terminal consensus sequence, three amino acids on the carboxyl-side of the cysteine residue are cleaved, and the carboxylic acid of the cysteine residue is methyl-esterified.

The physiological significance of these lipid–protein linkages has yet to be fully elucidated. Absolute correlation between lipid-linkage and membrane association has been established only for Pl-glycan anchor structures (type iii) (Ferguson and Williams, 1988) and the polyisoprenyl structure attached to the carboxyl-terminus of p21Ras (type iv) (Hancock *et al.*, 1989). Palmitoylation (type ii) of p21Ras (Buss *et al.*, 1989) increases the membrane-binding avidity of this protein and enhances transforming activity. On the other hand, it has been suggested that myristoylation on the amino-terminus (type i) exerts a regulatory role by directing the interaction of acylated proteins with other cellular components such as membranes and other peptides (Towler *et al.*, 1988). In fact, the myristoylated amino-terminal portion of $p60^{src}$ is required for interaction with a 32-kDa protein which then enables $p60^{src}$ to be localized to the intracellular side of the plasma membrane (Resh and Ling, 1990). Other proposed functions for lipid-linkage include localizing a certain protein to the correct cellular compartment, directing membrane fusion events, influencing the correct folding of a protein, and, for type iii, achieving faster lateral movement and faster release from the membrane.

B. Biosynthetic Labeling of CD45 with Fatty Acids

Earlier studies of CD45 demonstrated (1) little incorporation of palmitic acid into CD45 by biosynthetic radiolabeling of lymphocytes (Omary and Trowbridge, 1981), and (2) resistance to Pl-specific PLC treatment which should release membrane proteins anchored to the plasma membrane by Pl-glycans (Stiernberg *et al.*, 1987). Thus, the second and the third types of lipid-linkage described above seemed unlikely. In our own study (Takeda and Maizel, 1990), a biosynthetic radiolabeling experiment was carried out by employing a Moloney leukemia virus-induced murine $CD4^{+}8^{-}$ T lymphoma cell line, YAC-1 (Cikes *et al.*, 1973; Le Corre *et al.*, 1987). The cells were cultured overnight in the presence of [^{3}H]myristic acid or [^{3}H]palmitic acid and the incorporation of both labels into CD45 was compared to their incorporation into the transferrin receptor, a representative fatty acylated protein [thioester-linkage as described in type (ii) above] (Omary and Trowbridge, 1981). The transferrin receptor incorporated both myristic acid and palmitic acid efficiently. In contrast, CD45 incorporated myristic acid to the same level as the transferrin receptor, while palmitic acid was poorly incorporated. The fatty acid label incorporated into the transferrin receptor was released from the protein under mild alkaline methanol treatment, while the label incorporated into CD45 was resistant to the same treatment. Alkaline methanolysis cleaves thio- as well as hydroxy-ester linkages without affecting peptide bonds (Dittmer and Wells, 1969).

A significant degree of fatty acid turnover into amino acids appears unlikely under the conditions employed, because the label incorporated into the transferrin receptor was mild alkali-labile, indicating that the label was indeed incorporated in the fatty acyl-thioester form. In order to confirm that the [^{3}H]myristic acid label incorporated into CD45 was not due to metabolic conversion of the fatty acid label into amino acids, tryptic peptides of CD45 were analyzed by two-dimensional (2D) "mapping" (Fig. 3). CD45 was prepared from YAC-1 cells which were cultured in the presence of either [^{3}H]myristic acid, a ^{14}C-labeled amino acid mixture, or [^{35}S]methionine, or were surface-radioiodinated with ^{125}I (Cone and Brown, 1976). CD45 was then digested with trypsin and subjected to electrophoresis in the first dimension and chromatography in the second dimension (Gibson, 1974; Omary and Trowbridge, 1980). Virtually all ^{3}H-labeled peptides were found along the electrophoresis axis, whereas only one-third of ^{14}C-labeled peptides remained on the electrophoresis axis, with the rest being widely distributed. These results clearly indicate that the fatty acid label was not incorporated in the form of amino acids, since

---

FIG. 3. Two-dimensional tryptic peptide mapping of CD45. YAC-1 cells were labeled with [9,10-^{3}H]myristic acid, an L-[U-^{14}C]-labeled amino acid mixture (consisting of Ala, Arg, Asp, Glu, Gly, His, Ile, Leu, Lys, Phe, Pro, Ser, Thr, Tyr, and Val), or [^{35}S]methionine. Cell-surface labeling was carried out by lactoperoxidase-catalyzed radioiodination with ^{125}I. CD45 was prepared by immunoprecipitation and electroeluted from SDS-PAGE gels. After thiol-reduction and alkylation treatments, each sample was digested with tosylamidophenylethyl chloromethyl ketone (TPCK)–trypsin for 23 hours and spotted on thin-layer cellulose plates. Electrophoresis in the first dimension (E, the direction toward the cathode, as shown by arrows) was carried out in *n*-butanol : pyridine : acetic acid : H_2O (2 : 1 : 1 : 18, v/v/v/v) at 1000 V for 1 hour. Ascending chromatography in the second dimension (C, the direction as indicated by arrows) was carried out with a mixture of *n*-butanol : pyridine : acetic acid : H_2O (97 : 75 : 15 : 60, v/v/v/v). [Modified with permission from Takeda and Maizel (1990). Copyright 1990 by the AAAS.]

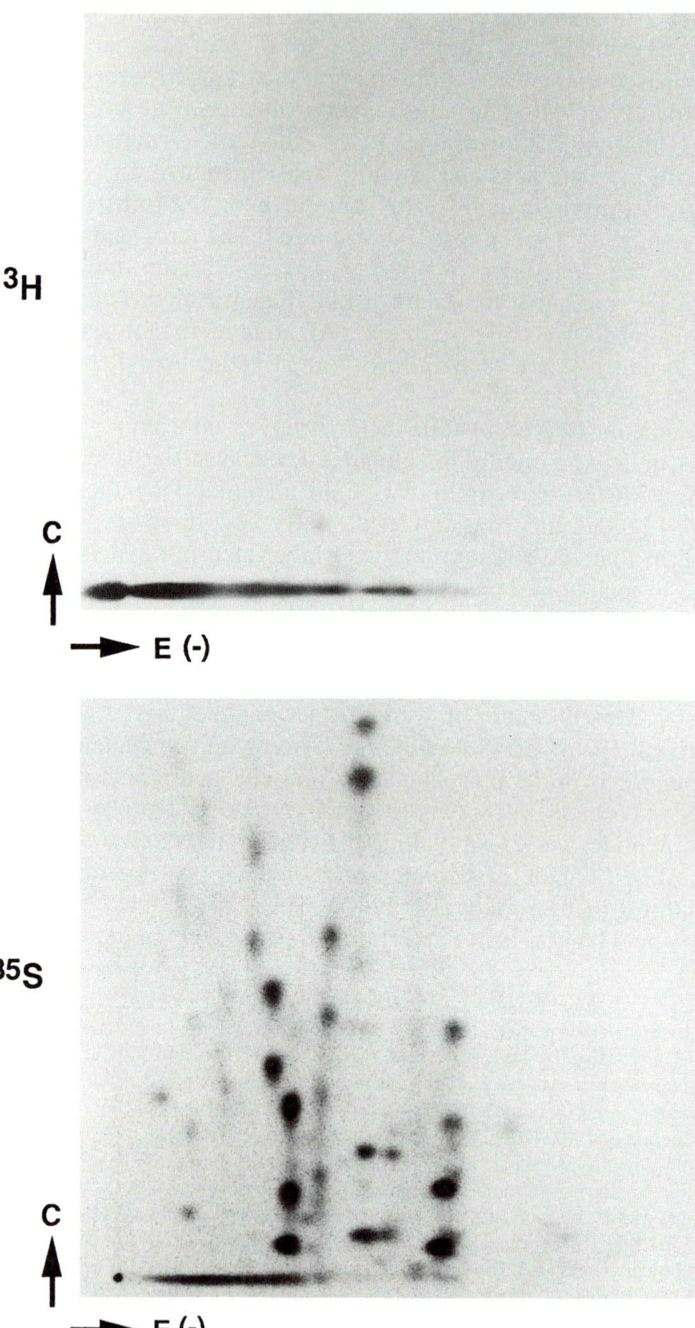

³H
C
E (-)
³⁵S
C
E (-)

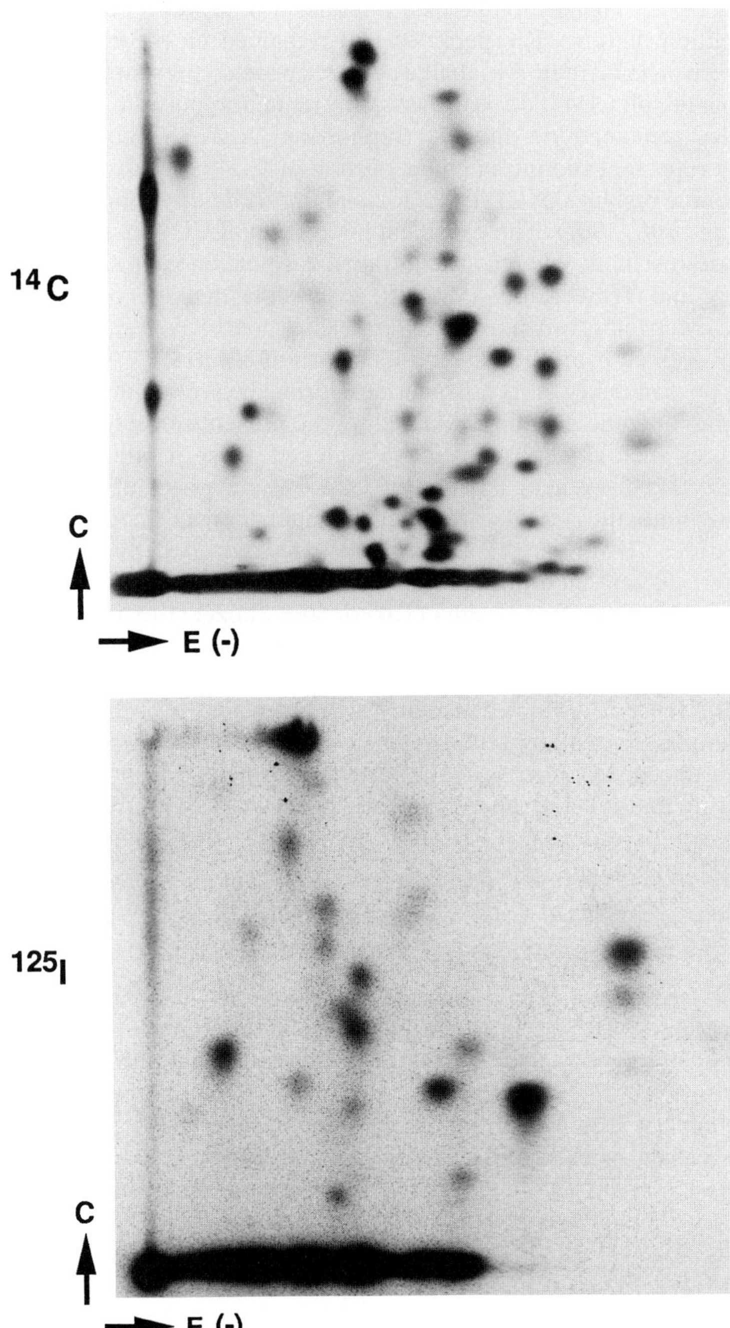

^{14}C
C
E (-)
^{125}I
C
E (-)

in that event ^{3}H- and ^{14}C-labeled peptides would have shared a similar distribution pattern. The peptides that remained on the electrophoresis axis were prominent in ^{125}I-labeled peptides which represent the extracellular domain of CD45. In contrast, only a minor portion of ^{35}S-labeled peptides remained on the electrophoresis axis. ^{35}S-Labeled peptides largely represent the intracellular portion of CD45, because 17 of the 19 methionine residues of CD45 are located intracellularly (Saga *et al.,* 1986; Thomas *et al.,* 1987). The extracellular portion of CD45 is reported to be protease-resistant, whereas the intracellular portion is protease-sensitive (Omary and Trowbridge, 1980). It is possible that the peptides found on the electrophoresis axis were large-sized, partially digested peptides, mostly derived from the extracellular portion of CD45.

The poor mobility of tryptic peptides during the second-dimension chromatography could also be explained by the presence of large hydrophilic structures such as oligosaccharides attached to the peptides. Thus, it is possible that fatty acid labeling of CD45 was a result of myristic acid turnover into the acetyl form, which might then be incorporated into saccharides. However, this possibility was excluded by the fact that [^{3}H]myristic acid incorporated into CD45 of YAC-1 cells was resistant to various glycosidase treatments (Takeda and Maizel, 1990). Endoglycosidase F as well as *N*-glycanase (Elder and Alexander, 1982) treatments of CD45 reduced the size of the molecule by a few thousand daltons by SDS-PAGE, but did not cause any significant loss of radioactive label. Neuraminidase treatment followed by *O*-glycanase (Umemoto *et al.,* 1977) did not appear to affect CD45 at all, in agreement with previous results indicating that most, if not all, oligosaccharides are in the N-linked form in the isotype of CD45 (CD45 RO) present in YAC-1 cells (Saga *et al.,* 1986; Thomas *et al.,* 1987).

C. Putative Identification of the Lipid Linked to CD45

CD45 labeled by [^{3}H]myristic acid incorporation was subjected to hydrolysis with 6 *N* HCl at 110°C for 22 hours in an attempt to characterize the lipid moiety attached to CD45. Theoretically, under these hydrolysis conditions, all lipids, saccharides, and peptides will be broken down to their simplest structural components (Dittmer and Wells, 1969). The hydrolysate was extracted in a chloroform : methanol : H_2O extraction mixture (Folch *et al.,* 1957), and both the upper and the lower phases of the extract were analyzed by thin-layer chromatography (TLC) (Skipski and Barclay, 1969) (Fig. 4). Lipids of high polarity are resolved better under the TLC conditions employed here than most neutral lipids which migrate near or with the solvent front. The TLC of the upper aqueous phase

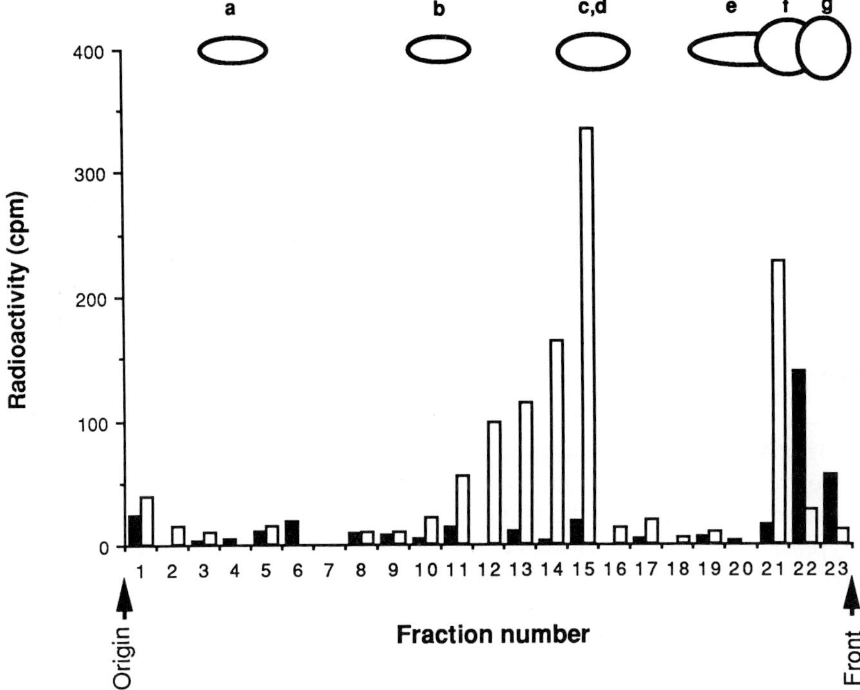

FIG. 4. TLC analysis of acid hydrolysis product derived from CD45 labeled with [^{3}H]myristic acid. CD45 labeled with [^{3}H]myristic acid was prepared as described in Fig. 3. Hydrolysis was carried out in 6 *N* HCl at 110°C for 22 hours. The hydrolysate (6000 cpm) was extracted by adding chloroform, methanol, and H_2O in a final ratio of 8 : 4 : 3 (v/v/v). The silica gel G plate was developed first in acetone : petroleum ether (1 : 3, v/v), and then in chloroform : methanol : acetic acid : H_2O (25 : 15 : 4 : 2, v/v/v/v). The standards (all from Sigma) are (a) sphingomyelin (from bovine brain), (b) psychosine (1-β-D-galactosylsphingosine), (c) sphingosine (derived from bovine brain sphingomyelin), (d) dihydrosphingosine (DL-1,3-dihydroxy-2-aminooctadecane), (e) cerebroside (from bovine brain), (f) myristic acid (tetradecanoic acid), (g) methyl esters of fatty acids (methyl esters of palmitic acid, stearic acid, oleic acid, linolenic acid, and arachidic acid). Open bars represent the upper phase, and solid bars represent the lower phase. [Reprinted with permission from Takeda and Maizel (1990). Copyright 1990 by the AAAS.]

revealed two main radioactive peaks. One of them comigrated with myristic acid, and the other with sphingosine–dihydrosphingosine standards. It is not surprising to find sphingosine in the upper phase, since the high polarity and the presence of the amino group in sphingosine are known to make its extraction by the typical chloroform : methanol : H_2O mixture into the chloroform phase difficult, especially in highly acidic conditions (Dittmer and Wells, 1969). The material "tailing" behind the sphingosine–dihydrosphingosine standards may represent a group of sphingosines with hydrocarbon chains shorter than those of the standards, which are either derived from brain sphingomyelin or chemically synthesized (C_{18}). Analysis of the lower phase gave only one peak which comigrated with myristic acid and the solvent front. The material at the solvent front is probably ester forms of fatty acids, since the presence of methanol during extraction and storage of lipid preparations are expected to cause the partial conversion of free fatty acids into their methyl ester forms (Dittmer and Wells, 1969). When similar hydrolysis products were subjected to another TLC condition suited for the separation of neutral lipids (Skipski and Barclay, 1969), the radioactivity was again found in three locations where sphingosine–dihydrosphingosine, fatty acid, and methyl ester of fatty acid standards migrated (A. Takeda, unpublished results).

The radioactive material recovered from TLC is only a fraction of the original radioactivity subjected to HCl hydrolysis. The cause for this poor recovery is not clear. The HCl hydrolysis conditions employed are certainly harsher than the standard conditions employed for hydrolyzing typical sphingolipids. However, hydrolysis at 70 or 95°C in lower concentrations of HCl (1–2 *N*) in methanol–H_2O (Gaver and Sweeley, 1965) instead of the conditions described above did not improve the recovery, and it often resulted in unidentifiable materials which probably were products of partial hydrolysis (A. Takeda, unpublished results).

Enzymes known to degrade sphingolipids were employed in an attempt to elucidate the unusual form of lipid attached to CD45. Well-characterized and commercially available enzymes for sphingolipid breakdown are rather limited. Endoglycoceramidase isolated from *Rhodococcus* (Genzyme) degrades the linkage between oligosaccharide and ceramide in both neutral and acidic glycosphiongolipids, but not in glycoglycerolipids (Ito and Yamagata, 1989). Ceramide glycanase isolated from leech (Boehringer Mannheim) has similar activity to that of the endoglycoceramidase of *Rhodococcus* and cleaves the binding between glucose and ceramide, but not between galactose and ceramide (Li *et al.*, 1986). Sphingomyelinase isolated from *Bacillus cereus* (EC 3.1.4.12; Boehringer Mannheim) hydrolyzes sphingomyelin to ceramide and phosphorylcholine (Ikezawa *et al.*, 1978). These enzymes failed to cause a decrease of radioactivity or a detectable shift in SDS-PAGE mobility of CD45 labeled by [^{3}H]myristic acid incorporation, when whole cells, immunoprecipitates, or electroeluted materials were treated (A. Takeda, unpublished results). However, if the sphingolipid-like structure forms a covalent bond with the CD45 peptide through its fatty acyl or sphingosine moiety, these enzymes will not release the [^{3}H]myristic acid label from the peptide, and may not result in any detectable shift in SDS-PAGE mobility. Therefore, [^{3}H]myristic acid-labeled CD45 was treated, at first, with pronase for "complete" proteolysis, and was then subjected to treatments by these enzymes in an attempt to detect any radioactive material whose patterns of organic–aqueous phase partition and/or TLC migration are altered by the enzyme treatments. None of the enzymes employed exhibited such effects (A. Takeda, unpublished results).

Evidence for covalent linkage between sphingolipids and peptides is limited. Slomiany *et al.* (1984) analyzed a glycolipid fraction derived from

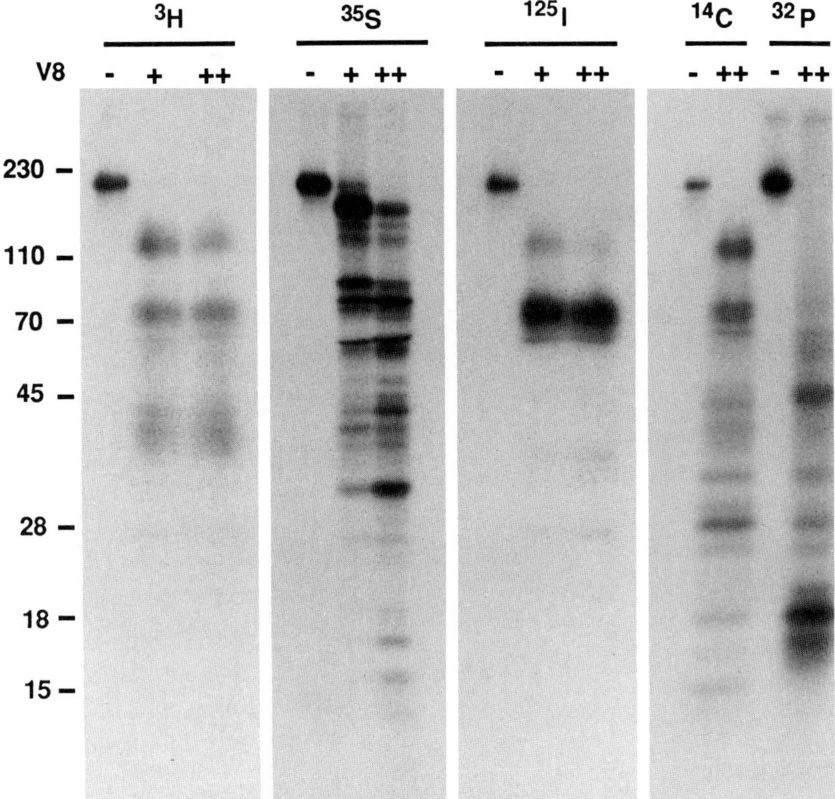

FIG. 5. One-dimensional peptide mapping analysis of CD45 after partial proteolysis. YAC-1 cells were labeled by incorporation of [^{3}H]myristic acid, [^{35}S]methionine, $H_3{}^{32}PO_4$, or a ^{14}C-labeled amino acid mixture, or by ^{125}I-iodination reaction on the cell surface. CD45 was prepared as described in Fig. 3. The samples were incubated without (−) or with V8 protease 37.5 μg/ml (+), or 75 μg/ml (+ +) for 1 hour, followed by SDS-PAGE in an 8–15% acrylamide gradient. The positions of molecular mass markers (expressed in kilodaltons) are shown on the left.

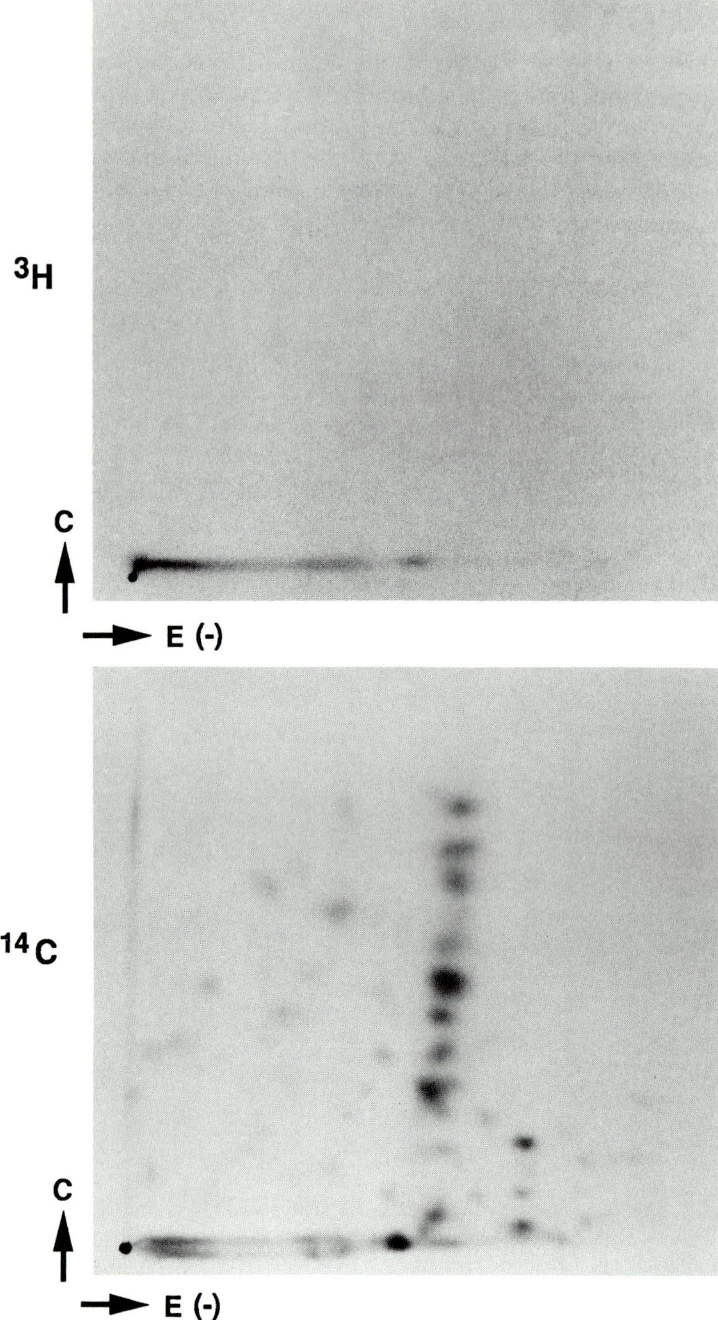

³H
C
E (-)
¹⁴C
C
E (-)

dog gastric epithelium after pronase digestion and detected a glycosphingolipid–glycoprotein complex of a molecular mass of about 14 kDa which consisted mainly of carbohydrates and small amounts of protein, sphingosine, and fatty acid. Extensive deglycosylation by trifluoromethanesulfonic acid left a core structure which contained sphingosine, glucose (Glc), mannose, *N*-acetylglucosamine (GlcNAc), and amino acids (aspartic acid, glutamic acid, serine, threonine, and lysine). Slomiany *et al.* (1984) proposed that the glycosphingolipid and the glycoprotein components are covalently linked through an amide linkage involving the amino group of sphingosine and a carboxyl group of an acidic amino acid, for example, Glc-sphingosine-NH-OC-peptide-GlcNAc. The nature of the intact peptide is not known, since the glycolipid–glycoprotein complex was isolated after extensive proteolysis. The proposed structure does not contradict our preliminary results regarding the lipid-linkage of CD45. On the other hand, one may envision the sphingolipid-like structure linked to CD45 to be similar to that of the Pl-glycan structure (Fig. 2), i.e., a sphingolipid attached to the peptide through an oligosaccharide-like structure.

D. Localization of the Lipid-Linkage in the CD45 Peptide

The two-dimensional tryptic peptide mapping pattern of [^{3}H]myristic acid-labeled peptides described above suggests that the fatty acid label was incorporated either into or in the proximity of the extracellular portion of CD45. This is, in fact, in agreement with the results obtained by peptide mapping of CD45 after partial digestion with V8 protease (Takeda and Cone, 1984) (Fig. 5). CD45 was prepared from YAC-1 cells labeled by incorporation of [^{3}H]myristic acid, [^{35}S]methionine, ^{14}C-labeled amino acid mixture, or [^{32}P]phosphoric acid, or by ^{125}I-labeling on the cell surface. The ^{3}H- and the ^{125}I-labeled samples exhibited a remarkable resemblance. Both of them produced two large peptides which were prominent and comigrated. The diffuse migration pattern of these bands is indicative of glycosylated peptides. The ^{14}C-labeled sample produced not only these dominant bands but also many smaller peptides. On the other hand, the ^{35}S- or ^{32}P-labeled samples produced a pattern quite different from the ^{3}H-

Fig. 6. Two-dimensional peptide mapping of CD45 after chymotryptic digestion. CD45 labeled by [^{3}H]myristic acid or ^{14}C-labeled amino acid mixture incorporation was prepared as described in Fig. 3. Samples were digested with chymotrypsin for 23 hours. Electrophoresis in the first dimension (E, the direction toward the cathode, as shown by arrows) and ascending chromatography in the second dimension (C, the direction as indicated by arrows) were carried out as described in Fig. 3.

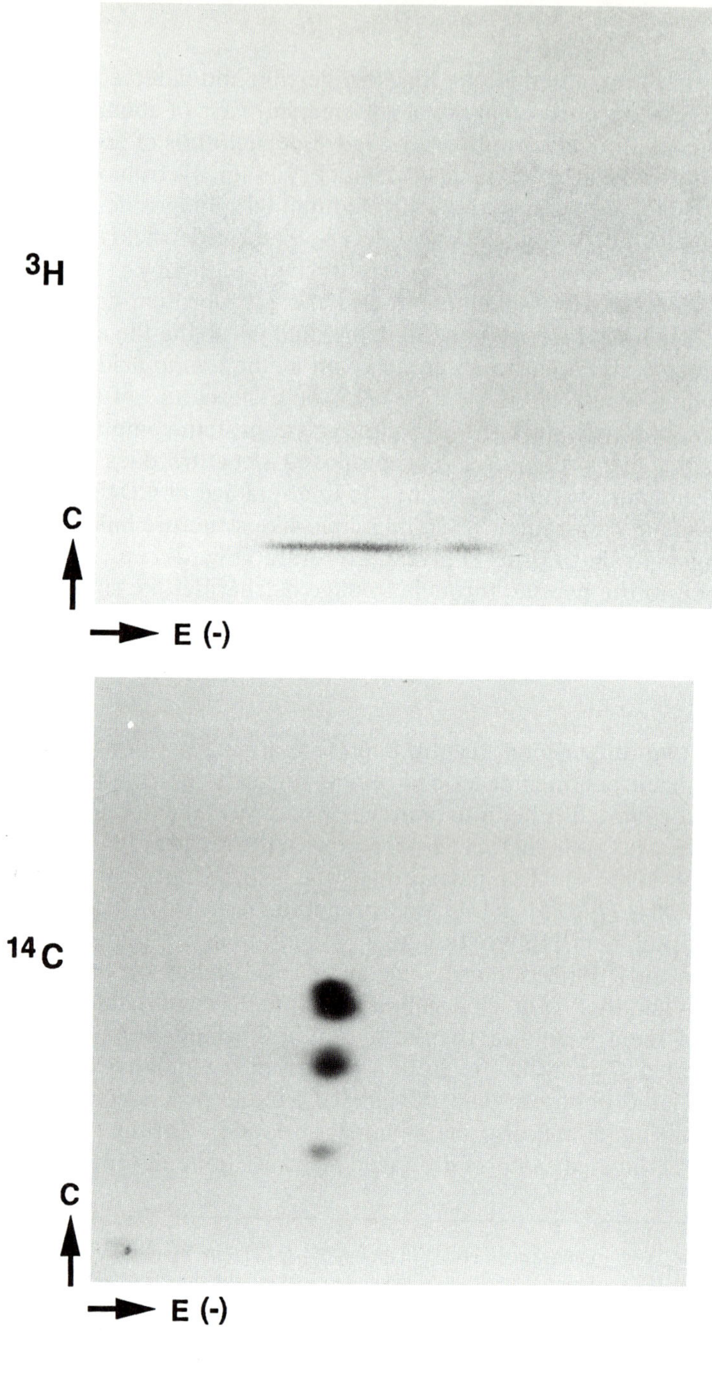

^3H
C
E (-)
^{14}C
C
E (-)

and the ^{125}I-labeled samples. They were devoid of the prominent large fragments and exhibited many peptides of smaller sizes, some of which comigrated with those of the ^{14}C-labeled sample. The resemblance between the ^{3}H- and the ^{125}I-labeled samples further supports the notion that the myristic acid label was incorporated either into or in the proximity of the extracellular portion of CD45.

Amino acid sequencing of the CD45–lipid linkage site and more definitive analysis of the lipid structure linked to the peptide require purification of a distinct proteolytic oligopeptide containing the myristic label. The objective of purifying such a fragment has been rather elusive. As in the case of tryptic digestion (Fig. 3), chymotrypsin also failed to generate a distinct fragment from [^{3}H]myristic acid-labeled CD45 (Fig. 6). The pattern was similar to that of tryptic digestion, although chymotrypsin produced various distinct fragments in the ^{14}C-labeled amino acid sample. Surprisingly, even the use of pronase did not succeed in producing a distinct fragment from the [^{3}H]myristic acid-labeled sample (Fig. 7). The results indicate that the environment in the vicinity of the lipid-linkage is highly resistant to proteolysis. Another less likely, but possible, explanation is that there is a structural polymorphism in the vicinity of the lipid structure which is responsible for the diverse mobility during electrophoresis and the immobility in the second dimension. For example, various oligosaccharide structures might be linked to the lipid structure.

Finally, another unusual characteristic of the lipid structure was revealed when tryptic peptide mapping was carried out on CD45 prepared from YAC-1 cells labeled by incorporation of either [^{3}H]myristic acid or [^{32}P]phosphoric acid (Fig. 8). The first dimension electrophoresis was carried out in pH 1.9, instead of pH 4.7, as was the case for Figs. 3, 6, and 7, and samples were spotted halfway in the electric field on TLC plates so that highly acidic fragments migrating toward the anode could also be observed. Under these highly acidic conditions, even the peptides containing phosphorylated amino acid residues will mostly migrate toward the cathode. However, a considerable portion of the myristic acid-labeled sample was seen migrating toward the anode. This result suggests the presence of a highly acidic group(s) consisting partly of the lipid structure

---

FIG. 7. Two-dimensional peptide mapping of CD45 after pronase digestion. CD45 labeled by [^{3}H]myristic acid or ^{14}C-labeled amino acid mixture incorporation was prepared as described in Fig. 3. Samples were digested with pronase for 21 hours. Electrophoresis in the first dimension (E, the direction toward the cathode, as shown by arrows), and ascending chromatography in the second dimension (C, the direction as indicated by arrows) were carried out as described in Fig. 3.

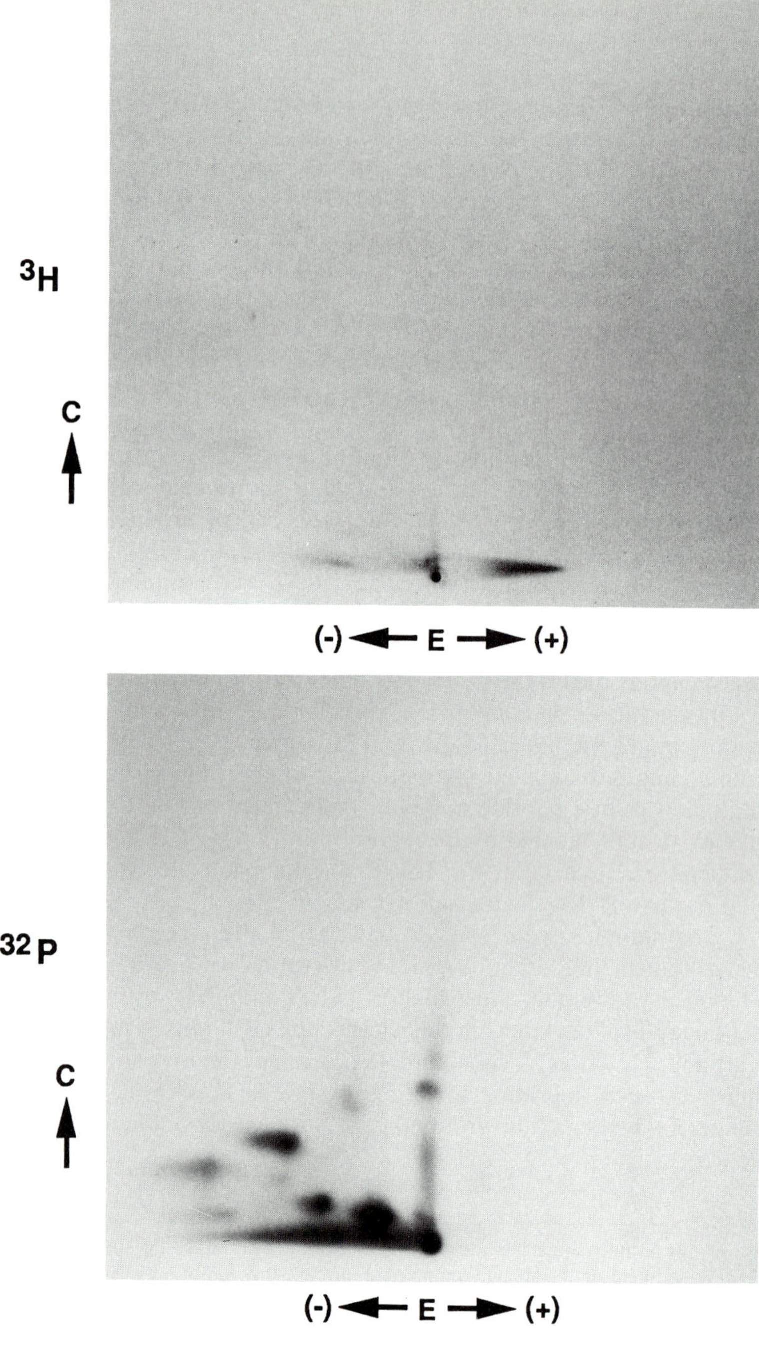

^3H
C
(-) E (+)
^{32}P
C
(-) E (+)

linked to CD45. A sulfate or a phosphate group might be responsible for the high acidity.

III. Conclusions and Hypothesis

Sphingolipids and their derivatives are known to exert numerous biological effects as described in the other articles of this volume. The analysis of the sphingolipid-like structure linked to CD45 is still too preliminary to allow speculation on its physiological significance. However, some aspects may deserve discussion at this stage. Several possible scenarios could be envisioned by which the lipid structure linked to CD45 might affect the overall role of CD45 and, therefore, signal transduction mediated by CD45. The lipid attachment may affect (1) other enzymes which are influenced by sphingolipids, (2) the PTP activity of CD45, (3) the half-life of CD45, (4) the intracellular localization of CD45, (5) the quarternary structure of CD45 and association with other proteins.

CD45 has been reported to affect directly at least one kinase, $p56^{lck}$, by dephosphorylating one of its tyrosine residues (Mustelin *et al.,* 1989; Ostergaard *et al.,* 1989). Because of the known effect of sphingolipids in downregulating protein kinases (Hannun *et al.,* 1986), it is tempting to speculate a role for CD45 and its "sphingolipid" in regulating protein kinases. Is it possible that CD45 exerts its effect by adjusting the amount of free sphingosine through formation and breakdown of sphingolipid–peptide linkages? This does not seem very likely, since none of the known forms of covalent lipid-linkages has been shown to exert any effect via released lipid moiety. However, the possibility should not be entirely ignored, since molecules similar to the Pl–glycan structure that anchors membrane proteins have been implicated as partly responsible for signal transduction mediated by insulin (Gaulton *et al.,* 1988) as well as nerve growth factor (Chan *et al.,* 1989). The origins or exact structures of the Pl–glycans are not known, and it is possible that they are derived from anchors for some membrane proteins.

FIG. 8. Two-dimensional tryptic peptide mapping of fatty acid-labeled CD45 peptides compared to phosphate-labeled peptides. CD45 labeled by [^{3}H]myristic acid or $H_3{}^{32}PO_4$ incorporation was prepared as described in Fig. 3. Samples were digested with TPCK-trypsin for 22 hours. Electrophoresis in the first dimension [E, the directions toward the cathode (−) and the anode (+) are shown by arrows] was carried out in formic acid : acetic acid : H_2O (11 : 39 : 450, v/v/v) at 1000 V for 40 minutes. Ascending chromatography in the second dimension (C, the direction as indicated by arrows) was carried out in a mixture of *n*-butanol : pyridine : acetic acid : H_2O (15 : 10 : 3 : 12, v/v/v/v).

There is no report to indicate that sphingolipids and related products affect the PTP activity of CD45. Our preliminary experiments also did not show any significant effect on CD45 PTP activity when sphingolipids were added up to about 50 μM (A. Takeda, unpublished results). A small inhibitory effect was seen at higher concentrations of sphingosines, but this effect was also seen with other non-sphingosine-containing lipids as well.

Our limited survey of human and mouse cell lines indicates that not all leukocytes form the lipid-linkage with the CD45 peptide. For example, a mouse T cell line, EL 4 (Old *et al.*, 1965), does not appear to incorporate [^{3}H]myristic acid into CD45, although EL 4 synthesizes CD45 just as actively as YAC-1 does (A. Takeda, unpublished results). If such a deficiency in the lipid-linkage formation could be confirmed, by comparing EL 4 with YAC-1, some insights to the physiological role of the sphingolipid-like structure attached to CD45 may be gained. In addition, there appears to be a subpopulation of CD45 which lacks the lipid-linkage within a given cell. In YAC-1 cells, this population of CD45 migrates slightly but distinctly faster than the population containing the lipid-label in SDS-PAGE, and appears mostly after long-term labeling (Takeda and Maizel, 1990). Furthermore, surface-radioiodinated preparations of CD45 do not reveal the fast-migrating band (A. Takeda, unpublished results). Combined with the data indicating that the lipid-linkage to CD45 occurs in its extracellular portion as discussed above, these results are consistent with the hypothesis that the lipid-linked form of CD45 is the one expressed on the plasma membrane. Given that various glycosphingolipids serve as recognition sites for cellular interactions (Hakomori, 1990), it is possible that the lipid linked to CD45 is part of a receptor–ligand recognition site(s).

The data currently available regarding the role of amino-terminal myristoylation on various proteins appear to indicate that myristoylation is critical but not sufficient for membrane association (Towler *et al.*, 1988). It has been proposed that myristoylation is required for correct localization of proteins in cellular compartment(s), probably by affecting the physical association of the protein with other molecules. A similar role might be assigned to the lipid linked to CD45. A need for such a modulator seems justified, since CD45 has been reported to form physical and functional associations with various other cellular proteins, notably those involved in signal transduction (Thomas, 1989; Fischer *et al.*, 1991; Trowbridge, 1991). Furthermore, T cell activation by either anti-TCR or an activating anti-Thy 1 antibody has been found to cause a rapid redistribution of CD45 in the cells (Minami *et al.*, 1991). On activation, CD45 is cleared from the Golgi, where 25–30% of the total CD45 in resting cells is stored, and is redistributed to an unknown location in the cell. In order to understand the mechanism of signal transduction in leukocytes, it is essential to learn what factors influence the physical associations and intracellular localization of CD45 under given circumstances. Understanding the nature of the lipid structure linked to CD45 may add another hue to the intricate picture of signal transduction.

ACKNOWLEDGMENTS

Supported in part by American Cancer Society grant ACS-IN 45-30. I thank Nabeel Yaseen for critical reading of the manuscript, and Abby Maizel and James Wu for their intellectual and experimental contributions.

References

Aitken, A., Cohen, P., Santikarn, S., Williams, D. H., Calder, A. G., Smith, A., and Klee, C. B. (1982). *FEBS Lett.* **150,** 314–318.

Berridge, M. J. (1989). *JAMA, J. Am. Med. Assoc.* **262,** 1834–1841.

Bourguignon, L. Y. W., Suchard, S. J., Nagpal, M. L., and Glenney, J. R., Jr. (1985). *J. Cell Biol.* **101,** 477–487.

Buss, J. E., Solski, P. A., Schaeffer, J. P., MacDonald, M. J., and Der, C. J. (1989). *Science* **243,** 1600–1603.

Carr, S. A., Biemann, K., Shoji, S., Parmelee, D. C., and Titani, K. (1982). *Proc. Natl. Acad. Sci. U.S.A.* **79,** 6128–6131.

Chan, B. L., Chao, M. V., and Saltiel, A. R. (1989). *Proc. Natl. Acad. Sci. U.S.A.* **86,** 1756–1760.

Charbonneau, H., Tonks, N. K., Walsh, K. A., and Fischer, E. H. (1988). *Proc. Natl. Acad. Sci. U.S.A.* **85,** 7182–7186.

Cikes, M., Friberg, S., Jr., and Klein, G. (1973). *J. Natl. Cancer Inst.* (*U.S.*) **50,** 347–362.

Clark, E. A., and Ledbetter, J. A. (1989). *Immunol. Today* **10,** 225–228.

Cone, R. E., and Brown, W. C. (1976). *Immunochemistry* **13,** 571–579.

Dittmer, J. C., and Wells, M. A. (1969). *In* "Methods in Enzymology" (J. Lowenstein, ed.), Vol. 14, pp. 482–530. Academic Press, New York.

Elder, J. H., and Alexander, S. (1982). *Proc. Natl. Acad. Sci. U.S.A.* **79,** 4540–4544.

Farnsworth, C. C., Gelb, M. H., and Glomset, J. A. (1990). *Science* **247,** 320–322.

Ferguson, M. A. J., and Williams, A. F. (1988). *Annu. Rev. Biochem.* **57,** 285–320.

Fischer, E. H., Charbonneau, H., and Tonks, N. K. (1991). *Science* **253,** 401–406.

Folch, J., Lees, M., and Sloane-Stanley, G. H. (1957). *J. Biol. Chem.* **226,** 497–509.

Gaulton, G. N., Kelly, K. L. Pawlowski, J., Mato, J. M., and Jarett, L. (1988). *Cell (Cambridge, Mass.)* **53,** 963–970.

Gaver, R. C., and Sweeley, C. C. (1965). *J. Am. Oil Chem. Soc.* **42,** 294–298.

Gibbs, J. B. (1991). *Cell (Cambridge, Mass.)* **65,** 1–4.

Gibson, W. (1974). *Virology* **62,** 319–336.

Gilman, A. G. (1989). *JAMA, J. Am. Med. Assoc.* **262,** 1819–1825.

Guan, K-L., and Dixon, J. E. (1991). *J. Biol. Chem.* **266,** 17026–17030.

Hakomori, S. I. (1990). *J. Biol. Chem.* **265,** 18713–18716.

Hancock, J. F., Magee, A. I., Childs, J. E., and Marshall, C. J. (1989). *Cell (Cambridge, Mass.)* **57,** 1167–1177.

Hannun, Y. A., Loomis, C. R., Merrill, A. H., Jr., and Bell, R. M. (1986). *J. Biol. Chem.* **261,** 12604–12609.

Hunter, T. (1987). *Cell (Cambridge, Mass.)* **50,** 823–829.

Ikezawa, H., Mori, M., Ohyabu, T., and Taguchi, R. (1978). *Biochim. Biophys. Acta* **528,** 247–256.

Ito, M., and Yamagata, T. (1989). *J. Biol. Chem.* **264,** 9510–9519.

Kaufman, J. F., Krangel, M. S., and Strominger, J. L. (1984). *J. Biol. Chem.* **259,** 7230–7238.

Kiener, P. A., and Mittler, R. S. (1989). *J. Immunol.* **143,** 23–28.

Klausner, R. D., and Samelson, L. E. (1991). *Cell (Cambridge, Mass.)* **64,** 875–878.

Koretzky, G. A., Picus, J., Thomas, M. L., and Weiss, A. (1990). *Nature (London)* **346,** 66–68.

Koretzky, G. A., Picus, J., Schyltz, T., and Weiss, A. (1991). *Proc. Natl. Acad. Sci. U.S.A.* **88,** 2037–2041.

Krebs, E. G. (1989). *JAMA, J. Am. Med. Assoc.* **262,** 1815–1818.

LaForgia, S., Morse, B., Levy, J., Barnea, G., Cannizzaro, L. A., Li, F., Nowell, P. C., Boghosian-Sell, L., Glick, J., Weston, A., Harris, C. C., Drabkin, H., Patterson, D., Croce, C. M., Schlessinger, J., and Huebner, K. (1991). *Proc. Natl. Acad. Sci. U.S.A.* **88,** 5036–5040.

Le Corre, R., Gerlier, D., Martin, A., Le Garrec, Y., Corradin, G., Bron, C., and Toujas, L. (1987). *Eur. J. Immunol.* **17,** 327–333.

Ledbetter, J. A., Tonks, N. K., Fischer, E. H., and Clark, E. A. (1988). *Proc. Natl. Acad. Sci. U.S.A.* **85,** 8628–8632.

Ledbetter, J. A., Schieven, G. L., Uckun, F. M., and Imboden, J. B. (1991). *J. Immunol.* **146,** 1577–1583.

Li, S.-C., DeGasperi, R., Muldrey, E., and Li, Y.-T. (1986). *Biochem. Biophys. Res. Commun.* **141,** 346–352.

Minami, Y., Stafford, F., Lippincott-Schwartz, J., Yuan, L., and Klausner, R. D. (1991). *J. Biol. Chem.* **266,** 9222–9230.

Mustelin, T., Coggeshall, K. M., and Altman, A. (1989). *Proc. Natl. Acad. Sci. U.S.A.* **86,** 6302–6306.

Nishibe, S., Wahl, M. I., Hernandez-Sotomayor, S. M. T., Tonks, N. K., Rhee, S. G., and Carpenter, G. (1990). *Science* **250,** 1253–1256.

Nishizuka, Y. (1989). *JAMA, J. Am. Med. Assoc.* **262,** 1826–1833.

Old, L. J., Boyse, E. A., and Stockert, E. (1965). *Cancer Res.* **25,** 813–819.

Omary, M. B., and Trowbridge, I. S. (1980). *J. Biol. Chem.* **255,** 1662–1669.

Omary, M. B., and Trowbridge, I. S. (1981). *J. Biol. Chem.* **256,** 4715–4718.

Ostergaard, H. L., and Trowbridge, I. S. (1990). *J. Exp. Med.* **172,** 347–350.

Ostergaard, H. L., and Trowbridge, I. S. (1991). *Science* **253,** 1423–1425.

Ostergaard, H. L., Shackelford, D. A., Hurley, T. R., Johnson, P., Hyman, R., Sefton, B. M., and Trowbridge, I. S. (1989). *Proc. Natl. Acad. Sci. U.S.A.* **86,** 8959–8963.

Pingell, J. T., and Thomas, M. L. (1989). *Cell (Cambridge, Mass.)* **58,** 1055–1065.

Resh, M. D., and Ling, H-P. (1990). *Nature (London)* **346,** 84–86.

Rilling, H. C., Breunger, E., Epstein, W. W., and Crain, P. F. (1990). *Science* **247,** 318–320.

Roberts, W. L., and Rosenberry, T. L. (1985). *Biochem. Biophys. Res. Commun.* **133,** 621–627.

Rudd, C. E., Trevillyan, J. M., Dasgupta, J. D., Wong, L. L., and Schlossman, S. F. (1988). *Proc. Natl. Acad. Sci. U.S.A.* **85,** 5190–5194.

Saga, Y., Tung, J.-S., Shen, F.-W., and Boyse, E. A. (1986). *Proc. Natl. Acad. Sci. U.S.A.* **83,** 6940–6944.

Samelson, L. E., Fletcher, M. C., Ledbetter, J. A., and June, C. H. (1990a). *J. Immunol.* **145,** 2448–2454.

Samelson, L. E., Phillips, A. F., Luong, E. T., and Klausner, R. D. (1990b). *Proc. Natl. Acad. Sci. U.S.A.* **87,** 4358–4362.
Schlessinger, J. (1986). *J. Cell Biol.* **103,** 2067–2072.
Schraven, B., Samstag, Y., Altevogt, P., and Meuer, S. C. (1990). *Nature (London)* **345,** 71–74.
Schraven, B., Kirchgessner, H., Gaber, B., Samstag, Y., and Meuer, S. (1991). *Eur. J. Immunol.* **21,** 2469–2477.
Sefton, B. M., Trowbridge, I. S., Cooper, J. A., and Scolnick, E. M. (1982). *Cell (Cambridge, Mass.)* **31,** 465–474.
Skipski, V. P., and Barclay, M. (1969). *In* "Methods in Enzymology" (J. Lowenstein, ed.), Vol. 14, pp. 530–598. Academic Press, New York.
Slomiany, A., Takagi, A., and Slomiany, B. L. (1984). *Biochem. Biophys. Res. Commun.* **125,** 211–217.
Stamenkovič, I., Sgroi, D., Aruffo, A., Sy, M. S., and Anderson, T. (1991). *Cell (Cambridge, Mass.)* **66,** 1133–1144.
Stiernberg, J., Low, M. G., Flaherty, L., and Kincade, P. W. (1987). *J. Immunol.* **138,** 3877–3884.
Stover, D. R., Charbonneau, H., Tonks, N. K., and Walsh, K. A. (1991). *Proc. Natl. Acad. Sci. U.S.A.* **88,** 7704–7707.
Streuli, M., Krueger, N. X., Tsai, A. Y. M., and Saito, H. (1989). *Proc. Natl. Acad. Sci. U.S.A.* **86,** 8698–8702.
Takeda, A., and Cone, R. E. (1984). *Biochem. Biophys. Res. Commun.* **122,** 932–937.
Takeda, A., and Maizel, A. L. (1990). *Science* **250,** 676–679.
Takeda, A., Wu, J. J., and Maizel, A. L. (1992). *J. Biol. Chem.* **267,** 16651–16659.
Telfer, J. C., and Rudd, C. E. (1991). *Science* **254,** 439–441.
Thomas, M. L. (1989). *Annu. Rev. Immunol.* **7,** 339–369.
Thomas, M. L., Reynolds, P. J., Chain, A., Ben-Neriah, Y., and Trowbridge, I. S. (1987). *Proc. Natl. Acad. Sci. U.S.A.* **84,** 5360–5363.
Todderud, G., Wahl, M. I., Rhee, S. G., and Carpenter, G. (1990). *Science* **249,** 296–298.
Tonks, N. K., Charbonneau, H., Diltz, C. D., Fischer, E. H., and Walsh, K. A. (1988). *Biochemistry* **27,** 8695–8701.
Tonks, N. K., Diltz, C. D., and Fischer, E. H. (1990). *J. Biol. Chem.* **265,** 10674–10680.
Torimoto, Y., Dang, N. H., Vivier, E., Tanaka, T., Schlossman, S. F., and Morimoto, C. (1991). *J. Immunol.* **147,** 2514–2517.
Towler, D. A., Gordon, J. I., Adams, S. P., and Glaser, L. (1988). *Annu. Rev. Biochem.* **57,** 69–99.
Trowbridge, I. S. (1991). *J. Biol. Chem.* **266,** 23517–23520.
Tse, A. G. D., Barclay, A. N., Watts, A., and Williams, A. F. (1985). *Science* **230,** 1003–1008.
Ullrich, A., Coussens, L., Hayflick, J. S., Dull, T. J., Gray, A., Tam, A. W., Lee, J., Yarden, Y., Libermann, T. A., Schlessinger, J., Downward, J., Mayes, E. L. V., Whittle, N., Waterfield, M. D., and Seeburg, P. H. (1984). *Nature (London)* **309,** 418–425.
Umemoto, J., Bhavanandan, V. P., and Davidson, E. A. (1977). *J. Biol. Chem.* **252,** 8609–8614.
Veillette, A., Bookman, M. A., Horak, E. M., and Bolen, J. B. (1988). *Cell (Cambridge, Mass.)* **55,** 301–308.
Volarevič, S., Burns, C. M., Sussman, J. J., and Ashwell, J. D. (1990). *Proc. Natl. Acad. Sci. U.S.A.* **87,** 7085–7089.
Yarden, Y., and Ullrich, A. (1988). *Annu. Rev. Biochem.* **57,** 443–478.

Part IV

COMPARATIVE AND DEVELOPMENTAL BIOLOGY

ADVANCES IN LIPID RESEARCH, VOL. 26

Glycosphingolipids of the Invertebrata as Exemplified by a Cestode Platyhelminth, *Taenia crassiceps*, and a Dipteran Insect, *Calliphora vicina*

ROGER D. DENNIS AND HERBERT WIEGANDT

Physiologisch–Chemisches Institut
Philipps Universität—Marburg
W-3550 Marburg, Germany

I. Introduction

A study of phylogeny dictates that bacteria preceded plants, which in their turn, preceded the animals, with a certain amount of overlap into the invertebrates and vertebrates, respectively. Their corresponding chemical evolution can be studied in terms of their plasma membrane-located glycolipids. The bacteria are characterized by variations on the theme of glycoglycerolipids, which may be phosphorylated, e.g., those of *Streptococcus* spp. (Fischer *et al.*, 1973), as well as of the lipopolysaccharides of gram-negative bacteria (Rietschel *et al.*, 1991). Gram-negative bacteria exhibit more exotic glycolipids, such as the glucuronic acid-containing glycopeptidolipids of *Mycobacterium* spp. (Chatterjee *et al.*, 1987) and the, thus far, unique prokaryotic glucuronic acid-containing glycosphingolipids of

Sphingomonas paucimobilis (Yamamoto *et al.*, 1978; Kawahara *et al.*, 1991). The thylakoid membrane of the chloroplast of dicotyledonous plants is again distinguished by the presence of glycoglycerolipids, mainly mono- and digalactosyldiacylglycerol (as reviewed by Curatolo, 1987). In addition, glycophosphosphingolipids have been analyzed from plants, yeast, and fungi (Laine and Hsieh, 1987). For the vertebrates, as of 1989 (Stults *et al.*, 1989), the structures of some 210 glycosphingolipids had been elucidated: 117 neutral and 93 acidic species, the latter being subdivided into 79 sialic acid-, 2 glucuronic acid-, and 12 sulfate-containing types. In contrast and for the corresponding period, 36 glycosphingolipids were listed for the invertebrates: 24 neutral and 12 acidic species, the latter being subdivided into 10 sialic acid- and 2 glucuronic acid-containing types (Stults *et al.*, 1989).

As regards their biological function(s), perhaps the one universal property of bacterial and plant glycolipids, and the simple vertebrate glycosphingolipids, is the stabilization of plasma membranes through interlipid hydrogen bonding. Such stability is of particular relevance in the thylakoid membrane of chloroplasts and the myelin and membranes of the brush border epithelium, because of the functional necessity of maintaining transmembrane ionic gradients. The putative roles of vertebrate glycosphingolipids may be categorized under regulation of cell growth, cellular interaction/communication, and differentiation (Hakomori, 1981). When these regulatory processes go awry in oncogenesis, the consequence may be tumor-distinctive glycolipids that can operate as tumor-associated antigens/markers resulting from precursor accumulation or neoglycolipid biosynthesis. The capability to glycosylate specified proteins and lipids is a characteristic of all eukaryotic organisms, including the invertebrates. With regard to glycolipids, the inference from acceptor specificity studies of various vertebrate glycosyltransferases using oligosaccharides, glycolipids, and glycoproteins (Makita and Taniguchi, 1985; Nagai *et al.*, 1987) is that at least some of these enzymes are capable of generalized glycoconjugate glycosylation.

Of the limited studies undertaken on the glycosphingolipids of invertebrates, the main effort on their structural elucidation has been centred on various species of the Platyhelminthes, Insecta, Mollusca (Hori *et al.*, 1981, 1983), and Echinodermata (Kochetkov *et al.*, 1976; Sugita, 1979a,b). This comparative review pertains mainly to the glycosphingolipids of the Platyhelminthes and Insecta as so far "dissected" at the biological, biochemical, chemical, and molecular level.

Because of the presumed location of glycosphingolipids in the outer leaflet of the plasma membrane, a study of these putative entities in parasitic helminths has the added dimension of host–parasite interaction/relationship. The interface between parasitic helminth (Cestoda/Trematoda) and vertebrate host is the tegumental surface membrane. The basic questions are, What is the component composition of this tegument, and What are the characteristics that allow most helminths to avoid the cellular and humoral responses of the host's immune defense system? The tegument of members of both helminth classes is invested with a glycocalyx of largely unknown chemical composition (Naiki *et al.*, 1985; Schmidt and Peters, 1987; Schmidt, 1988). Evidence for its involvement in immunological recognition has been obtained from schistosomula of the trematode *Schistosoma mansoni,* i.e., the glycanic epitope of the 38,000 M_r antigen (Capron *et al.*, 1987). The same functional properties are elicited by a carbohydrate epitope present on the oligosaccharide moiety of keyhole limpet (*Megathura crenulata*) hemocyanin (Grzych *et al.*, 1987).

As our model of an integrative host–helminth parasite system for the study of the role(s) of glyco(sphingo)lipids, we chose the metacestodes of the fox tapeworm, *Taenia crassiceps* (Cestoda: Cyclophyllidea; Freeman, 1962). This prototype has been utilized for various reasons, not the least of which being the high yield of larvae per infected mouse and its close phylogenetic relationship to species of medical and/or veterinary importance, e.g., *Taenia solium* (pork tapeworm) and *Taenia saginata* (beef tapeworm; M. Katz *et al.*, 1988).

In anticipation that some of the basic conundra concerning vertebrate glycosphingolipid function(s) could be approached using a "simplified" animal model, 10 years ago we turned our attention to the blowfly *Calliphora vicina* (Insecta: Diptera) prototype, for a systematic investigation of glycosphingolipid chemistry, biochemistry, and biology. Of relevance to this analysis was that these insects have a year-round, short life cycle of approximately 18 days at 23°C; they can be bred to yield large numbers of larvae, pupae, or adults, i.e., kilogram quantities; and they undergo a metamorphosis which is triggered by a single morphogenetic hormone, ecdysterone (Highnam and Hill, 1977).

II. Glycosphingolipids of the Platyhelminthes

A. Introduction: Glycosphingolipids of the Platyhelminthes and Nematoda

The presence of glycolipids in parasitic helminths has been reported for members of the phyla Platyhelminthes (Cestoda and Trematoda) and Nematoda. For the Cestoda, these have included *Hymenolepis diminuta* and *H. citelli* (Harrington, 1965; Webb and Mettrick, 1971, 1973), *Echino-*

coccus granulosus and *Moniezia benedeni* (Hrzenjak and Ehrlich, 1976; Richards *et al.*, 1987), *Echinococcus multilocularis* (Persat *et al.*, 1988, 1990a), *Spirometra mansonoides* (Singh *et al.*, 1987a), *T. crassiceps* (Mills *et al.*, 1981) and *T. taeniaeformis* (Lesuk and Anderson, 1941); for the Trematoda, *Fasciola hepatica* and *Paramphistomum microbothrium* (Hrzenjak and Ehrlich, 1975) and *S. mansoni* (Naiki *et al.*, 1985; Weiss *et al.*, 1986; Maloney *et al.*, 1990); for the Nematoda, *Angiostrongylus cantonensis* (Kwong *et al.*, 1990), *Ascaris lumbricoides* (Ehrlich and Hrzenjak, 1975), *Onchocerca gibsoni* (Maloney and Semprevivo, 1991), and *Trichuris globulosa* (Sarwal *et al.*, 1989). In many of these studies, the unverified glycolipid components have been detected by thin-layer chromatographic techniques and spray reagents. The observation of serine incorporation into the sphingosine moiety of the cerebroside fraction (Webb and Mettrick, 1973) had indicated the capability of a cestode to synthesize sphingolipid(s). Structural analyses of parasitic helminth glycosphingolipids have been performed on the mono- to tetraosylceramides of *E. multilocularis* (Persat *et al.*, 1990b, 1992), the ceramide mono- to tetrasaccharides of the cestode *Metroliasthes coturnix* (Nishimura *et al.*, 1991), the ceramide monosaccharide of *S. mansonoides* (Singh *et al.*, 1987b), and the ceramide mono- and disaccharides of *S. mansoni* (Makaaru *et al.*, 1992). The identification of highly fucosylated glycosphingolipids in *S. mansoni* (Levery *et al.*, 1992) in association with the ceramide disaccharide (Makaaru *et al.*, 1992) represents a new monosaccharide sequence, the "schisto"-series (core structure: GlcNAc3GalNAc3GalNAcβ4GlcCer).[1]

[1] Glycosphingolipids are designated $(X)_{size}$Cer, whereby (X) stands for root name, e.g., arthro-series (Ap, sequence GalNAcβ4GlcNAcβ3Manβ4Glcβ<), globo-series (Gb, sequence GalNAcβ3Galα4Galβ4Glcβ<), lacto-series (Lc, sequence Galβ3GlcNAcβ3Galβ4Glcβ<), neolacto-series (nLc, sequence Galβ4GlcNAcβ3Galβ4Glcβ<), mollu-series (Ml, sequence GlcNAcβ2Manα3Manβ4Glcβ<), xylomollu-series (XMl, GlcNAcβ2Manα3[Xylβ2]Manβ-4Glcβ<), neogala-series (nGa, Galβ6Galβ6Galβ6Galβ<), and the size is given by the subscript number of the monosaccharide constituents. Monosaccharide residues carrying substitutions are indicated by roman numerals (counting from the ceramide end), a superscript indicating the substitution position. Short-hand designations used only for glycosphingolipids of the arthro-series are N, Nz, A, and Az, where N stands for GSL with a neutral oligosaccharide chain and A for those with an additional acidic substituent (glucuronic acid). To this, z is added for members that carry the zwitterionic substituent 2-aminoethylphosphate; arabic numerals 1 to 9 indicate the number of constituent neutral monosaccharides. *P*Etn, Phosphoethanolamine (2-aminoethyl-phosphate).

B. GLYCOSPHINGOLIPIDS OF *Taenia crassiceps* METACESTODES

The acidic glycolipid fraction was negative for sialic acid, free amino groups, and phosphate esters, as determined by spray reagent (Kunz *et al.*, 1991). The neutral glycolipid fraction, at a yield of 20 mg/g lyophilized larvae, was similarly negative for sialic acid, free amino groups, and phosphate esters (Baumeister *et al.*, 1992) and served as the source for further structural investigations. Structural analyses were performed on the four major glycolipid components. From thin-layer chromatographic migration properties, they corresponded to glycolipids with a sugar chain-length of mono-, di-, tri-, and tetrasaccharide [thin-layer chromatography-designated glycolipid components a–d, respectively (Baumeister *et al.*, 1992)], respectively (Fig. 1).

These carbohydrate chains have been structurally elucidated as belonging to the neogala-series with the carbohydrate core structure Galβ6Galβ6-Galβ6Galβ1 Cer for the neogalatetraosylceramide, nGa_4Cer (Table I; Dennis *et al.*, 1992).

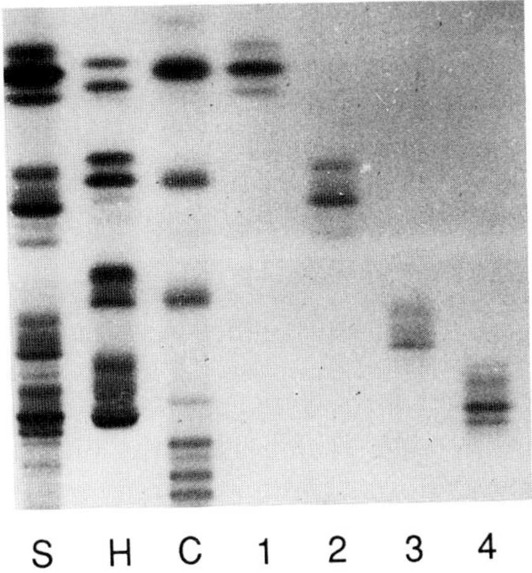

FIG. 1. Thin-layer chromatogram of *T. crassiceps* neutral-fraction glycosphingolipids. 1, 2, 3, 4, Isolated *T. crassiceps* neutral-fraction GSL corresponding to ceramide mono-, di-, tri-, and tetrasaccharides, respectively; S, *T. crassiceps* total neutral-fraction GSL; H, human spleen neutral-fraction GSL, C, *C. vicina* neutral-fraction GSL. High-performance thin-layer chromatographic (HPTLC) Silica gel-60 plates (Merck, Darmstadt, Germany), running solvent was chloroform–methanol–water (65:25:4 v/v/v); visualized with orcinol–sulfuric acid spray reagent.

Table I

CHEMICAL CONSTITUTION OF THE NEOGALA-SERIES GLYCOSPHINGOLIPIDS AND THEIR FUCOSYLATED DERIVATIVES IN THE CESTODA *T. crassiceps*, *M. coturnix*, AND *E. multilocularis*

GSL[a]	*T. crassiceps*[b]	Systematic abbreviation	*M. coturnix*[c] *E. multilocularis*[d]
a	Galβ Cer		Galβ1Cer
		Ga_2Cer	Galα4GalCer[c]
b	Galβ6Galβ1Cer	nGa_2Cer	Galβ6Galβ1Cer[d]
		II^3Fucα-nGa_2Cer	Fucα3Galβ6GalCer[d]
c	Galβ6Galβ6Galβ1Cer	nGa_3Cer	Galβ6Galβ6GalCer[c,d]
		II^3Fucα-nGa_3Cer	Galβ6[Fucα3]Galβ6GalCer[d]
d[e]	Galβ6Galβ6Galβ6Galβ1Cer	nGa_4Cer	Galβ6Galβ6Galβ6GalCer[c,d]
	Galα4Galβ6Galβ6Galβ1Cer	III^4Galα-nGa_3Cer	

[a] Glycosphingolipid components as designated in Baumeister *et al.* (1992).
[b] Dennis *et al.* (1992).
[c] Nishimura *et al.* (1991).
[d] Persat *et al.* (1992).
[e] Present in approximately equimolar amounts.

Each thin-layer chromatography-resolved glycolipid component was characterized by multiple banding, a direct result of ceramide moiety heterogeneity. The long-chain bases were typified by the predominance of dihydrosphingosine (sphinganine; main constituent) and phytosphingosine (4-hydroxysphinganine; major constituent). The major ceramide fatty acids were particularly long-chained and saturated, with C26 : 0 hexacosanoic and C28 : 0 octacosanoic acids as the dominant species, however, 2-hydroxy fatty acids were not detected (Dennis *et al.*, 1992). These data closely resemble those of the two other systematic, structural analyses of cestode neutral-fraction glycosphingolipids to date, i.e., *M. coturnix* (Nishimura *et al.*, 1991) and *E. multilocularis* (Persat *et al.*, 1992), despite being derived from different parasitic helminths, different life-cycle stages, and different vertebrate hosts. This homology refers to the neogala-series carbohydrate moiety, the very long-chained saturated fatty acids, and sphinganine and 4-hydroxysphinganine as sphingoid bases. The ceramide disaccharide of *M. coturnix* corresponded to the gala-series (Galα4GalCer; Nishimura *et al.*, 1991). A fraction of the ceramide disaccharide and trisaccharide components of *E. multilocularis* were extant as their fucosylated derivative (II^3Fucα-nGa_2Cer and II^3Fucα-nGa_3Cer, respectively; Persat *et al.*, 1992). The possibility therefore exists that the neogala-series is of more prevalent occurrence than first thought, at least in the larval and adult life-cycle stages of the Cestoda. The data for the neutral-fraction glycosphingolipids of *T. crassiceps* (Dennis *et al.*, 1992) did not substantiate the previous identification of 14 different sugars from 5 glycolipid species (Mills *et al.*, 1981).

From a phylogenetic viewpoint, neogala-series glycosphingolipids are not confined to the Cestoda but have been identified in the phyla Gastropoda (Matsubara and Hayashi, 1981, 1986; Hayashi and Matsubara, 1982, 1989) and Annelida. Phosphocholine homologs of ceramide mono- and disaccharides have been characterized in the annelid species *Marphysa sanguinea* (Noda *et al.*, 1992) and *Pheretima hilgendorfi* (Sugita *et al.*, 1992), as well as in the occurrence of major glycolipids of Galβ1Cer, Galβ6Galβ1Cer, and Galβ6Galβ6Galβ1Cer in the latter. The disaccharide structure Galβ6Galβ⟨ has surprisingly been found in a primitive deuterostome, the sea urchin *Hemicentrotus pulcherrimus* (Kubo *et al.*, 1992). This provokes the speculation that glycosphingolipids with this terminal oligosaccharide are correlated in some way with life in a marine environment, i.e., parasitic helminths, marine annelids and gastropods.

C. Immunological and Immunochemical Investigation of Platyhelminthic Glycosphingolipids

The tegumental surface expression of glycolipids has been shown by oxidation with galactose oxidase or periodate treatment followed by reduction with tritiated borohydride for *S. mansoni* schistosomula (Samuelson and Caufield, 1982) and adults (Naiki *et al.*, 1985). Are the expressed glycolipids immunologically functional, i.e., immunogenic and/or antigenic, as a first step toward studying biological function? Parasitic helminth-derived glycolipids can be readily demonstrated to be antigenic by manifesting immunoreactivity with the corresponding infection serum, e.g., the anti-neutral-fraction glycolipid antibody activity of *E. multilocularis* metacestode-induced alveolar hydatid sera and *E. granulosus* metacestode-induced hydatid cyst sera (Persat *et al.*, 1991), but are they immunogenic? This question is obscured by the fact that the same or very similar carbohydrate epitopes are frequently present on both glycoproteins and glycolipids, as is the case with monoclonal antibodies directed against glycoconjugate components of various *S. mansoni* life-cycle stages (Weiss *et al.*, 1986; Levery *et al.*, 1992). Glyco(sphingo)lipids are immunogenic, being capable of inciting an immune response (the polyhexosamine ceramide complex of *E. granulosus* hydatid cyst fluid; Hrzenjak *et al.*, 1977), although the antibodies generated regularly cross-react with the same or very similar carbohydrate antigenic determinants present on proteins (Marcus, 1984).

D. Immunological and Immunochemical Investigation of *Taenia crassiceps* Glycosphingolipids

Employing ELISA and HPTLC-immunostaining, anti-neutral-fraction glycolipid antibody activity has been detected in *T. crassiceps* metacestode-induced infection sera (Kunz *et al.*, 1991). Mice developed a rapid and increasing titer to parasite-derived neutral glycolipids (ELISA: IgM and IgG specificity), being significantly elevated from 5 days postinfection onward. Immunostaining revealed that, by 3 days postinfection, immunoreactivity to certain tetra- and greater than tetrasaccharide glycolipids was evident, and, by 7 days, all but 2 of the 24 thin-layer chromatography-designated bands were serologically active. The question as to whether the parasite glycolipids themselves are immunogenic remains open. As an initial step in addressing this question, mice were immunized as in preparation for the generation of monoclonal antibodies against parasite-specific, lipid-bound carbohydrate epitopes, by repeated injection of pretreated, bacterial carrier *Salmonella minnesota* R595-bound *T. crassiceps* metacestode total neutral-fraction glycolipids [at 1 : 1 (w/w) with 25 μg of glycolipid/dose; Galanos *et al.*, 1971]. This yielded an ELISA-determined, anti-glycolipid antibody titer of 1 : 6400, equivalent to 1 : 100 for the controls (uninjected or pretreated bacterial carrier only; Höllerer *et al.*, 1993). The stimulus for the generation of specific, anti-carbohydrate antibodies, i.e., parasite-derived neutral-fraction glycolipids, is assumed not to be due to cross-reacting *T. crassiceps* metacestode-originated glycoproteins and/or polysaccharides.

Immunochemically, the carbohydrate nature of the antigenic determinant(s) of *T. crassiceps* neutral-fraction glyco(sphingo)lipids has been demonstrated by the abolition of immunoreactivity, either irreversibly with mild sodium metaperiodate oxidation or reversibly with peracetylation, followed by deacetylation on the thin-layer plate (Baumeister *et al.*, 1992). The minor antigenic determinant of the structurally determined glycosphingolipids (Galα4Galβ6Galβ⟨; Dennis *et al.*, 1992) is apparently nonimmunogenic in mice, at least under the conditions applied, because no anti-glycoconjugate antibody activity was detected by HPTLC-immunostaining with the homologous infection sera. However, enzymatic removal of the terminal galactose residue by α-galactosidase cleavage restored immunoreactivity to the homologous infection sera (Baumeister *et al.*, 1993). The major carbohydrate epitope of the structurally determined glycosphingolipids (Galβ6Galβ6Galβ⟨; Dennis *et al.*, 1992) has been established serologically in the murine system, with normal serum, homologous infection serum, and a monospecific, polyclonal antibody directed to neogalatriaosylceramide (Baumeister *et al.*, 1992). The latter antibody was isolated from homologous infection serum by affinity chromatography on a nGa_3Cer-bound, octyl-Sepharose Cl-4B column (Hirabayashi *et al.*, 1983). Immunochemically, the major common epitope expressed by *T. crassiceps* metacestode neutral-fraction glyco(sphingo)lipids is found in various marine, archaeogastropod mollusks, i.e., *Turbo cornutus, Chlorostoma argyrostoma turbinatum,* and *Monodonta labio* (Galβ6Galβ6Gal-β1Cer; Matsubara and Hayashi, 1981, 1986; Hayashi and Matsubara, 1982, 1989).

Serologically, the HPTLC patterns of *T. crassiceps* metacestode, *T. solium* metacestode, and *T. saginata* metacestode and adult neutral-fraction glycosphingolipids, in both the homologous and heterologous infection sera, were virtually identical (Baumeister *et al.*, 1992). This coincidence of immunological reactivity included the monospecific, polyclonal antibody directed to the neogala epitope (nGa_3Cer). The implication is that the neogala-carbohydrate series is not confined to neutral-fraction glyco(sphingo)lipids of *T. crassiceps* metacestodes, but may occur in at least two closely related species, *T. solium* and *T. saginata*. This analogy can apparently be further extended in the Cestoda, e.g., the immunological cross-reactivity of *E. multilocularis* neutral-fraction glycolipids with alveolar hydatid- and hydatid cyst-infection sera (Persat *et al.*, 1991), and the heterologous immunoreactivity of *E. granulosus* metacestode neutral-fraction glycolipids with *T. crassiceps* metacestode-induced infection sera and the monospecific, polyclonal antibody directed against nGa_3Cer (Fig. 2; Irmer *et al.*, 1993).

An unforeseen complication has been the weak immunoreactivity of *T. crassiceps* neutral-fraction glycolipids with noninfected mice sera (Kunz *et al.*, 1991), whereby 50% of those tested proved positive toward three thin-layer chromatography-designated component bands, whose migration properties corresponded to possessing sugar chains of a tetrasaccharide and two greater than tetrasaccharides, respectively. The phenomenon of seemingly naturally occurring anti-carbohydrate antibodies is not an isolated one, being detected in normal human sera and directed against such epitopes as GalCer (Avila and Rojas, 1990), Galα3Gal (anti-Gal antibody; Towbin *et al.*, 1987), various glycosphingolipids including Forssman (Yasuda *et al.*, 1982), and insect neutral, zwitterionic, and acidic glycosphingolipids (Nores *et al.*, 1991). A possible explanation(s) for the occurrence of heterophile antibodies could be either the presence of gram-negative bacteria in the gastrointestinal and/or pulmonary tracts that carry the same or very similar epitopes on their cell-envelope lipopolysaccharides and capsular polysaccharides (e.g., blood group substances and terminal Galα3Gal$\langle$ residues) (Galili, 1988), or the prior infection with parasitic protozoa or helminths of other species that express the same or

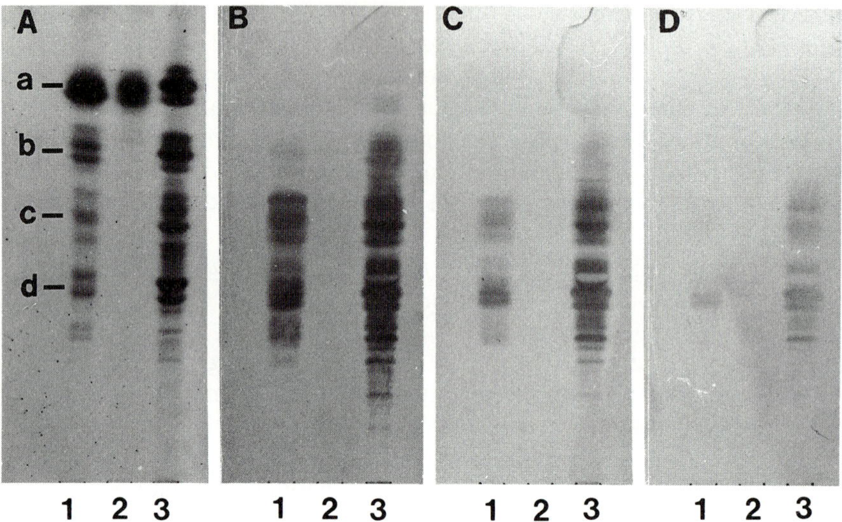

FIG. 2. Immunoreactivity of the neutral-fraction glycolipids from the laminated and germinal layers of *E. granulosus* hydatid cysts and *T. crassiceps* metacestodes with mouse normal serum, mouse *T. crassiceps*-infection serum, and infection serum-derived, monospecific, polyclonal antibody directed to *n*Ga_3Cer. Silica-gel-60 HPTLC plates (Merck, Darmstadt, Germany); running solvent was chloroform–methanol–0.2% aqueous $CaCl_2$ (60 : 35 : 8 v/v/v). Glycolipids were visualized either chemically with (A) orcinol–sulfuric acid or serologically (dilution 1 : 400) with (B) mouse infection serum, (C) mouse monospecific, polyclonal anti-*n*Ga_3Cer antibody, or (D) mouse normal serum. 1, *E. granulosus* germinal layer, 2, *E. granulosus* laminated layer; 3, *T. crassiceps* metacestodes. Letters a–d represent the number of constituent monosaccharides from 1 to 4, respectively.

very similar glycoconjugates, which is an interpretation for the elevated anti-Gal antibody titers of kinetoplast protozoan infections (Avila *et al.*, 1988a,b). In support of the former hypothesis, the immunodominant epitope of lipopolysaccharide O-chain specificity of various *Bacteroides fragilis* strains, a human colonic anaerobe, was the *β*6-linked D-galactose oligomer of the neogala-series (Weintraub *et al.*, 1985a,b; Lindberg *et al.*, 1990). The immunodominant epitope, *β*6-linked D-galactosyl disaccharide, was also present on the capsular polysaccharide of *Streptococcus agalactiae* (group B) type II bacteria (Jennings *et al.*, 1983). This molecular mimicry has been observed recently in the lipooligosaccharides of various human bacterial pathogens of mucosa of the genera *Haemophilus* and *Neisseria,* which expressed carbohydrate epitopes similar to those present on mammalian glycosphingolipids (Mandrell *et al.*, 1992).

E. Hypotheses of Glycocalyx Biological Function

Among the relevant, putative biological function(s) of the tegument and tegumental-bound glycocalyx of cestodes and trematodes the following may be categorized: evasion of the host immune response by sloughing-off of the tegument or structural segments; evasion of the host immune response resulting from inherent properties of the tegument; and stabilization of the tegument as an absorptive surface for nutrient uptake.

In order to evade the host's immune response, the tissue-migrating juvenile of the trematode *F. hepatica* appears to counter by sloughing-off the accumulating antibody–glycocalyx complex (Hanna and Trudgett, 1983) and secreting a replacement. This process is mediated by the soluble form of a M_r 28,000 parasite-derived protease secreted during the transformation of *S. mansoni* to the schistosomulum (as reviewed by Arnon, 1991). The relative resistance of the developing cestode, *H. diminuta,* to presumed immunological, but complement-independent, expulsion from the intestinal lumen of the mouse (Schmidt and Ruppel, 1988) is correlated with the asymmetric distribution of the tegument glycoconjugates. The scolex and strobilation zone are characterized by a dense glycocalyx sheath (Schmidt, 1988) that is virtually absent from the discarded strobila segment.

The previous hypothesis, that the tegumental adsorption of host blood-derived glycolipids enabled the helminth parasite to avoid the immune reaction by molecular mimicry, e.g., *S. mansoni* schistosomula (McLaren *et al.,* 1975; Goldring *et al.,* 1976), has proved untenable (as reviewed by Arnon, 1991). Other parasite-intrinsic evasion mechanism(s) must be involved to explain the extant binding of antibodies but resistance to complement-dependent lysis. Of pertinence to this question may be the finding of a tegument-located IgG (Fc_γ)-binding protein from the metacestode of *T. crassiceps* (Kalinna and McManus, 1992), whereby antibody binding may occur but be immunologically inoperative. Another tegument-dependent, immunological evasion maneuver may be exhibited by *S. mansoni* schistosomula, which display an acquired refractoriness to antibody-dependent, complement-induced lysis within 1 hour on *in vitro* incubation at 37°C. This schistosomular resistance to complement is believed to be due in part to a membrane form of the M_1 28,000 parasite-derived protease (glycoprotein) that is able to cleave the complement components C3, C3b, and C9 (as reviewed by Arnon, 1991).

Through evolutionary pressure of functional adaptation to the common environmental constraints of the vertebrate small intestine, the host's epithelial cell plasma membrane, e.g., that of rat (Breimer *et al.,* 1982), and the helminth parasite tegument membrane, e.g., that of *S. mansonoides*

(Singh *et al.*, 1987a,b), may have evolved the same or similar mechanism(s) of stabilization. These are thought to be ceramide monohexoside (galactose) as the dominant neutral-fraction glycosphingolipid, enhanced hydrogen bonding through increased sphingoid base hydroxylation and/or fatty acid components, and enhanced hydrophobic bonding through increased fatty acid chain-length.

III. Glycosphingolipids of the Insecta

A. Introduction: Glycosphingolipids of the Protostomia

Initial reports on the presence of glycolipids in the Invertebrata relied on thin-layer chromatographic identification: Central nervous system tissue of *Periplaneta americana* (Insecta), *Homarus* sp. (Crustacea), and *Helix pomatia* (Mollusca; Honegger and Freyvogel, 1963); nervous tissue of *P. americana* and *Astacus fluvialitis* (Crustacea; Reinišová and Michaleč, 1966); nervous tissue of *Aplysia kurodai* (Mollusca) and *Cambarus clarkii* (Crustacea; Komai *et al.*, 1973); as well as from a mosquito cell line (*Aedes albopictus;* Luukkonen *et al.*, 1973). Although ceramide-2-aminoethyl phosphonate had been chemically identified and its distribution restricted to the molluscan *A. kuroda* (Komai *et al.*, 1973), systematic structural analyses of protostomial glycosphingolipids have only been performed on members of the Platyhelminthes, Mollusca (freshwater and marine snails), and the Insecta (Diptera: flies).

The glycosphingolipids of *Hyriopsis schlegelii* can be taken as typical of freshwater mollusks in deriving the mollu- and xylomollu-carbohydrate series, corresponding to the core structures Manα3Manβ4Glc$\beta\langle$ and Manα3[Xylβ2]Manβ4Glc$\beta\langle$, respectively (Higashi and Hori, 1968; Hori *et al.*, 1977a; Sugita *et al.*, 1981, 1984a). One acidic fraction GSL component of this mollu-carbohydrate series, with a more complex chemical constitution, carries a terminal 4-*O*-methylglucuronic acid (Hori *et al.*, 1977b, 1981, 1983). The side-chain moiety, 2-aminoethylphosphate in a hexose 6-position, is also characteristic of freshwater bivalves. In contrast, the side-chain 2-aminoethylphosphonate in hexose 6-position is characteristic of various marine mollusk glycosphingolipids that do not belong to the mollu-series, but with a carbohydrate core sequence GalNAcα3[Galα2-]Galβ4Glc$\beta\langle$ [suggested designation: GalGp, galagastro(pod)-series] (Araki *et al.*, 1986, 1987a,b). Pyruvylation of the terminal galactose was found in glycosphingolipids from *A. kurodai* (Araki *et al.*, 1989).

A systematic survey of larval glycosphingolipids of the dipteran *Lucilia caesar* has shown their arthro-carbohydrate series (Ap) core structure to be derived from mactose: GlcNAcβ3Manβ4Glcβ1Cer (Sugita *et al.*, 1982a). Linear elongation of the oligosaccharide chain was achieved by hexoses and hexosamines of the neutral-type glycosphingolipids (Sugita *et al.*, 1982b, 1990); hexoses and hexosamines on the 2-amino ethylphosphate-containing Ap_3Cer of the zwitterionic-type (Itonori *et al.*, 1991); and embellishment with glucuronic acid (with/without $III^6$2-aminoethylphosphate) in the acidic-type glycosphingolipids (Sugita *et al.*, 1988a, 1989).

Not all the glycolipids of the insects have proved to be sphingolipids, as exhibited by gentiobiose-based glycoglycerolipids, i.e., blaberosides, of the cockroach *Blaberus colloseus* (Stoskopf *et al.*, 1989). As for the non-insect arthropods (and insects), GlcCer has been chemically identified in nerve tissue (Shimomura *et al.*, 1983; Okamura *et al.*, 1985, 1986).

B. Glycosphingolipids of *Calliphora vicina* Pupae

Calliphora vicina was chosen as an example of dipteran insects, because they are among the most highly evolved of the invertebrata. The glycosphingolipid chemical structures from *C. vicina* pupae are closely paralleled in their composition by those of the phylogenetically closely related *L. caesar,* despite being derived from different dipteran species and different developmental stages (Table II).

1. *Ceramide Moiety*

The ceramide moiety has been designated dipteran ceramide because of the characteristic fatty acid and long-chain base composition. The fatty acids are predominantly straight-chained, even-numbered, and saturated: C18 : 0 (stearic acid), C20 : 0 (arachidic acid; dominant species), and C22 : 0 (behenic acid). The long-chain bases are dominated by C14 : 1 (tetradecasphing-4-enine) and C16 : 1 (hexadecasphing-4-enine) (Dennis *et al.*, 1985a). From consideration of the insect glycosphingolipid structures (Table II), they form a biogenetic series, whereby synthesis occurs on a dipteran ceramide moiety by stepwise addition of the respective monosaccharide residue.

2. *Carbohydrate Moiety*

The arthro-carbohydrate series is defined by its triaose oligosaccharide chain of characteristic sequence, linkage position, and anomeric configuration (GlcNAcβ3Manβ4Glcβ(). Its longest species, arthrononaosylceramide (Ap_9Cer, with nine monosaccharide residues), represents what is perhaps the longest unbranched, nonrepetitive carbohydrate structure

Table II

CHEMICAL STRUCTURES OF ARTHRO-SERIES GLYCOSPHINGOLIPIDS OF PUPAL *Calliphora vicina* AND LARVAL *Lucilia caesar*

Laboratory designation	Shorthand designation	Chemical structure	Refs[a]
N, Neutral			
N1	GlcCer	GlcβCer	1,2
N2	MacCer	Manβ4GlcβCer	1,2
N3	Ap_3Cer	GlcNAcβ3Manβ4GlcβCer	1,2
N4	Ap_4Cer	GalNAcβ4GlcNAcβ3Manβ4GlcβCer	1,3
N5a	Ap_5Cer	GalNAcα4GalNAcβ4GlcNAcβ3Manβ4GlcβCer	3,4
N5b	IV^3Galα-Ap_4Cer	Galα3GalNAcβ4GlcNAcβ3Manβ4GlcβCer	3
N5c	IV^3Galβ-Ap_4Cer	Galβ3GalNAcβ4GlcNAcβ3Manβ4GlcβCer	5
N6	Ap_6Cer	Galβ3GalNAcα4GalNAcβ4GlcNAcβ3Manβ4GlcβCer	3,4
N7	Ap_7Cer	GlcNAcβ3Galβ3GalNAcα4GalNAcβ4GlcNAcβ3Manβ4GlcβCer	3,4
N8	Ap_8Cer	GalNAcβ3GlcNAcβ3Galβ3GalNAcα4GalNAcβ4GlcNAcβ3Manβ4GlcβCer	6
N9	Ap_9Cer	Galβ3GalNAcβ3GlcNAcβ3Galβ3GalNAcα4GalNAcβ4GlcNAcβ3Manβ4GlcβCer	6
Nz, Neutral-zwitterionic			
Nz3	III^6*P* Etn-Ap_3Cer	(*P* Etn-6)GlcNAcβ3Manβ4GlcβCer	7,8
Nz4	III^6*P* Etn-Ap_4Cer	GalNAcβ4(*P* Etn-6)GlcNAcβ3Manβ4GlcβCer	7–9
Nz5a	III^6*P* Etn-Ap_5Cer	GalNAcα4GalNAcβ4(*P* Etn-6)GlcNAcβ3Manβ4GlcβCer	7–9
Nz5c	IV^3Galβ-,III^6*P* Etn-Ap_4Cer	Galβ3GalNAcβ4(*P* Etn-6)GlcNAcβ3Manβ4GlcβCer	8
Nz6	III^6*P* Etn-Ap_6Cer	Galβ3GalNAcα4GalNAcβ4(*P* Etn-6)GlcNAcβ3Manβ4GlcβCer	7,8
Nz7	III^6*P* Etn-Ap_7Cer	GlcNAcβ3Galβ3GalNAcα4GalNAcβ4(*P* Etn-6)GlcNAcβ3Manβ4GlcβCer	7,8
A, Acidic			
A5c	IV^3GlcAβ3Galβ-Ap_4Cer	GlcAβ3Galβ3GalNAcβ4GlcNAcβ3Manβ4GlcβCer	5
A6	VI^3GlcAβ-AP_6Cer	GlcAβ3Galβ3GalNAcα4GalNAcβ4GlcNAcβ3Manβ4GlcβCer	10
Az, Acidic-zwitterionic			
Az5c	IV^3GlcAβ3Galβ-,III^6 *P* Etn-Ap_4Cer	GlcAβ3Galβ3GalNAcβ4(*P* Etn-6)GlcNAcβ3Manβ4GlcβCer	5
Az6	VI^3GlcAβ-,III^6 *P* Etn-Ap_6Cer	GlcAβ3Galβ3GalNAcα4GalNAcβ4(*P* Etn-6)GlcNAcβ3Manβ4GlcβCer	10

[a] References: 1, Sugita *et al.*, 1982a; 2, Dennis *et al.*, 1985a; 3, Dennis *et al.*, 1985b; 4, Sugita *et al.*, 1982b; 5, Weske *et al.*, 1990; 6, Sugita *et al.*, 1990; 7, Itonori *et al.*, 1991; 8, Helling *et al.*, 1991; 9, Dabrowski *et al.*, 1990; 10, Sugita *et al.*, 1989.

detected so far: Galβ3GalNAcβ3GlcNAcβ3Galβ3GalNAcα4GalNAcβ4-GlcNAcβ3Manβ4Glcβ⟨ (Sugita *et al.*, 1990). The Ap-series glycosphingolipids are further classified into three types according to attributes of their carbohydrate moiety: neutral species (N1–7), containing from one to seven neutral monosaccharides in the arthro-series sequence (Dennis *et al.*, 1985a,b); neutral-zwitterionic species (Nz3–7), consisting of three to seven neutral sugars of the arthro-oligosaccharide chain and a *N*-acetylglucosamine-bound phosphoethanolamine at the third monosaccharide residue (III^6*P*Etn–Ap$_3$Cer; Dabrowski *et al.*, 1990; Helling *et al.*, 1991); and acidic species (A), distinguished by a terminal glucuronic acid residue of the arthro-series sequence and the absence or presence (Az) of a phosphoethanolamine side chain (Dennis *et al.*, 1987; Weske *et al.*, 1990). The glucuronic acid-containing glycosphingolipids have been given the generic name of arthrosides in implying putative chemical and functional parallelism to the sialic acid-containing glycosphingolipids (gangliosides) of the vertebrates. The chemical properties of the phosphoethanolamine moiety were investigated following cleavage with hydrofluoric acid (Fischer *et al.*, 1973; Itasaka and Hori, 1979). N/Nz as well as A/Az glycosphingolipids can be readily resolved by two-dimensional HPTLC and visualized by carbohydrate-detecting spray reagents. Nz and Az components may be additionally identified by their positive reaction with ninhydrin (Fig. 3).

The carbohydrate moiety of the three insect glycosphingolipid types is defined by the presence of mannose, glucuronic acid, phosphoethanolamine side chain, and high *N*-acetylhexosamine content. In addition, the monosaccharide sequence of the arthro-series oligosaccharide chain (Table II) displays a partial structural coincidence with various vertebrate glycosphingolipids, as exemplified by the similarity of the "iso-Forssman"[2] epitope, GalNAcα4GalNAcβ⟨, of Ap$_5$Cer and III^6-*P*Etn–Ap$_5$Cer to that of terminal GalNAcα3GalNAcβ⟨ of Forssman glycolipids; Ap$_3$Cer to that of Lc$_3$Cer; Ap$_4$Cer to that of nLc$_4$Cer; Ap$_5$Cer with the human erythrocyte P1 antigen, IV^{4}Galα–nLc$_4$Cer; IV^{3}Galα–Ap$_4$Cer (N5b) with the rabbit erythrocyte B-active glycosphingolipid, IV^3-Galα–nLc$_4$Cer; as well as IV^{3}Galβ–Ap$_4$Cer (N5c) to the human erythrocyte glycolipid, IV^{3}Galβ–nLc$_4$Cer. This reciprocity is to some degree the consequence of extension of the Ap$_4$Cer core oligosaccharide into three isomeric structural sequences of the pentaosylceramide, IV^{4}GalNAcα–Ap$_4$Cer (N5a, dominant component), IV^{3}Galα–Ap$_4$Cer, and IV^3-Galβ–Ap$_4$Cer. These correlations are indicative of the structural conserva-

[2] Because of its strong heterophilic immunogenicity and chemical isomerism to the "Forssman" epitope GalNAcα3GalNAcβ⟨, the Ap$_5$Cer terminus GalNAcα4GalNAcβ< may be designated as "iso-Forssman".

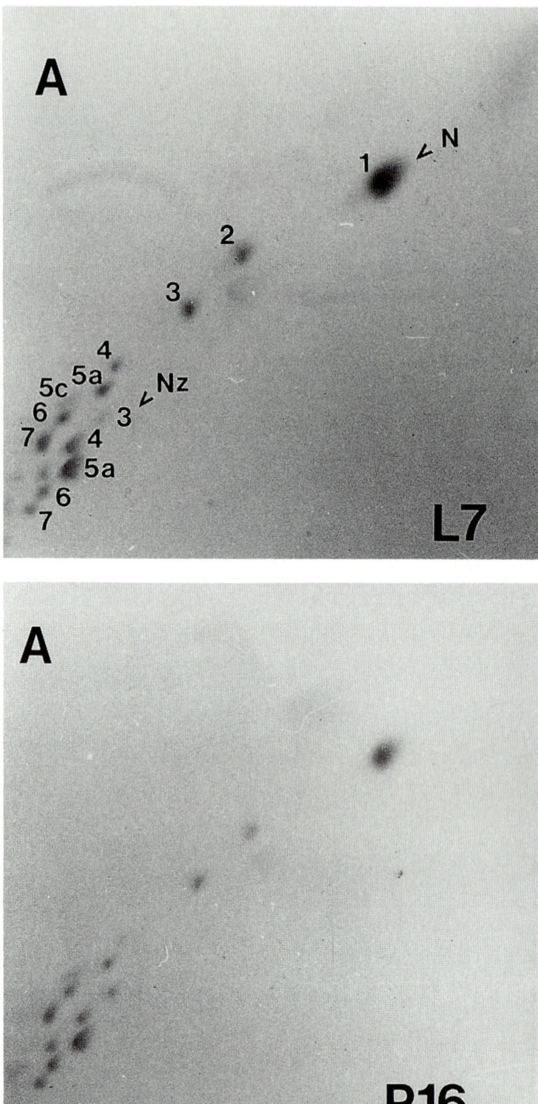

FIG. 3. Comparative two-dimensional HPTLC of the neutral- (A) and the acidic- (B) fraction GSL of *C. vicina* 7-day larvae (L7) and 16-day pupae (P16). Silica gel-60 HPTLC plates (Merck, Darmstadt, Germany); running solvents for (A) first dimension, chloroform–methanol–0.2% aqueous $CaCl_2$ (50 : 40 : 10 v/v/v); second dimension, chloroform–methanol–2.5% aqueous ammonia (50 : 40 : 10 v/v/v); for (B), first dimension, 2-propanol–0.2% aqueous $CaCl_2$ (7 : 3, v/v); second dimension, *n*-butanol–acetic acid–water (2 : 1 : 1, v/v/v). Glycolipids were visualized with orcinol–sulfuric acid spray reagent.

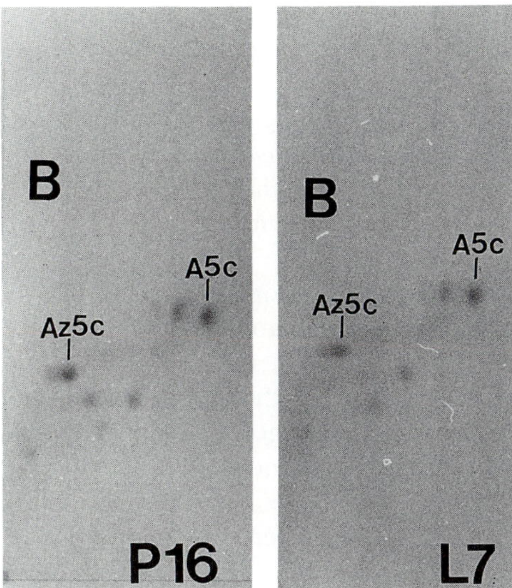

FIG. 3. Continued

tion of "ancestral" glycosphingolipid–oligosaccharide core structures, such that minor mutation-induced alterations in transferase monosaccharide substrate specificity at position II, mannose/galactose, and position IV, *N*-acetylgalactosamine/galactose, of insects and vertebrates, respectively, may account for the differences between these two phylogenetically distant groups. The significance of these correlations must await the availability of data as to the information content of these glycosphingolipids.

C. Organ Distribution of *Calliphora vicina* Larvae Glycosphingolipids

A result of structural analyses of glycosphingolipids in selected organs of invertebrates is the accumulation of indirect data as to their putative biological activities. Their localization in molluscan organs has indicated (1) the proposed stabilization of absorptive plasma membrane surfaces by oligogalactosylceramides in the rat intestinal, epithelial cell (Breimer *et al.,* 1982), and cestode tegument (Nishimura *et al.,* 1991; Dennis *et al.,* 1992) as relevant to the significance of viscerally located neogala-series glycosphingolipids of *T. cornutus* (Matsubara and Hayashi, 1981); (2) phosphonoglycosphingolipids of *A. kurodai* skin and central and peripheral nervous systems (Araki *et al.,* 1989) as functional entity replacements for the analogous gangliosides of vertebrates; and (3) the gametic restriction of the acidic component, $V^4GlcA4Me\beta$-,$V^3GalNAc3Me\alpha$-XMl_6Cer, to the spermatozoa of *H. schlegelii* (Hori *et al.*, 1983) as proposed involvement in sperm–egg recognition. For arthropodal organs, the glycoglycerolipids of *B. colosseus* have been isolated from the head (Stoskopf *et al.*, 1989), while the glucosylceramide of the crustacean *P. aztecus* showed a neural tissue distribution correlated with nerve conduction (Shimomura *et al.*, 1983).

To accumulate indirect information as to their putative biological activities in the Insecta, the glycosphingolipid–larval organ distribution of *C. vicina* was studied (Sickmann *et al.*, 1992) by documenting and correlating stadium-associated differences with their possible parallel evolution in vertebrate-equivalents, i.e., analogous molecules having evolved under similar environmental constraints and performing the same tasks. Analysis of organ-derived neutral glycolipids demonstrated qualitative similarities but quantitative dissimilarities in terms of total glycolipid–carbohydrate content and component glycosphingolipid distribution. Acidic glycolipids exhibited pronounced qualitative and quantitative organ-related variation. The neutral glycolipids were most characteristic of the fat body, the insect equivalent of the vertebrate liver. The zwitterionic glycolipids were most representative of the central nervous system and imaginal discs, i.e., organs persisting throughout metamorphosis. The acidic glycolipids were most typical of the imaginal discs, a reflection of their assumed role in regulated cell reorganization during pupal development.

D. Immunological, Immunochemical, and Immunohistochemical Investigation of Protostomial Glycoconjugates

The embryogenesis of all invertebrate organ systems has demonstrated the cardinal importance of cell–cell recognition and specific cell–cell interaction in the morphogenetic processes of development, differentiation, maturation, and regeneration. The cellular mechanisms must be highly conserved, as exemplified by the insect central nervous system, with its common embryonic plan that is extrapolated throughout the Arthropoda and the philosophy of the "identified" neuron, as characterized by location, shape, electrical activity, and axoneurite pattern (Goodman and Spitzer, 1979). This infers that differentiation of the identified neuron involves conserved patterns of cell–cell interaction, growth cone projection, and specific fasciculation (Goodman *et al.*, 1984; Thomas *et al.*, 1984). The working hypothesis, to explain the molecular mechanisms involved, is that the specificity of recognition throughout ontogeny is the consequence of sequential expression, both temporally and spatially, of different functional molecules at the plasma membrane interface (Goodman and Bastiani, 1984). The identification of these putative molecules as developmental/functional markers can most readily be accomplished by the generation of monoclonal antibodies or monospecific polyclonal antibodies. The antigenic determinants of these developmental antigens are largely unknown, but, because of their cell surface location, many have been found to be glycoconjugates, i.e., protein- and lipid-bound carbohydrate chains. An alternative approach is to generate monoclonal antibodies or monospecific, polyclonal antibodies against components of suspected morphogenetic activity and of known structure, and use them as immunochemical reagents.

Monoclonal antibodies have been applied to reveal the antigens of unknown antigenic determinant(s) in subsets of leech (*Haemopis marmorata*) sensory and motor neurons (Zipser and McKay, 1981); to determine the degree of molecular conservation in the phylogeny of the central nervous system between *Drosophila melanogaster* and human brains (Miller and Benzer, 1983) and categorize the cells according to four types of specificity, i.e., tissue, regional, cell-type and neuron-subset specificities (Fujita *et al.*, 1982); and to investigate the working hypothesis of specific cell–cell recognition during neuronal regeneration to reform the innervation pattern of *P. americana* axotomized leg muscles (Denburg *et al.*, 1986).

Monoclonal antibodies have identified a number of glycoprotein antigens as presumed effectors for cell–cell recognition events in the developing insect central nervous system. The sequential appearance of antigens in the neuronal differentiation of the *D. melanogaster* retina has been correlated with pattern formation in the eye imaginal disc (Zipursky *et al.*, 1984; Tomlinson, 1988). Fasciclins I–III of *Schistocerca nitens* and *D. melanogaster* nervous systems are expressed on axon fascicle subsets and in a spatiotemporal pattern during development which is consistent with the "labeled pathway hypothesis," i.e., axon fascicles in the embryonic neuropil are differentially labeled in terms of recognition molecules utilized by growth cones for selective fasciculation and guidance clues (Bastiani *et al.*, 1987; Patel *et al.*, 1987; Zinn *et al.*, 1988). Antibodies against horseradish peroxidase recognize a neuronal and developmental antigen in the nervous system of primitive (orthopteran, *S. nitens*) and advanced (dipteran, *D. melanogaster*) insects (Jan and Jan, 1982; Snow *et al.*, 1987). The neural tissue-specific carbohydrate is expressed on a complex set of developmentally regulated proteins (F. Katz *et al.*, 1988). The antigen is expressed on all growth cones as well as selected "guidepost" cells believed to assist in pioneer neuronal pathway formation (Murray *et al.*, 1984). The carbohydrate epitope is present on pineapple stem bromelain as $XylMan_2FucGlcNAc_2$ (Ishihara *et al.*, 1979). This xylose-containing glycan antigenic determinant has been highly conserved, as indicated by its presence on glycoproteins of plants, mollusks, and insects (Faye and Chrispeels, 1988).

Immunochemical and immunohistochemical investigations of molluscan glyco- and phosphonoglycosphingolipids have been initiated to determine the extent of taxonomic order homology, as well as tissue localization as a first step in the study of biological function. MI_4Cer- and cross-reacting XMI_5Cer-immunoreactivity was restricted in its taxonomic order distribution in the Mollusca (Sugita *et al.*, 1988b). The antigenic determinant GlcA4Meβ4Fucα⟨, of the *H. schlegelii* acidic glycosphingolipid V^4GlcA4Meβ-,V^3GalNAc3Meα-XMI_6Cer, is located on the spermatozoan plasma membrane and is again restricted in its taxonomic order distribution (Sugita *et al.*, 1984b). In the central nervous system of *A. kurodai*, phosphonoglycosphingolipid reactivity was restricted to the ganglionic neuropil and glia (Abe *et al.*, 1985, 1988).

Immunochemical and immunohistochemical analyses have been performed for the detection of lipid-bound carbohydrate epitopes in the Arthropoda. Mactosylceramide, Manβ4Glcβ1Cer, isolated from spermatozoa of the mollusk *H. schlegelii* was identified and characterized in two crustacean species (Itonori *et al.*, 1992). In *P. americana*, an embryonic, neuronal developmental stage-specific antigen was reexpressed in the adult as a neuronal growth-associated antigen and was correlated with axonal regeneration for specific reformation of leg muscle innervation following axotomy (Denburg, 1989). Preliminary data suggest the molecule(s) mediating the cell–cell recognition process is a glycolipid(s) (Denburg *et al.*, 1989).

E. Immunological, Immunochemical, and Immunohistochemical Investigation of *Calliphora vicina* Glycosphingolipids

1. *Serology*

Following determination of terminal glucuronic acid in the acidic glycosphingolipid fraction of *C. vicina* (Dennis *et al.*, 1987), and in the search for a useful immunological reagent and adjunct, *C. vicina* glycolipids were screened for immunoreactivity with terminal sulfated, glucuronic acid-containing, epitope-specific, monclonal antibodies. The antigenic determinant is represented by two acidic glycosphingolipids of the mammalian peripheral nervous system, i.e., $IV^3GlcA3SO_3\beta$–nLc_4Cer and $VI^3GlcA3SO_3\beta$-nLc_6Cer (Chou *et al.*, 1986, 1987; Ariga *et al.*, 1987).

Insect acidic glycolipids that contained no sulfate ester were only recognised by L2.334 and human IgM M-protein (LT), and not by other L2 or HNK-1 monoclonal antibodies. This result was consistent with the findings of Ilyas *et al.* (1986) and Noronha *et al.* (1986): only L2.334 and certain human IgM M-proteins do not have an absolute sulfate requirement for reactivity toward this glucuronic acid-containing moiety. The antigenic determinant of insect acidic glycolipids was assumed to be a terminal nonsulfated, glucuronic acid-containing moiety (Dennis *et al.*, 1988). As *C. vicina* acidic glycolipids reacted with a diseased-state IgM M-protein, equivalent to a monoclonal antibody, it was relevant to determine if normal human serum contained antibodies directed against insect lipid-bound carbohydrate epitopes.

Enzyme-linked immunosorbent assay (ELISA) of all human sera exhibited the presence of IgG- and IgM-specific antibodies against insect total neutral–zwitterionic (average titer 1 : 1000–2000) and acidic (average titer 1 : 500–1000) glycolipid fractions, respectively (Dabrowski *et al.*, 1988; Wiegandt, 1992). This titer was considerably elevated in IgM M-protein anti-acidic glycolipid fraction reactivity. Closer inspection revealed that these heterophile antibodies (IgM specificity) reacted with components of the three types of insect glycosphingolipid, classified as neutral, neutral–zwitterionic, and acidic (Table II). Antibody activities were segregated by sequential immunoaffinity chromatography into three populations of differing epitope specificity. Population I reacted with epitopes on the neutral oligosaccharide chain backbone common to all three glycolipid types, the main antigenic determinant containing a terminal HexNAc residue. Immunoreactivities were separable into at least four subpopulations of differing epitope specificity. Population II identified phosphoethanolamine-containing epitopes of neutral–zwitterionic and acidic–zwitterionic glycolipids. Of the two subpopulations of antibody present, the majority required the free amino group of phosphoethanolamine for reactivity. Population III detected terminal β-glucuronic acid-containing epitopes that were restricted to the acidic glycolipids (Nores *et al.*, 1991).

Analogous to the specificity of antigen–antibody interaction and antigenic determinant recognition is the ligand–receptor system. The hypothesis to be tested was that a glycoconjugate(s) on the plasma membrane of susceptible insect cells acts as a receptor and/or modulator for specific interaction with the protoxin/activated toxin ligand of the biological insecticide δ-endotoxin of *Bacillus thuringiensis* var. *kurstaki* (Knowles *et al.*, 1984; Knowles and Ellar, 1986). The thin-layer chromatographic technique signified $IV^3Gal\alpha$–Ap_4Cer (Dennis *et al.*, 1986) and $IV^3GlcA\beta3Gal\beta$–Ap_4Cer (Weske *et al.*, 1990) as the main protoxin/activated toxin-binding neutral and acidic glycosphingolipids, respectively. Because two glycosphingolipids of different terminal structures were specified, this would suggest that binding is either the result of a common subterminal sequence, as proposed for the attachment of bacteria to glycosphingolipid receptor sites (Karlsson, 1989), or more than one molecular determinant is required as a functional toxin-binding site of relevant insect plasma membrane glycoproteins.

2. *Immunochemistry*

As potential immunochemical and immunohistochemical instruments, monoclonal antibodies were generated against *C. vicina* lipid-bound carbohydrate epitopes by immunization of Balb/c mice with pretreated, bacterial carrier *S. minnesota* R595-bound total neutral/neutral–zwitterionic- and acidic-fraction glycolipids, respectively [at 1 : 1 (w/w) with 100 μg of glycolipid/dose; Galanos *et al.,* 1971]. Three cloned, hybridoma cell lines were established that secreted the monoclonal antibodies CAF-I (from an acidic glycolipid-stimulated spleen cell fusion), and CNF-I and CNF-III (from neutral/neutral–zwitterionic glycolipid-stimulated spleen cell fusions). CAF-I was an IgG3 isotype. The acidic glycolipid-bound carbohydrate epitope of GlcAβ3Galβ⟨ was exemplified by IV^{3}GlcAβ3-Galβ-Ap$_4$Cer and IV^{3}GlcAβ3Galβ–,III^6*P*Etn–Ap$_4$Cer (Table II; Keller *et al.,* 1990). Under the conditions used, Western blot analysis failed to detect specifically glycoproteins expressing the CAF-I epitope. It is worth noting that component A6 exhibited only minimal binding to the CAF-I antibody. CNF-I was an IgG3 isotype. The neutral glycolipid-bound "iso-Forssman" epitope of GalNAcα4GalNAcβ⟨ was exemplified by Ap$_5$Cer and III^6*P*Etn–Ap$_5$Cer (Table II; Keller *et al.,* 1993). Under the conditions used, Western blot analysis failed to detect specifically *C. vicina*-derived glycoproteins expressing the CNF-I epitope. CNF-III was of the IgM class. The neutral–zwitterionic glycolipid-bound carbohydrate epitope of GalNAcβ4(*P*Etn–6)GlcNAcβ3Manβ⟨ was exemplified by III^6*P*Etn–Ap$_4$Cer and IV^{3}Galβ–,III^6*P*Etn–Ap$_4$Cer (Table II; Weske *et al.,* 1993). Western blot analysis demonstrated the determinant CNF-III to be present on a number of *C. vicina*-derived glycoproteins during larval (88 kDa major, and 52 kDa minor) and pupal (88, 66, and 52 kDa major, and 33 kDa minor) development, but it was not expressed on proteins in the adult. This would indicate that protein-bound CNF-III can be considered as a stadium-associated antigen.

3. *Immunohistochemistry*

In interpreting the result of the immunohistochemical staining of insect tissue during development, the following criterion was applied: spatially and temporally altered patterns of epitope expression that were correlatable with ephemeral morphogenetic events, e.g., cell–cell recognition and interaction occurring in cells containing the antigen, were regarded as being of functional relevance to ontogeny.

The main expression of the CAF-I epitope was in the brain and imaginal discs of *D. melanogaster* late third-instar larvae and retinular cells of the adult compound eye retina (Keller *et al.*, 1990). The central nervous systems of *C. vicina* and the coleopteran *Tenebrio molitor* displayed a coordinated temporal and spatial regulation of CAF-I expression during metamorphosis (Breidbach *et al.*, 1990). In *T. molitor,* this sequence of immunoreactivity in the midbrain central complex glomeruli and ventral nerve cord has been defined as glia-dependent (Breidbach *et al.*, 1992; Wegerhoff and Breidbach, 1992). These cells ensheath the glomeruli of the developing central complex's fan-shaped body and serially homologous neuronal somata in the cell body layer of thoracic ganglia, as well as in the ventral associative neuropil (Meyer *et al.*, 1987; Carlson and Saint Marie, 1990; Tolberg and Oland, 1990). Reexpression of the CAF-I epitope can be induced on nerve-extirpation, with immunoreactivity demonstrable in the severed sensory nerve neurons, ventral-associative neuropil, cellular cortex, and nerve stump (Breidbach *et al.*, 1992). The cellular distribution of CAF-I expression in the central nervous system of an insect bears a superficial resemblance to that for the phosphonoglycosphingolipids of the mollusk *A. kurodai* (Abe *et al.*, 1985, 1988) and may indicate the involvement of glia in the resultant immunostaining pattern. The reexpression of the CAF-I epitope following axotomy in *T. molitor* could correspond to the reappearance of the growth-associated antigen of adult *P. americana* following the postulated involvement of cell–cell recognition and interaction in axonal regeneration (Denburg *et al.*, 1989).

The functional carbohydrate epitopes L2/HNK-1 and L3, characterizing two overlapping glycoprotein families of vertebrate neural cell adhesion molecules (Kruse *et al.*, 1984; Kücherer *et al.*, 1987; Fahrig *et al.*, 1990), have been immunohistochemically detected in the late third-instar larvae and imagoes of *C. vicina* and *D. melanogaster* (Dennis *et al.*, 1991). The L2/HNK-1 cell surface-associated and L3 extracellular matrix-localized distribution of immunoreactivity was not restricted to neural structures. The functional significance of these carbohydrate epitopes has not been investigated in the invertebrates. As the L2/HNK-1 carbohydrate of vertebrates is itself functionally active in cell–cell interaction (Künemund *et al.*, 1988), the reason for phylogenetic conservation could be due to provision of the same or similar vital functions in the developing insect.

IV. Conclusions

There are strong indications from experimental data for the involvement of vertebrate glycosphingolipids in morphogenetic events of ontogeny, e.g., cell–cell recognition and interaction, modulation of cell growth, and induction of differentiation. Because comparable processes occur in invertebrate organisms, as a first approach the most reasonable assumption would be that the regulatory mechanisms are comparable and also involve glycosphingolipids. However, prior to the investigation of biological and/or functional activities of invertebrate glycosphingolipids in relevant, developmental systems, there is an essential requirement for the systematic analysis of their chemical structures, immunochemical properties, and immunohistochemical localization. Information on the monosaccharide residue sequences of the sphingolipid-bound oligosaccharide chains is the basis for all future chemical and immunological identification. Their immunological properties, as determined by the specificity of generated monoclonal antibodies, include antigenic determinant characterization, occurrence of protein- and/or sphingolipid-bound carbohydrate epitopes, and modulation of reactivity in functional assays of development. The immunohistochemical localization of the tissue and cellular distribution is an initial step in the study of function, to correlate differential, spatial, and temporal patterns of antigenic determinant(s) expression in cells undergoing particular morphogenetic events of development. The analysis of these properties of invertebrate glycosphingolipids has been initiated in a cestode platyhelminth (*T. crassiceps*) and dipteran insect (*C. vicina*).

Cestode neutral-fraction glycosphingolipids are specified by the neogala-series oligosaccharide chain (nGa_4Cer core structure: Galβ6Galβ6Galβ6-Galβ1Cer). Although an additional tetraosylceramide component has been isolated (Galα4Galβ6Galβ6Galβ1Cer), elongation may yield variants of different monosaccharide sequences, and new structures probably exist in the acidic glycolipid fraction. The cestode ceramide moiety is distinguished by particularly long-chained, saturated fatty acids and saturated, long-chain bases with or without supplementary hydroxylation. Immunochemically, the neutral-fraction glycolipid expressed a major carbohydrate epitope (Galβ6Galβ6Galβ⟨), as determined by an infection serum-derived, immunoaffinity-purified, monospecific polyclonal antibody and a minor antigenic determinant (Galα4Galβ6Galβ⟨), which is apparently nonimmunogenic in the mouse. Serologically, the lipid-bound epitope(s) displayed a more prevalent taxonomic distribution in the Cestoda than the species under study. The working hypothesis to elucidate the occurrence in control sera of anti-cestode neutral-fraction glycolipid antibody activity is of an immune response to gastrointestinal and/or pulmonary tract-located, gram-negative bacteria manifesting the same or very similar epitopes on lipopolysaccharides or capsular polysaccharides of the cell envelope. Immunohistochemical localization of the cestode-derived, neogala-series carbohydrate epitope will be an initial strategy in the functional analysis of tapeworm glycosphingolipid interaction at the host–parasite interface.

Insect glycosphingolipids are typified by the arthro-series oligosaccharide chain (Ap_4Cer core structure: GalNAcβ4GlcNAcβ3Manβ4Glc-β1Cer), which occurs as neutral (no additional substituents or side chains), neutral–zwitterionic (GlcNAc-bound phosphoethanolamine side chain), and acidic/acidic–zwitterionic (terminal glucuronic acid residue with or without GlcNAc-bound phosphoethanolamine side chain) variants. The dipteran ceramide moiety consists of a C20 : 0 (arachidic acid) fatty acid and a C14 : 1 (tetradecasphing-4-enine) long-chain base. Organ distribution of the glycosphingolipids was asymmetric, whereby various larval organs were defined by particular proportions of components of the three glycolipid variants. Neutral glycolipids were most characteristic of the fat body, neutral–zwitterionic glycolipids most representative of the central nervous system and imaginal discs (persist throughout metamorphosis), and acidic glycolipids most typified the imaginal discs (undergo regulated cell reorganization during pupal development). Immunochemically, antigenic determinants of the three arthro-series variants have been defined by generation of the anti-insect oligosaccharide monoclonal antibodies CNF-I (GalNAcα4GalNAcβ⟨), CNF-III (GalNAcβ4[PEtn–6]GlcNAcβ3Manβ⟨), and CAF-I (GlcAβ3Galβ⟨), respectively. Serologically, in human blood three populations of circulating heterophile antibodies were detected that were directed against the corresponding carbohydrate variants of insect glycosphingolipids. Immunohistochemical visualization of the CAF-I epitope in immunoreactive larval and adult organs demonstrated an apparent plasma membrane-restricted pattern of expression. In the two holometabolous insects investigated, CAF-I reactivity during metamorphosis of the central nervous system displayed different but coordinated regulation of spatial and temporal patterns. In the midbrain and ventral nerve cord, this demarcated glia tissue ensheathed cellular and neuropilar structures. The induction of CAF-I epitope reexpression on nerve extirpation agrees with the postulated cell–cell recognition and interaction events on initiation of axonal regeneration. From the accumulated descriptive data on structure, immunochemistry, and immunohistochemistry, it is evident that systems of neural tissue and imaginal disc origin should form the basis for functional assays on insect glycosphingolipid involvement in the morphogenetic processes of ontogeny.

References

Abe, S., Kumanishi, T., Araki, S., and Satake, M. (1985). *Brain Res.* **327,** 259–267.

Abe, S., Watanabe, Y., Araki, S., Kumanishi, T., and Satake, M. (1988). *J. Biochem.* (*Tokyo*) **104,** 220–226.

Araki, S., Satake, M., Ando, S., Hayashi, A., and Fujii, N. (1986). *J. Biol. Chem.* **261,** 5138–5144.

Araki, S., Abe, S., Ando, S., Fujii, N., and Satake, M. (1987a). *J. Biochem.* (*Tokyo*) **101,** 145–152.

Araki, S., Abe, S., Odani, S., Ando, S., Fujii, N., and Satake, M. (1987b). *J. Biol. Chem.* **262,** 14141–14145.

Araki, S., Abe, S., Ando, S., Kon, K., Fujiwara, N., and Satake, M. (1989). *J. Biol. Chem.* **264,** 19922–19927.

Ariga, T., Kohriyama, T., Freddo, L., Latov, N., Saito, M., Kon, K., Ando, S., Suzuki, M., Hemling, M. E., Rinehart, K. L., Kusunoki, S., and Yu, R. K. (1987). *J. Biol. Chem.* **262,** 848–853.

Arnon, R. (1991). *Vaccine* **9,** 379–394.

Avila, J. L., and Rojas, M. (1990). *Am. J. Trop. Med. Hyg.* **43,** 52–60.

Avila, J. L., Rojas, M., and Towbin, H. (1988a). *J. Clin. Microbiol.* **26,** 126–132.

Avila, J. L., Rojas, M., and Garcia, L. (1988b). *J. Clin. Microbiol.* **26,** 1842–1847.

Bastiani, M. J., Harrelson, A. L., Snow, P. M., and Goodman, S. C. (1987). *Cell* (*Cambridge, Mass.*) **48,** 745–755.

Baumeister, S., Dennis, R. D., Kunz, J., Wiegandt, H., and Geyer, E. (1992). *Mol. Biochem. Parasitol.* **53,** 53–62.

Baumeister, S., Dennis, R. D., Geyer, R., Wiegandt, H., and Geyer, E. (1993). In preparation.

Breidbach, O., Dennis, R. D., Keller, M., and Wiegandt, H. (1990). *Neurosci. Lett.* **109,** 265–270.

Breidbach, O., Dennis, R. D., Marx, J., Görlach, C., Wiegandt, H., and Wegerhoff, R. (1992). *Neurosci. Lett* **147,** 5–8.

Breimer, M. E., Hansson, G. C., Karlsson, K. A., and Leffler, H. (1982). *J. Biol. Chem.* **257,** 557–568.

Capron, A., Dessaint, J. P., Capron, M., Ouma, J. H., and Butterworth, A. E. (1987). *Science* **238,** 1065–1072.

Carlson, S. D., and Saint Marie, R. L. (1990). *Annu. Rev. Entomol.* **35,** 597–621.

Chatterjee, D., Aspinall, G. O., and Brennan, P. J. (1987). *J. Biol. Chem.* **262,** 3528–3533.

Chou, D. K. H., Ilyas, A. A., Evans, J. E., Costello, C., Quarles, R. H., and Jungalwala, F. B. (1986). *J. Biol. Chem.* **261,** 11717–11725.

Chou, D. K. H., Schwarting, G. A., Evans, J. E., and Jungalwala, F. B. (1987). *J. Neurochem.* **49,** 865–873.

Curatolo, W. (1987). *Biochim. Biophys. Acta* **906,** 137–160.

Dabrowski, U., Dabrowski, J., Dennis, R. D., Egge, H., Helling, F., Keller, M., Peter-Katalinič, J., Weske, B., and Wiegandt, H. (1988). *Proc. Jpn. Conf. Biochem. Lipids* **27,** 54–55.

Dabrowski, U., Dabrowski, J., Helling, F., and Wiegandt, H. (1990). *J. Biol. Chem.* **265,** 9737–9743.

Denburg, J. L. (1989). *J. Neurosci.* **9,** 3491–3504.

Denburg, J. L., Caldwell, R. T., and Marner, J. A. M. (1986). *J. Comp. Neurol.* **245,** 123–136.

Denburg, J. L., Norbeck, B. A., Caldwell, R. T., and Marner, J. A. M. (1989). *Dev. Biol.* **132,** 1–13.

Dennis, R. D., Geyer, R., Egge, H., Menges, H., Stirm, S., and Wiegandt, H. (1985a). *Eur. J. Biochem.* **146,** 51–58.

Dennis, R. D., Geyer, R., Egge, H., Peter-Katalinič, J., Li, S.-C., Stirm, S., and Wiegandt, H. (1985b). *J. Biol. Chem.* **260,** 5370–5375.

Dennis, R. D., Haustein, D., Knowles, B. H., Ellar, D. J., and Wiegandt, H. (1986). *Biomed. Chromatogr.* **1,** 31–37.

Dennis, R. D., Geyer, R., Egge, H., Peter-Katalinič, J., Keller, M., Menges, H., and Wiegandt, H. (1987). *In* "Gangliodides and Modulation of Neuronal Functions" (H. Rahmann, ed.), pp. 351–358. Springer-Verlag, Berlin.

Dennis, R. D., Antonicek, H., Wiegandt, H., and Schachner, M. (1988). *J. Neurochem.* **51,** 1490–1496.

Dennis, R. D., Martini, R., and Schachner, M. (1991). *Cell Tissue Res.* **265,** 589–600.

Dennis, R. D., Baumeister, S., Geyer, R., Peter-Katalinič, J., Hartmann, R., Egge, H., Geyer, E., and Wiegandt, H. (1992). *Eur. J. Biochem.* **207,** 1053–1062.

Ehrlich, I., and Hrzenjak, T. (1975). *Vet. Arh.* **45,** 129–133.

Fahrig, T., Schmitz, B., Weber, D., Kücherer-Ehret, A., Faissner, A., and Schachner, M. (1990). *Eur. J. Neurosci.* **2,** 153–161.

Faye, L., and Chrispeels, M. J. (1988). *Glycoconjugate J.* **5,** 245–256.

Fischer, W., Ishizuka, I., Landgraf, H. R., and Herrmann, J. (1973). *Biochim. Biophys. Acta* **296,** 527–545.

Freeman, R. S. (1962). *Can. J. Zool.* **40,** 969–990.

Fujita, S. C., Zipursky, S. L., Benzer, S., Ferrus, A., and Shotwell, S. L. (1982). *Proc. Natl. Acad. Sci. U.S.A.* **79,** 7929–7933.

Galanos, C., Lüderitz, O, and Westphal, O. (1971). *Eur. J. Biochem.* **24,** 116–122.

Galili, U. (1988). *Transfus. Med. Rev.* **2,** 112–121.

Goldring, O. L., Clegg, J. A., Smithers, S. R., and Terry, R. J. (1976). *Clin. Exp. Immunol.* **26,** 181–187.

Goodman, C. S., and Bastiani, M. J. (1984). *Sci. Am.* **251,** 50–58.

Goodman, C. S., and Spitzer, N. C. (1979). *Nature (London)* **280,** 208–214.

Goodman, C. S., Bastiani, M. J., Doe, C. Q., Dulac, S., Helfand, S. L., Kuwada, J. Y., and Thomas, J. B. (1984). *Science* **225,** 1271–1279.

Grzych, J.-M., Dissous, C., Capron, M., Torres, S., Lambert, P.-H., and Capron, A. (1987). *J. Exp. Med.* **165,** 865–878.

Hakomori, S.-I. (1981). *Annu. Rev. Biochem.* **50,** 733–764.

Hanna, R. E. B., and Trudgett, A. G. (1983). *Parasite Immunol.* **5,** 409–425.

Harrington, G. W. (1965). *Exp. Prasitol.* **17,** 287–295.

Hayashi, A., and Matsubara, T. (1982). *Adv. Exp. Med. Biol.* **152,** 103–114.

Hayashi, A., and Matsubara, T. (1989). *Biochim. Biophys. Acta* **1006,** 89–96.

Helling, F., Dennis, R. D., Weske, B., Nores, G., Peter-Katalinič, J., Dabrowski, U., Egge, H., and Wiegandt, H. (1991). *Eur. J. Biochem.* **200,** 409–421.

Higashi, S., and Hori, T. (1968). *Biochim. Biophys. Acta* **152,** 568–575.

Highnam, K. C., and Hill, L. (1977). "The Comparative Endocrinology of the Invertebrates," pp. 104–130. Edward Arnold, London.

Hirabayashi, Y., Suzuki, T., Suzuki, Y., Taki, T., Matsumoto, M., Higashi, H., and Kato, S. (1983). *J. Biochem. (Tokyo)* **94,** 327–330.

Höllerer, Dennis, R. D., and Wiegandt, H. (1993). In preparation.

Honegger, C. G., and Freyvogel, T. A. (1963). *Helv. Chim. Acta* **46,** 2265–2270.

Hori, T., Sugita, M., Kanbayashi, J., and Itasaka, O. (1977a). *J. Biochem. (Tokyo)* **81,** 107–114.

Hori, T., Takeda, H., Sugita, M., and Itasaka, O. (1977b). *J. Biochem. (Tokyo)* **82,** 1281–1285.

Hori, T., Sugita, M., Ando, S., Kuwahara, M., Kumauchi, K., Sugie, E., and Itasaka, O. (1981). *J. Biol. Chem.* **256,** 10979–10985.

Hori, T., Sugita, M., Ando, S., Tsukada, K., Shiota, K., Tsuzuki, M., and Itasaka, O. (1983). *J. Biol. Chem.* **258,** 2239–2245.

Hrzenjak, T., and Ehrlich, I. (1975). *Vet. Arh.* **45,** 299–309.

Hrzenjak, T., and Ehrlich, I. (1976). *Vet. Arh.* **46,** 9–15.

Hrzenjak, T., Ehrlich, I., Muič, V., and Debogovič, Z. (1977). *Vet. Arh.* **47,** 317–322.

Ilyas, A. A., Dalakas, M. C., Brady, R. O., and Quarles, R. H. (1986). *Brain Res.* **385,** 1–9.

Irmer, Baumeister, S., Dennis, R. D., and Geyer, (1993). In preparation.

Ishihara, H., Takahashi, N., Oguri, S., and Tejima, S. (1979). *J. Biol. Chem.* **254,** 10715–10719.

Itasaka, O., and Hori, T. (1979). *J. Biochem. (Tokyo)* **85,** 1469–1481.

Itonori, S., Nishizawa, M., Suzuki, M., Inagaki, F., Hori, T., and Sugita, M. (1991). *J. Biochem. (Tokyo)* **110,** 479–485.

Itonori, S., Hiratsuka, M., Sonku, N., Tsuji, H., Itasaka, O., Hori, T., and Sugita, M. (1992). *Biochim. Biophys. Acta* **1123,** 263–268.

Jan, L. Y., and Jan, Y. N. (1982). *Proc. Natl. Acad. Sci. U.S.A.* **79,** 2700–2704.

Jennings, H. J., Rosell, K. G., Katzenellenbogen, E., and Kasper, D. L. (1983). *J. Biol. Chem.* **258,** 1793–1798.

Kalinna, B., and McManus, D. P. (1992). *Parasitology* (submitted for publication).

Karlsson, K. A. (1989). *Annu. Rev. Biochem.* **58,** 309–350.

Katz, F., Moats, W., and Jan, Y. N. (1988). *EMBO J.* **7,** 3471–3477.

Katz, M., Despommier, D. D., and Gwadz, R. W. (1988). "Parasitic Diseases," 2nd ed., pp. 65–89. Springer-Verlag, New York.

Kawahara, K., Seydel, U., Matsuura, M., Danbara, H., Rietschel, E. T., and Zähringer, U. (1991). *FEBS Lett.* **292,** 107–110.

Keller, M., Dennis, R. D., Weske, B., and Wiegandt, H. (1990). *Hybridoma* **9,** 295–307.

Keller, M., Weske, B., Dennis, R. D., and Wiegandt, H. (1993). *Hybridoma* (in press).

Knowles, B. H., and Ellar, D. J. (1986). *J. Cell Sci.* **83,** 89–101.

Knowles, B. H., Thomas, W. E., and Ellar, D. J. (1984). *FEBS Lett.* **168,** 197–202.

Kochetkov, N. K., Smirnova, G. P., and Chekareva, N. V. (1976). *Biochim. Biophys. Acta* **424,** 274–283.

Komai, Y., Matsukawa, S., and Satake, M. (1973). *Biochim. Biophys. Acta* **316,** 271–281.

Kruse, J., Mailhammer, R., Wernecke, H., Faissner, A., Sommer, I., Gloridis, C., and Schachner, M. (1984). *Nature (London)* **311,** 153–155.

Kubo, H., Jiang, G. J., Irie, A., Morita, M., Matsubara, T., and Hoshi, M. (1992). *J. Biochem.* **111,** 726–731.

Kücherer, A., Faissner, A., and Schachner, M. (1987). *J. Cell Biol.* **104,** 1597–1602.

Künemund, V., Jungalwala, F. B., Fischer, G., Chou, D. K. H., Keilhauer, G., and Schachner, M. (1988). *J. Cell Biol.* **106,** 213–223.

Kunz, J., Baumeister, S., Dennis, R. D., Kuytz, B., Wiegandt, H., and Geyer, E. (1991). *Parasitol. Res.* **77,** 443–447.

Kwong, A. Y. H., Wong, P. C. L., and Ko, R. C. (1990). *Comp. Biochem. Physiol. B* **95B,** 193–197.

Laine, R. A., and Hsieh, T. C.-Y (1987). *In* "Methods in Enzymology" (V. Ginsburg, ed.), Vol. 138, pp. 186–195. Academic Press, Orlando, FL.

Lesuk, A., and Anderson, R. J. (1941). *J. Biol. Chem.* **139,** 457–469.

Levery, S. B., Weiss, J. B., Salyan, M. E. K., Roberts, C. E., Hakomori, S.-I., Magnani, J. L., and Strand, M. (1992). *J. Biol. Chem.* **267,** 5542–5551.

Lindberg, A. A., Weintraub, A., Zähringer, U., and Rietschel, E. T. (1990). *Rev. Infect. Dis.* **12,** 133–141.

Luukkonen, A., Brummer-Korvenkontio, M., and Renkonen, O. (1973). *Biochim. Biophys. Acta* **326,** 256–261.

Makaaru, C. K., Damian, R. T., Smith, D. F., and Cummings, R. D. (1992). *J. Biol. Chem.* **267,** 2251–2257.

Makita, A., and Taniguchi, N. (1985). *In* "Glycolipids" (H. Wiegandt, ed.), pp. 1–99. Elsevier, Amsterdam.

Maloney, M. D., Semprevivo, L. H., and Coles, C. C. (1990). *Int. J. Parasitol.* **20,** 1091–1093.

Mandrell, R. E., McLaughlin, R., Kwaik, Y. A., Lesse, A., Yamasaki, R., Gibson, B., Spinola, S. M., and Apicella, M. A. (1992). *Infect. Immun.* **60,** 1322–1328.

Marcus, D. M. (1984). *Mol. Immunol.* **21,** 1083–1091.

Matsubara, T., and Hayashi, A. (1981). *J. Biochem.* (*Tokyo*) **89,** 645–650.

Matsubara, T., and Hayashi, A. (1986). *J. Biochem.* (*Tokyo*) **99,** 1401–1408.

McLaren, D. J., Clegg, J. A., and Smithers, S. R. (1975). *Parasitology* **70,** 67–75.

Meyer, M. R., Reddy, G. R., and Edwards, J. S. (1987). *J. Neurosci.* **7,** 512–521.

Miller, C. A., and Benzer, S. (1983). *Proc. Natl. Acad. Sci. U.S.A.* **80,** 7641–7645.

Mills, G. L., Taylor, D. C., and Williams, J. F. (1981). *Comp. Biochem. Physiol. B* **69B,** 553–557.

Murray, M. A., Schubiger, M., and Palka, J. (1984). *Dev. Biol.* **104,** 259–273.

Nagai, Y., Sanai, Y., and Nakaishi, H. (1987). *In* "Gangliosides and Modulation of Neuronal Functions" (H. Rahmann, ed.), pp. 275–292. Springer-Verlag, Berlin.

Naiki, M., Ramasamy, R., Ochanda, J. O., and Maina, G. (1985). *Jpn. J. Vet. Sci.* **47,** 777–786.

Nishimura, K., Suzuki, A., and Kino, H. (1991). *Biochim. Biophys. Acta* **1086,** 141–150.

Noda, N., Tanaka, R., Miyahara, K., and Kawasaki, T. (1992). *Tetrahedron Lett.* **33,** 7527–7530.

Nores, G., Dennis, R. D., Helling, F., and Wiegandt, H. (1991). *J. Biochem.* (*Tokyo*) **110,** 1–8.

Noronha, A. B., Ilyas, A. A., Antonicek, H., Schachner, M., and Quarles, R. H. (1986). *Brain Res.* **385,** 237–244.

Okamura, N., Stoskopf, M., Hendricks, F., and Kishimoto, Y. (1985). *Proc. Natl. Acad. Sci. U.S.A.* **82,** 6779–6782.

Okamura, N., Yamaguchi, H., Stoskopf, M., Kishimoto, Y., and Saida, T. (1986). *J. Neurochem.* **47,** 1111–1116.

Patel, N. H., Snow, P. M., and Goodman, S. C. (1987). *Cell* (*Cambridge, Mass.*) **48,** 975–988.

Persat, F., Mojon, M., and Patavy, A. F. (1988). *Comp. Biochem. Physiol. B.* **91B,** 133–136.

Persat, F., Bouhours, J. F., Mojon, M., and Petavy, A. F. (1990a). *Mol. Biochem. Parasitol.* **38,** 97–104.

Persat, F., Bouhours, J. F., Mojon, M., and Petavy, A. F. (1990b). *Mol. Biochem. Parasitol.* **41,** 1–6.

Persat, F., Vincent, C., Mojon, M., and Petavy, A. F. (1991). *Parasite Immunol.* **13,** 379–389.

Persat, F., Bouhours, J. F., Mojon, M., and Petavy, A. F. (1992). *J. Biol. Chem.* **267,** 8764–8769.

Reinišová, J., and Michaleč, C. (1966). *Biochim. Biophys. Acta* **19,** 581–588.

Richards, K. S., Ilderton, E., and Yardley, H. J. (1987). *Comp Biochem. Physiol.* **86,** 209–212.

Rietschel, E. T., Kirikae, T., Feist, W., Loppnow, H., Zabel, P., Brade, L., Ulmer, A. J., Brade, H., Seydel, U., Zähringer, U., Schlaak, M., Flad, H. D., and Schade, U. (1991). *In* "Molecular Aspects of Inflammation," Colloq. Mosbach No. 42, pp. 207–231. Springer-Verlag, Berlin.

Samuelson, J. C., and Caulfield, J. P. (1982). *J. Cell Biol.* **94,** 363–369.

Sarwal, R., Sanyal, S. N., and Khera, S. (1989). *J. Helminthol.* **63,** 287–297.

Schmidt, J. (1988). *Parasitol. Res.* **75,** 155–161.

Schmidt, J., and Peters, W. (1987). *Parasitol. Res.* **73,** 80–86.

Schmidt, J., and Ruppel, A. (1988). *Int. J. Parasitol.* **18,** 675–682.

Shimomura, K., Hanjura, S., Ki, P. F., and Kishimoto, Y. (1983). *Science* **220,** 1392–1393.

Sickmann, T., Weske, B., Dennis, R. D., and Wiegandt, H. (1992). *J. Biochem.* (*Tokyo*) **111,** 662–669.

Singh, B. N., Beach, D. H., Walenga, R. W., Mueller, J. F., and Holz, G. G. (1987a). "Molecular Paradigms for Eradicating Helminthic Parasites," pp. 493–506. Liss, New York.

Singh, B. N., Costello, C. E., Levery, S. B., Walenga, R. W., Beach, D. H., Mueller, J. F., and Holz, G. G. (1987b). *Mol. Biochem. Parasitol.* **26,** 99–112.

Snow, P. M., Patel, N. H., Harrelson, A. L., and Goodman, S. C. (1987). *J. Neurosci.* **7,** 4137–4144.

Stoskopf, M. K., Kishimoto, Y., Tanaka, T., Okamura, N. Kan, L.-S., Cotter, R., and Fenselau, C. (1989). *J. Biol. Chem.* **264,** 4964–4971.

Stults, C. L. M., Sweeley, C. C., and Macher, B. A. (1989). *In* "Methods in Enzymology" (V. Ginsburg, ed.), Vol. 179, pp. 167–214. Academic Press, San Diego.

Sugita, M. (1979a). *J. Biochem.* (*Tokyo*) **86,** 289–300.

Sugita, M. (1979b). *J. Biochem.* (*Tokyo*) **86,** 765–772.

Sugita, M., Yamamoto, T., Masuda, S., Itasaka, O., and Hori, T. (1981). *J. Biochem.* (*Tokyo*) **90,** 1529–1535.

Sugita, M., Nishida, M., and Hori, T. (1982a). *J. Biochem.* (*Tokyo*) **92,** 327–334.

Sugita, M., Iwasaki, Y., and Hori, T. (1982b). *J. Biochem.* (*Tokyo*) **92,** 881–887.

Sugita, M., Nakano, Y., Nose, T., Itasaka, O., and Hori, T. (1984a). *J. Biochem.* (*Tokyo*) **95,** 47–55.

Sugita, M., Inoue, T., Itasaka, O., and Hori, T. (1984b). *J. Biochem.* (*Tokyo*) **95,** 737–742.

Sugita, M., Itonori, S., Inagaki, F., Itasaka, O., and Hori, T. (1988a). *Proc. Jpn. Conf. Biochem. Lipids* **29,** 77–80.

Sugita, M., Sanai, Y., Itonori, S., and Hori, T. (1988b). *Biochim. Biophys. Acta* **962,** 159–165.

Sugita, M., Itonori, S., Inagaki, F., and Hori, T. (1989). *J. Biol. Chem.* **264,** 15028–15033.

Sugita, M., Inagaki, F., Naito, H., and Hori, T. (1990). *J. Biochem.* (*Tokyo*) **107,** 899–903.

Sugita, M., Fujii, H., Inagaki, F., Suzuki, M., Hayata, C., and Hori, T. (1992). *J. Biol. Chem.* **267,** 22595–22598.

Thomas, J. B., Bastiani, M. J., Bate, M., and Goodman, C. S. (1984). *Nature* (*London*) **310,** 203–207.

Tolberg, L. P., and Oland, L. A. (1990). *Exp. Neurol.* **109,** 19–28.

Tomlinson, A. (1988). *Development* (*Cambridge, UK*) **104,** 183–193.

Towbin, H., Rosenfelder, G., Wieslander, J., Avila, J. L., Rojas, M., Szarfman, A., Esser, K., Nowack, H., and Timpl, R. (1987). *J. Exp. Med.* **166,** 419–432.

Webb, R. A., and Mettrick, D. F. (1971). *Can. J. Biochem.* **49,** 1209–1212.

Webb, R. A., and Mettrick, D. F. (1973). *Int. J. Parasitol.* **3,** 47–58.

Wegerhoff, R., and Breidbach, O. (1992). *Cell Tissue Res.* **268,** 341–358.
Weintraub, A., Larsson, B. E., and Lindberg, A. A. (1985a). *Infect. Immun.* **49,** 197–201.
Weintraub, A., Zähringer, U., and Lindberg, A. A. (1985b). *Eur. J. Biochem.* **151,** 657–661.
Weiss, J. B., Magnani, J. L., and Strand, M. (1986). *J. Immunol.* **136,** 4275–4282.
Weske, B., Dennis, R. D., Helling, F., Keller, M., Nores, G., Peter-Katalinič, J., Egge, H., Dabrowski, U., and Wiegandt, H. (1990). *Eur. J. Biochem.* **191,** 379–388.
Weske, B., Dennis, R. D., Keller, M., and Wiegandt, H. (1993). In preparation.
Wiegandt, H. (1992). *Biochim. Biophys. Acta* **1123,** 117–126.
Yamamoto, A., Yano, I., Masui, M., and Yabuuchi, E. (1978). *J. Biochem.* (*Tokyo*) **83,** 1213–1216.
Yasuda, T., Ueno, J., Naito, Y., and Tsumita, T. (1982). *Adv. Exp. Med. Biol.* **152,** 457–465.
Zinn, K., McAllister, L., and Goodman, S. C. (1988). *Cell* (*Cambridge, Mass.*) **53,** 577–587.
Zipser, B., and McKay, R. (1981). *Nature* (*London*) **289,** 549–554.
Zipursky, S. L., Venkatesh, T. R., Teplow, D. B., and Benzer, S. (1984). *Cell* (*Cambridge, Mass.*) **36,** 15–26.

ADVANCES IN LIPID RESEARCH, VOL. 26

Developmental Changes of Glycosphingolipid Composition of Epithelia of Rat Digestive Tract

JEAN-FRANÇOIS BOUHOURS,* DANIÈLE BOUHOURS,* AND GUNNAR C. HANSSON†

* *Institut National de la Santé et de la Recherche Médicale*
Unité 76
F-75739 Paris, France
and
† *Department of Medical Biochemistry*
University of Göteborg
S-413 90 Göteborg, Sweden

I. Introduction

Epithelia of the gastrointestinal tract, namely, the stomach, the small intestine, and the colon, share a common embryonic origin, the primitive gut, which itself derives from the endodermal germ layer. It was of interest to investigate how the three epithelia differentiate, in terms of glycosphingolipids content, during the late fetal and early postnatal periods, which are accessible to biochemical analysis. This ontogenetic study was compared to the process of cell differentiation that takes place in the adult intestine as part of the constant renewal of the epithelium. In addition to specific modifications of the glycolipid composition, defined by qualitative and quantitative changes of their glycosidic moiety, it was noted that substantial alterations of the ceramide part also occur during these differentiation processes. Therefore, the developmental and maturational changes of the glycolipid composition of the three epithelia of the gastrointestinal tract of the rat are presented in this review, with special emphasis on the remodeling of ceramide.

II. Glycosphingolipid Composition of the Rat Stomach

A. Ceramide and Glycosphingolipids of the Adult Rat Stomach

Ceramide and glucosylceramide occur in the adult rat stomach at concentrations (3.5 and 5 μmol/g protein) higher than any other glycolipid seen on thin-layer chromatograms (Fig. 1, lanes 1 and 4). Glucosylceramide contains C_{18}-spingenine,[1] C_{18}-4*D*-hydroxysphinganine, and C_{20}4*D*-hydroxysphinganine (65, 10, and 25% of the bases, respectively) mainly linked to C_{22}, C_{23}, and C_{24} α-hydroxy fatty acids (D. Bouhours and Bouhours, 1985a). The second most abundant glycolipid, isogloboside (iGb$_4$Cer, Fig. 1), does not have the same ceramide composition as glucosylceramide. It contains both C_{18} bases but no C_{20}-4*D*-hydroxysphinganine, and nonhydroxy C_{16}, C_{20}, C_{22}, and C_{24} fatty acids. It has been found that glucosylceramide is mainly a glycolipid of the epithelium, whereas iGb$_4$Cer is mainly a component of the mesenchyme (D. Bouhours and Bouhours, 1985a). Therefore, the specific ceramide composition of each glycolipid corresponds to its specific tissue location.

The stomach contains four major gangliosides (Fig. 1, lane 4): G_{M3}, which is a component of the mesenchyme, and G_{M2}, G_{M1}, and a blood group B-active ganglioside, termed B-G_{M1}, which are components of the epithelium (J.-F. Bouhours *et al.*, 1987). The *N*-acetyl-*O*-trimethylsilylated sphingoid bases of the B-G_{M1} ganglioside of the adult rat stomach were analyzed by gas chromatography–mass spectrometry (GC-MS) Fig. 2). Peaks recorded on scans 18 and 22 are *threo* and native *erythro* isomers of C_{18}-sphingenine. Characteristic fragments are seen at *m/z* 174 and 311 for C_{18}-sphingenine (Fig. 2B), at *m/z* 299, 382, and 401 for C_{18}-4*D*-hydroxysphinganine (not shown), and at *m/z* 327, 410, and 429 for C_{20}-4*D*-hydroxysphinganine (Fig. 2C) (D. Bouhours and Bouhours, 1985a). Quantitative gas chromatography indicated that B-G_{M1} contains C_{18}-sphingenine, and C_{18}- and C_{20}-4*D*-hydroxysphinganine in the proportions of 24, 21, and 55%, respectively, whereas, surprisingly, G_{M1} contains C_{18}-sphingenine only.

[1] The nomenclature used in this article conforms to the recommendations of IUPAC-IUB (1977). Sphingoid bases are termed as sphingenine (sphingosine), sphinganine (dihydrosphingosine), and 4*D*-hydroxysphinganine (phytosphingosine). They are assumed to have 18 carbon atoms unless otherwise specified. Carbohydrate chains are abbreviated according to the same recommendations: Gb$_3$Cer, globotriaosylceramide.

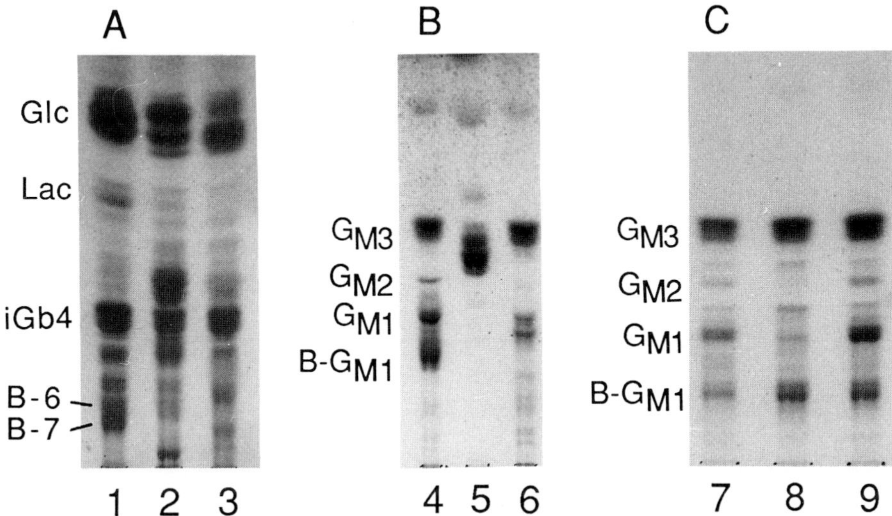

FIG. 1. Thin-layer chromatography of neutral and acid glycolipids of rat stomach, small intestine, and colon. (A) Neutral glycolipids analyzed on HPTLC plate in the solvent chloroform/methanol/water (60 : 35 : 8, v/v); vizualization: α-naphthol/H_2SO_4; (B) acid glycolipids analyzed in the solvent chloroform/methanol/water containing 0.2% $CaCl_2$ (50 : 40 : 10, v/v); (C) acid glycolipids analyzed as in B, except that the water contained 0.25% KCl and 2.5 *M* ammonia instead of $CaCl_2$; vizualization: resorcinol/HCl. Lanes 1, 4, and 7, total stomach; lanes 2 and 5, epithelium of the small intestine; lanes 3 and 6, total colon; lane 8, antrum; lane 9, fundus. The positions of the glycolipids of the stomach are indicated on the left side of each plate. B-6 and B-7 refer to blood group B-active ganglio-hexa- and isoglobo-heptaglycosylceramide, respectively (Hansson *et al.*, 1987). B-G_{M1} refers to a blood group B-active ganglioside (J.-F. Bouhours *et al.*, 1987).

From these data, it is obvious that the G_{M1} found in the ganglioside extract of the rat gastric epithelium is not a precursor of the synthesis of B-G_{M1}, although G_{M1} and B-G_{M1} are theoretically on the same pathway of biosynthesis. Separate extraction of gangliosides from the two functional parts of the stomach, namely, the antrum and fundus, showed that G_{M3} and B-G_{M1} are in both parts (Fig. 1C). In contrast, G_{M2} and G_{M1} are components of the fundus only (Fig. 1C, lane 9). Therefore, B-G_{M1} is expressed by epithelial cells that are present in both parts of the stomach, probably mucus-secreting cells, whereas G_{M1} and G_{M2} are likely to be expressed mainly in cells that have a location restricted to the fundus, such as oxyntic cells. In another species, the guinea pig, it has been shown that the different cell types constituting the gastric epithelium have specific glycolipid compositions (J.-F. Bouhours and Bouhours, 1978). These data

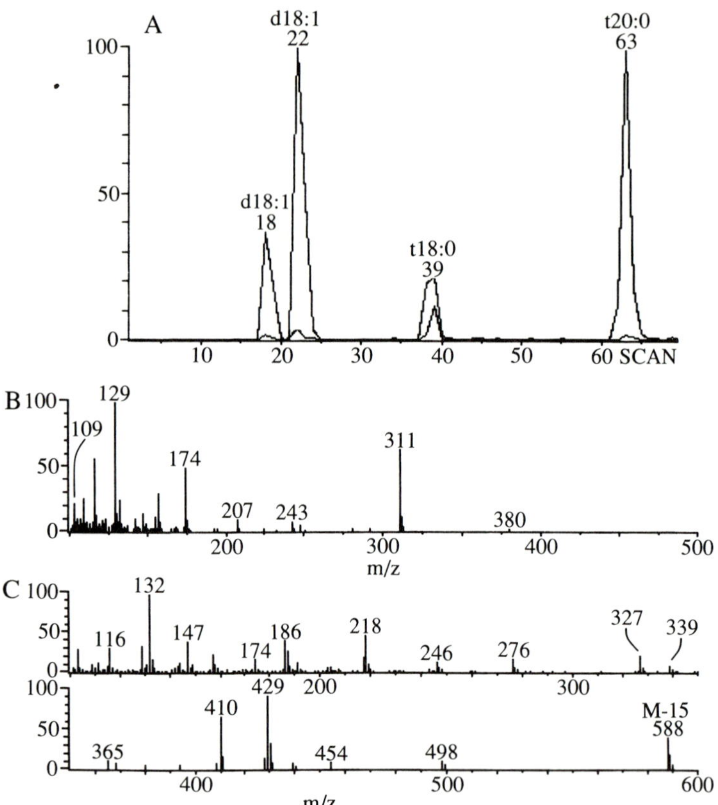

FIG. 2. GC-MS of *N*-acetyl-*O*-trimethylsilyl derivatives of the sphingoid bases of B-G_{M1} of the adult rat stomach. The chromatogram of the ion *m/z* 311 is the upper trace of peaks 18 and 22. The chromatogram of the ion *m/z* 132 is the upper trace of peaks 39 and 63. Mass spectrum B is from scan 22, showing C_{18}-sphingenine (*d*18 : 1). Spectrum C is from scan 63, showing C_{20}-4*D*-hydroxysphinganine (*t*20 : 0). A fused silica capillary column (10 m × 0.25 mm i.d.) coated with a 0.05-μm-thick cross-linked film of SE-54 was used in a Carlo Erba 4160 gas chromatograph which was directly interfaced to a VG ZAB-HF mass spectrometer. The oven temperature was programmed at a 10°C/minute rate from 70°C up to 300°C. For detailed operating conditions, see Hansson *et al.* (1989).

demonstrate that, within the same epithelium, the sphingoid base composition is part of the cell specificity of the glycolipid expression.

Comparison of the blood group B-active glycolipid based on an isoglobo core (B-7, Fig. 1) and iGb$_4$Cer shows that both have the same base composition (82 and 18% of C_{18}-sphingenine and C_{18}-4*D*-hydroxysphinganine, respectively, for isoGb$_4$Cer), but different fatty acid contents: C_{16} to C_{24}

hydroxy fatty acids are present on mass spectra of B-7, whereas they are absent from iGb_4Cer (D. Bouhours and Bouhours, 1985a; Hansson *et al.*, 1987). These differences in fatty acid composition can also be linked to the preferential tissue location of the glycolipids, as B-7 is in the epithelium and iGb_4Cer is in the mesenchyme.

Two free ceramide fractions, ceramide 1 and ceramide 2, were obtained by purifying ceramides which were present in the chloroform and chloroform–methanol (98 : 2, v/v) eluants, respectively, of the silicic acid column which was used for preparing glycolipids of the rat gastric epithelium (Hansson *et al.*, 1987). Both fractions were permethylated by the method of Ciucanu and Kerek (1984) and Larson *et al.* (1987) and analyzed by GC-MS. Ion chromatograms (*m/z* 88) are shown in Fig. 3A and B. Diagnostic fragments are found at M-253 for C_{18}-sphingenine-containing ceramides, and at M-285 and M-241 for C_{18}-4*D*-hydroxysphinganine-containing molecules (Fig. 3C and D). Free ceramide contains a major molecular species, *N*-palmitoylsphingenine, which is present in both ceramide fractions. Minor species are C_{18} to C_{24} nonhydroxyacylsphingenine (Fig. 3A), and C_{22}, C_{23}, and C_{24} nonhydroxyacyl-4-*D*-hydroxysphinganine and hydroxyacylsphingenine (Fig. 3B). The last eluted peaks contain C_{20} sphingoid bases.

B. Developmental Changes of Stomach Glycosphingolipids

Changes of the glycolipid composition of the stomach were studied from 2 days before to 60 days after birth. During all of that period, glucosylceramide remains the most abundant glycolipid and the only one accessible to structural analysis (D. Bouhours and Bouhours, 1985a). It was found that, 2 days before birth, 55% of the fatty acids are hydroxylated. This percentage rises to 70% at birth, increases moderately after, and reaches 82% by the age of 60 days (Fig. 4). The prenatal increase of fatty acid hydroxylation parallels an increase of the trihydroxylated base content (Fig. 5). Reversely, after birth, C_{18}-4*D*-hydroxysphinganine is partly replaced by C_{18}-sphingenine. The rapid changes of glucosylceramide structure which occur before and during the 2 weeks after birth are followed by a slow modification of its base composition that resembles the slow increase of C_{20}-sphingenine content which has been observed in aging human and rat brain gangliosides (Mansson *et al.*, 1978; Palestini *et al.*, 1990).

In fetus and newborn rats, 95% of the stomach gangliosides are G_{M3} and G_{D3}. After birth, the percentage of both gangliosides decreases and, between 10 and 20 days of age, a rapid change of the ganglioside profile is observed. G_{M1} and B-G_{M1}, which have not been seen before, are readily

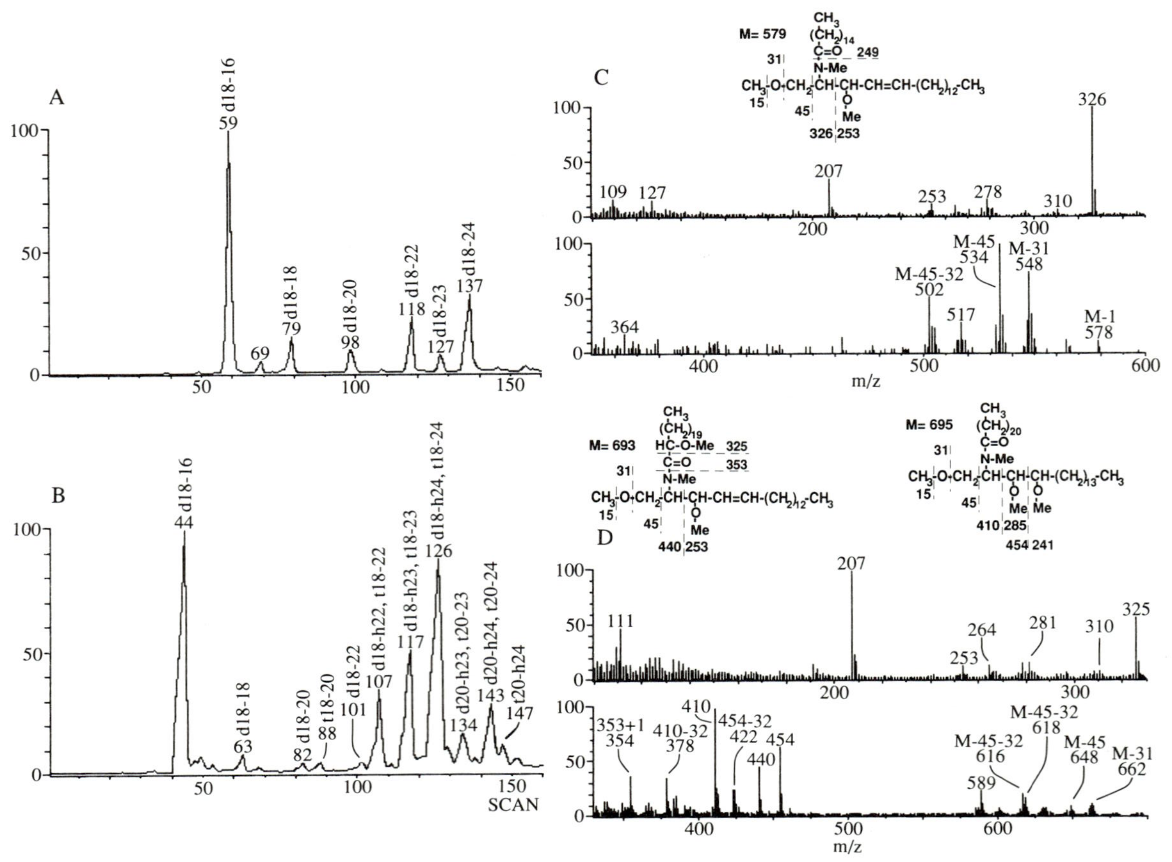
A
d18-16
59
69
d18-18
79
d18-20
98
d18-22
118
d18-23
127
d18-24
137
B
d18-16
44
d18-18
63
d18-20
82
t18-20
88
d18-22
101
d18-h22, t18-22
107
d18-h23, t18-23
117
d18-h24, t18-24
126
d20-h23, t20-23
134
d20-h24, t20-24
143
t20-h24
147
SCAN
C
M= 579
m/z
326
207
109
127
253
278
310
M-45
534
M-31
548
M-45-32
502
517
364
M-1
578
D
M= 693
M= 695
111
207
253
264
281
310
325
353+1
354
410-32
378
410
454-32
422
440
454
M-45-32
616
M-45-32
618
589
M-45
648
M-31
662

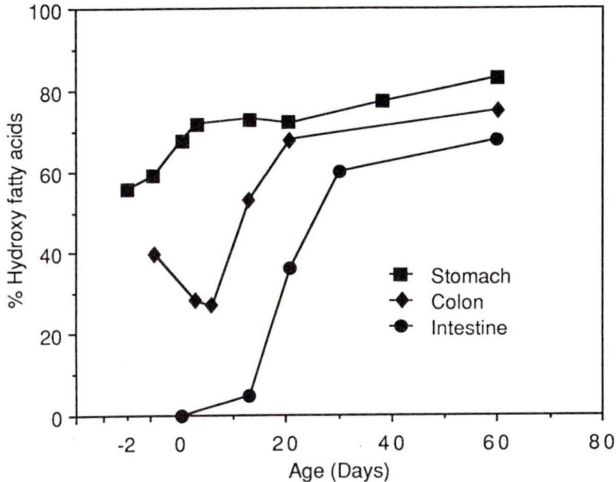

FIG. 4. Percentage of α-hydroxy fatty acids in the glucosylceramide of stomach, intestine, and colon of developing rats (D. Bouhours and Bouhours, 1981, 1983, 1985c). Age 0 means the day of birth.

detected on thin-layer chromatograms (J.-F. Bouhours *et al.*, 1987). Blood group B-active neutral glycolipids are detected by immunostaining in 13-day-old but not in younger rats. Therefore, the expression of G_{M1} and blood group B-active glycolipids of the ganglio- and isoglobo-series in the rat stomach is developmentally regulated and is triggered at the end of the second week after birth, after the completion of the first phase of glucosylceramide remodeling.

III. Glycosphingolipid Composition of the Rat Small Intestine

The epithelium of the small intestine is a dynamic cell population under constant renewal. A pool of proliferative cells is located at the bottom of the crypts. New cells move along the wall of the crypts and then along the villi, undergoing differentiation. From crypt to villus tip, the mean

---

FIG. 3. GC-MS of permethylated ceramides of rat stomach. Profiles are ion chromatograms recorded at *m/z* 88. (A) Ceramide 1, (B) ceramide 2. Peak identification: *d*18–16, C_{18}-sphingenine linked to nonhydroxy C_{16} fatty acid; *t*18–20, C_{18}-4*D*-hydroxysphinganine linked to hydroxy C_{20} fatty acid. Mass spectra recorded in scan 59 of profile A and in scan 107 of profile B are shown in C (spectrum of *d*18–16) and D (superimposed spectra of *d*18–*h*22 and *t*18–22), respectively. Ion at *m/z* 207 is a contaminant from the column coating. Operating conditions were as in Fig. 2.

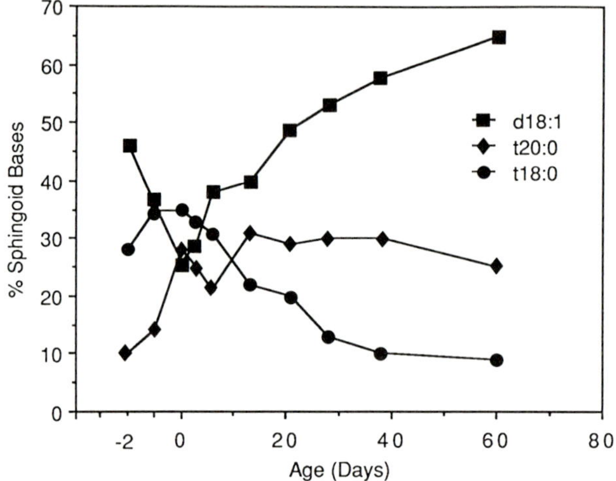

FIG. 5. Developmental changes of the sphingoid base composition of rat stomach glucosylceramide (D. Bouhours and Bouhours, 1985a).

transit time in rat intestine has been found to be 2–3 days (Lipkin, 1985). Isolation of cells according to their position along the crypt–villus axis, i.e., according to their degree of differentiation (Weiser, 1973), made possible the first study of the changes of the glycolipid composition of naturally differentiating cells, as opposed to transformed or cancerous cells (Glickman and Bouhours, 1976). The *differentiation–maturation* that takes place in all tissues undergoing renewal during the life span of an individual is distinct from the *ontogenic differentiation* which occurs during embryogenesis and development.

A. CERAMIDE AND GLYCOSPHINGOLIPIDS OF EPITHELIAL CELLS OF CRYPTS AND VILLI OF THE ADULT RAT INTESTINE

Most of the glycolipid fraction of the epithelial cells of adult rats (Fig. 1, lanes 2 and 5) consists of ceramide (17%), glucosylceramide (38%), globotriaosylceramide (Gb_3Cer, 15%), Galα1–3Gb_3Cer (5%), and G_{M3} (14%) (J.-F. Bouhours and Glickman, 1976), with variations depending on rat strains (Breimer *et al.*, 1981a) and on the proximal–distal position of the intestinal segment (Dahiya and Brasitus, 1986). The remaining 10% comprises at least 15 minor neutral glycolipids which are themselves unequally abundant (for reviews, see Breimer *et al.*, 1982c; Hansson, 1988). Among these minor glycolipids, some are either blood group H-active or blood group A-active, depending on the strain (Breimer *et al.*, 1980).

One-half of free ceramide contains sphingenine mainly linked to palmitic acid. 4*D*-Hydroxysphinganine and sphinganine are present in 45 and 5% of the ceramides, respectively. About 10% of 4*D*-hydroxysphinganine is linked to α-hydroxy fatty acids (J.-F. Bouhours and Guignard, 1979).

During the differentiation–maturation along the crypt–villus axis, quantitative changes of the glycolipid distribution occur, as well as modifications of the fatty acid composition. C_{18}-4*D*-Hydroxysphinganine and C_{18}-sphingenine account for 70 and 25% of the sphingoid bases, respectively, of all glycolipids of all cell types (J.-F. Bouhours and Glickman, 1977; Breimer *et al.,* 1981a). C_{20}-4*D*-Hydroxysphinganine is not a component of free ceramide and glycolipids of the small intestine.

Villus cells have more glucosylceramide and G_{M3}, but less ceramide and Gb_3Cer than crypt cells. The 40% decrease of Gb_3Cer and the 5- to 7-fold increase of G_{M3} concentration in villus compared to crypt cells are likely to originate in different activities of G_{M3} synthase and α-galactosyltransferase, which are competing for the same substrate, lactosylceramide. Indeed, it was found that the α-galactosyltransferase activity is reduced 3- to 5-fold (J.-F. Bouhours and Glickman, 1976), whereas the sialyltransferase activity is increased 30-fold between crypt and villus cells (Glickman and Bouhours, 1976).

However, G_{M3} and Gb_3Cer have different percentages of hydroxy fatty acids: 70 and 30%, respectively (J.-F. Bouhours and Glickman, 1977; Breimer *et al.,* 1982b; Dahiya and Brasitus, 1986). They are likely to be synthesized from different lactosylceramides. A probable hypothesis would be that Gb_3Cer is mainly synthesized in crypt cells in which glucosylceramide contains fewer hydroxy fatty acids than in villus cells (30 and 70%, respectively, according to Breimer *et al.,* 1982b). Gb_3Cer is likely to be partly degraded without full compensation by synthesis as cells mature. On the contrary, G_{M3} is actively synthesized in mature cells, and, therefore, acquires "mature" ceramide containing 70% of hydroxy fatty acids. This hypothesis is only partly acceptable, as the G_{M3} from crypt cells also has a high percentage of hydroxy fatty acids (J.-F. Bouhours and Glickman, 1977; Breimer *et al.,* 1982b). It is possible that precursors with specific ceramides are selected for the synthesis of each glycolipid.

In addition to the different levels of hydroxylation of fatty acids in crypt and villus cells, their chain length is also modified. C_{20} nonhydroxy fatty acids are more abundant in all glycolipids of crypt cells than of villus cells (J.-F. Bouhours and Glickman, 1977; Breimer *et al.,* 1982b). Glycolipids are mainly located in the brush border membrane of the intestinal epithelial cells (Forstner and Wherrett, 1973; J.-F. Bouhours, 1980; Hansson, 1983), and they are likely to have a turnover rate that is faster than the cellular turnover, as it is the case for brush border membrane proteins (Alpers,

1972). Therefore, the modification of the fatty acid composition probably results from a synthesis of new glycolipids during the migration of cells from crypt to villus.

B. Developmental Changes of Intestinal Glycosphingolipids

Three glycosphingolipids are present in the rat fetal intestine: glucosylceramide, G_{M3}, and G_{D3} (D. Bouhours and Bouhours, 1981; D. Bouhours *et al.,* 1982). After birth, G_{M3} becomes the major ganglioside of the small intestine and the only one found in intestinal epithelial cells of most strains of rats (D. Bouhours and Bouhours, 1988). Beside glucosylceramide, which remains the most abundant glycolipid during the whole life, other neutral glycolipids appear after birth. First, the lactosylceramide concentration rises from 5 to 10% of the intestinal glycolipid content between birth and 13 days of age. Then it declines and reaches the low adult level in 3-week-old rats. At the age of 21 days, Gb_3Cer becomes a major intestinal glycolipid, and blood group-active glycolipids, not seen in younger animals, are detected on thin-layer chromatograms. In addition, important structural modifications take place in glycolipids that are present during the entire developmental period.

1. *Modifications of the Ceramide Part of Glycolipids*

It was first observed that neonate and adult rats have a similar percentage of 4*D*-hydroxysphinganine in glucosylceramide, whereas they differ in their hydroxy fatty acid content (D. Bouhours and Bouhours, 1981). Therefore, the time of appearance of 4*D*-hydroxysphinganine was investigated before birth, and that of α-hydroxy fatty acids after birth (Fig. 6).

Five days before birth, sphingenine is the major base in all three glycolipids (D. Bouhours and Bouhours, 1984). The percentage of the other base, 4*D*-hydroxysphinganine, increases during the following days and reaches nearly the adult level in glucosylceramide and G_{M3} at birth (Fig. 6A). This was interpreted as a consequence of the prenatal organogenesis of the intestine (Mathan *et al.,* 1976) with increasing contribution of the epithelial cell mass. Therefore, 4*D*-hydroxysphinganine is present as early as 17 days of gestation, and is probably a component of glycolipids of the epithelium (mainly glucosylceramide and G_{M3}), whereas glycolipids of the mesenchyme (part of G_{M3} and most of G_{D3}) are likely to contain exclusively sphingenine, as in the adult small intestine (Breimer *et al.,* 1982d).

The data obtained from rats are consistent with the findings that meconium, obtained from human newborns, contains glycosphingolipids with 4*D*-hydroxysphinganine (Karlsson and Larson, 1979, 1981; Larson, 1986).

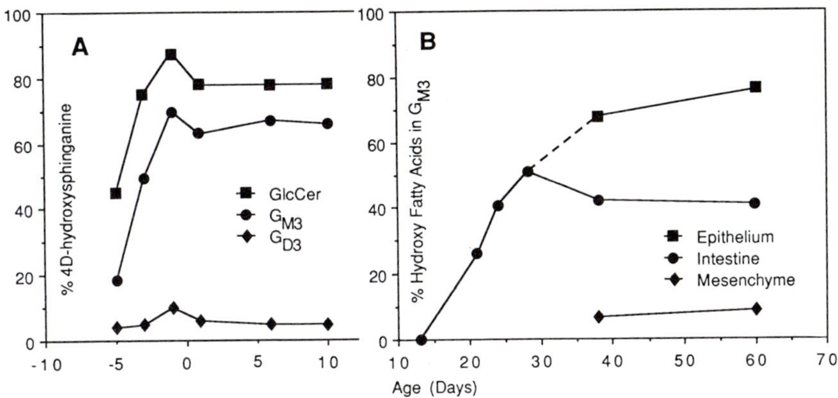

FIG. 6. Developmental changes of ceramide composition of intestinal glycolipids. (A) Percentage of 4*D*-hydroxysphinganine in intestinal glycolipids of fetus and neonate rats; (B) percentage of α-hydroxy fatty acids in intestinal G_{M3} as a function of age (D. Bouhours and Bouhours, 1983, 1984).

Hydroxy fatty acids appear in glucosylceramide after 13 days of age (D. Bouhours and Bouhours, 1981) (Fig. 4). The same changes take place in G_{M3} fatty acids. The α-hydroxylation of fatty acids concerns only the G_{M3} of the epithelium, as seen after separation of epithelial cells from the mesenchyme in 35- and 60-day-old rats (Fig. 6B) (D. Bouhours and Bouhours, 1983).

2. *Modification of* G_{M3}

G_{M3} of the rat small intestine undergoes three types of change during the ontogenic differentiation: a structural change affecting the amount of α-hydroxy fatty acids as mentioned above, another structural modification consisting of the hydroxylation of *N*-acetylneuraminic acid (NeuAc), and a huge variation of concentration.

a. Hydroxylation of the Sialic Acid of G_{M3}. G_{M3} is the only detectable ganglioside of the intestinal epithelium of the adult rat (Forstner and Wherrett, 1973), except in some inbred strains which also express small amounts of G_{D3} (D. Bouhours and Bouhours, 1988). It contains a large proportion of *N*-glycolylneuraminic acid (NeuGc), whereas stomach and colon gangliosides contain only NeuAc (Fig. 1B) (Glickman and Bouhours, 1976; Angström *et al.*, 1981). However, the sialic acid of G_{M3} is NeuAc before and immediately after birth (D. Bouhours *et al.*, 1982). The replacement of NeuAc by NeuGc in G_{M3} occurs after 13 days of age in a manner similar to the appearance of hydroxy fatty acids (Fig. 6B) (D. Bouhours and Bouhours, 1983). The expression of NeuGc in G_{M3} is controlled by the activity of the enzyme CMP-NeuAc hydroxylase (J.-F. Bouhours and Bouhours, 1989).

b. Concentration Changes of G_{M3}. The concentration of G_{M3} increases rapidly from a low level at 5 days before birth to a maximum at 6 days after birth, and then declines, reaching the adult level in 4-week-old rats (Fig. 7) (D. Bouhours *et al.*, 1982; D. Bouhours and Bouhours, 1983). The G_{M3} concentration of the epithelial cells of an adult animal of the same strain is shown for comparison in Fig. 7. Only extreme values are indicated, although the G_{M3} concentration increases continuously in cells from the crypt bottom to the villus tip. It appears that the dividing crypt cells of the adult animal have considerably less G_{M3} than the rapidly growing intestinal tissue of the immediate prenatal period. These results suggest that crypt cells are not surviving embryonic cells, but are more likely partially differentiated cells.

The total ganglioside content is only slightly higher than the G_{M3} content during the whole period of development (Fig. 7), except in 17- to 19-day-old fetuses in which G_{D3} is a major ganglioside (D. Bouhours *et al.*, 1982). A similar postnatal pattern of ganglioside concentration is observed in

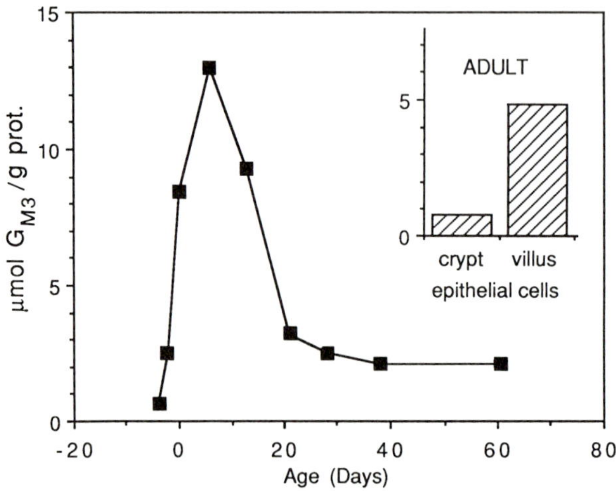

FIG. 7. Developmental changes of G_{M3} concentration in the small intestine (D. Bouhours *et al.*, 1982; D. Bouhours and Bouhours, 1983). G_{M3} concentrations of epithelial cells isolated from crypts and from villus tips of an adult rat intestine are shown in the insert.

mouse intestine (Sato *et al.*, 1982). The high concentration of G_{M3} during the first 3 weeks of life coincides with the high level of sialylation of brush border glycoproteins at this period of development (Kraml *et al.*, 1984; Srivastava *et al.*, 1987).

3. *Expression of Blood Group-Active Glycolipids*

In addition to glycolipids with short glycosidic chains, the intestine of the adult rat contains lacto-series blood group H- and also blood group A-active glycolipids, depending on the strain (Breimer *et al.*, 1982a,c,d). As the concentration of blood group ABH-active glycolipids of human erythrocytes is low at birth and increases during the first year of life (Watanabe and Hakomori, 1976; Fukuda and Levery, 1983), it was of interest to know whether the expression of similar glycolipids in the rat small intestine is also developmentally regulated. The concentration of blood group-active glycolipids actually rises abruptly around 21 days, although a low level of expression can be detected with anti-blood group antibodies in younger animals and neonates (D. Bouhours and Bouhours, 1985b; D. Bouhours *et al.*, 1987). As blood group A- or H-active glycolipids are expressed in crypt as well as in villus cells of adult animals (Breimer *et al.*, 1981a; D. Bouhours and Bouhours, 1985b), they are likely to appear in intestine, with a new generation of cells arising from the crypts at the time of weaning, when the epithelium initiates its adult-type renewal (Henning, 1985). However, the epithelium does not acquire full capacity to synthesize blood group-active glycolipids at once along the whole length of the small intestine. The earliest expression of blood group A-active glycolipids occurs in the distal part of the intestine, together with the adult form of glucosylceramide and G_{M3} (D. Bouhours *et al.*, 1990). It must be stressed also that a day-by-day analysis of blood group glycolipids between 14 and 22 days of age did not show a sequential appearance of precursor blood group H- before blood group A-active glycolipids (D. Bouhours *et al.*, 1987).

IV. Glycosphingolipid Composition of the Rat Colon

A. Glycosphingolipid Composition of the Adult Rat Colon

Glucosylceramide and tetraglycosylceramide are the major glycolipids appearing on thin-layer chromatograms (Fig. 1A, lane 3). The tetraglycosylceramide spot consists of a mixture of at least four glycolipids in which a blood group B-active tetraglycosylceramide is the main component

(Hansson *et al.*, 1980). Among many bands with lower intensity and slower mobility than the tetraglycosylceramides, at least two blood group A-active and two blood group B-active glycolipids are present (Hansson *et al.*, 1984). The latter glycolipids have been characterized as a monofucosylated lacto-series hexaglycosylceramide and a difucosylated neolacto-series heptoglycosylceramide (Angström *et al.*, 1987). Gangliosides account for only 5% of the total glycosphingolipids of the colon. G_{M3}(NeuAc) is the major ganglioside of the whole colon (Fig. 1B, lane 6) and the only one found in isolated epithelial cells. The mesenchyme mainly contains G_{M3} and G_{D3}.

B. Developmental Changes of the Ceramide Part of Glucosylceramide

In the whole colon, as well as in isolated epithelial colic cells, glucosylceramide and free ceramide account for 48 and 23% of the total glycolipid fraction, respectively (D. Bouhours and Bouhours, 1985c). These proportions are kept nearly unchanged from birth to adulthood. However, both the fatty acid composition and the base composition of glucosylceramide are modified. The most important changes of base composition occur between birth and 13 days of age, with the appearance of C_{18}-sphingenine and C_{20}-4*D*-hydroxysphinganine (Fig. 8) which, afterward, remain minor, by comparison with C_{18}-4*D*-hydroxysphinganine. The α-hydroxylation of fatty acids presents a biphasic pattern, with a decrease in the percentage of α-hydroxy acids immediately after birth and an increase after 6 days (Fig. 4).

The smaller amount of α-hydroxy fatty acids correlates with the higher percentage of C_{18}-sphingenine in 6-day-old rats, indicating that glucosylceramide is less hydroxylated at this age than in younger or older animals. This physical change may reflect a transient increase of the membrane fluidity for a physiological purpose that is not yet known, as the adult colon morphological structures are completed just before birth (Eastwood and Trier, 1974).

Between 6 and 60 days, free ceramide is less hydroxylated than the ceramide part of glucosylceramide. It does not contain hydroxy fatty acids, and its major base is C_{18}-sphingenine. C_{18}- and C_{20}-4*D*-hydroxysphinganine are minor components, contributing for 20 and 15% of the bases, respectively. The ceramide concentration does not change significantly during the postnatal period of development (D. Bouhours and Bouhours, 1985c).

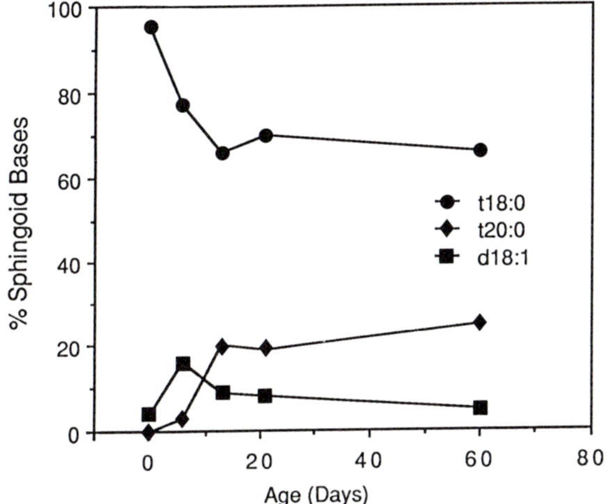

FIG. 8. Changes of the sphingoid base composition of rat colon glucosylceramide (D. Bouhours and Bouhours, 1985c).

V. Conclusion

Gangliosides of the Digestive Tract Epithelia

The three organs of the digestive tract contain mainly G_{M3} and G_{D3} during the end of the fetal life. Afterward, they differentiate and acquire distinct ganglioside compositions. Small and large intestines keep G_{M3} as the only ganglioside of the epithelium, although with differences of concentration and of sialic acid. The stomach epithelium loses most of its G_{M3} and expresses more complex gangliosides, i.e., G_{M2}, G_{M1}, and B-G_{M1}, in a cell-specific manner during postnatal differentiation. B-G_{M1} has also been described in rat as a tumor-associated antigen in hepatoma (Holmes and Hakomori, 1982) and PC12 pheochromocytoma cells (Ariga *et al.*, 1987). Therefore, the same glycolipid can be either an antigen of the normal ontogenic differentiation of a cell population or an oncogenic antigen in other cell types.

Blood Group Glycolipid Expression

It is remarkable that rat expresses different blood group ABH determinants in the same individual. The stomach expresses glycolipids carrying the blood group B determinant; the colon expresses glycolipids with blood group A or blood group B determinants. No genetic polymorphism has been described so far in these organs. In contrast, the small intestine expresses blood group A-active glycolipids in a restricted number of inbred strains, whereas most other inbred strains express blood group H-active glycolipids only (D. Bouhours *et al.*, 1991). Therefore, rats carry genes for both the α-galactosyltransferase and the α-*N*-acetylgalactosaminyltransferase, and express them in a tissue-specific manner.

Free Ceramide

Free ceramide is the sphingolipid second in concentration to glucosylceramide in the three organs of the digestive tract and at all ages. It has less α-hydroxy fatty acids, and less 4*D*-hydroxysphinganine than glucosylceramide. This composition is intermediate between the highly hydroxylated glycolipids and the nonhydroxylated sphingomyelin (J.-F. Bouhours and Guignard, 1979). It reflects the key metabolic position of ceramide, which is the initial acceptor substrate in the biosynthesis, and the end product of the degradation of all sphingolipids. As it is a molecule of quantitative and metabolic importance, free ceramide deserves characterization.

Purified ceramides can be analyzed as trimethylsilyl derivatives by direct-inlet electron-impact mass spectrometry (J.-F. Bouhours and Guignard, 1979) or by GC-MS (Samuelsson and Samuelsson, 1969; Hansson *et al.*, 1979). Spectra are interpreted by referring to known fragmentation patterns (Samuelsson and Samuelsson, 1969; Hammarström, 1970; Hammarström *et al.*, 1970). Free ceramides can also be analyzed as permethylated derivatives (Fig. 3). These derivatives are useful because of their lower molecular masses and lesser fragmentation than trimethylsilyl derivatives. Their GC-MS analysis permits the identification of different molecular species when the complexity of the sample is limited, as in Fig. 3A or 3B. When a combined ceramide sample of the rat gastric mucosa was analyzed, peaks were overlapping in the high mass region, making the interpretation of the GC-MS chromatogram difficult.

In addition to free ceramides naturally occurring in tissues, ceramides from sphingomyelin and from most glycosphingolipids can be analyzed as permethylated derivatives by GC-MS. First, they have to be hydrolyzed from sphingomyelin by phospholipase C (Karlsson, 1968) or from glycolipids by ceramide glycanase (S.-C. Li *et al.*, 1986; Y.-T. Li *et al.*, 1987; Ito and Yamagata, 1989). This latter technique gives the opportunity to analyze permethylated ceramides and oligosaccharides successively on the same column (Hansson *et al.*, 1989).

Developmental Changes of Glucosylceramide

The study of the developmental changes of the ceramide part of glycolipids is limited to the study of glucosylceramide, the only glycolipid abundant in all three organs of the digestive tract during the entire developmental period. In addition, G_{M3} of the small intestine is also accessible to analysis because of its exceptional high concentration during the perinatal period. Other glycolipids characteristic of the adult-type organ-specific glycolipid composition are more difficult to analyze because they appear at (small intestine) or after (stomach) the time when the major changes in the lipid part of glucosylceramide take place. It seems reasonable to assume that their fatty acid and base compositions follow the same trend as that of the glucosylceramide of their respective tissues, provided that they do not belong to cell subpopulations having different ceramides (for example, G_{M1} and B-G_{M1} in the rat stomach).

An intriguing finding is that α-hydroxylation of glucosylceramide fatty acids of stomach and colon does not follow the developmental pattern observed in the small intestine, as hydroxy fatty acids are already present at birth. The regulation of α-hydroxylase activity during the development appears to be tissue-specific, even in epithelia with the same embryonic origin.

It is known that, in rat brain, the α-hydroxylation of cerebroside fatty acids is activated at 13 days after birth, and reaches a maximum at 21–23 days (Murad *et al.,* 1976), in a manner similar to the appearance of α-hydroxy fatty acids in the small intestine (Figs. 4 and 7B). It is likely that the onset of α-hydroxylation of fatty acids is under a similar developmental control in intestine and brain.

The delayed appearance of α-hydroxy fatty acids in the small intestine might have a functional significance. The absence of α-hydroxy fatty acids in glycolipids before 21 days may partly explain why the fluidity of the brush border membrane is higher in suckling than in weaned rats (Brasitus *et al.,* 1984; Hübner *et al.,* 1988). It is interesting to note that the absorption of nutrients at the time of higher fluidity of the brush border membrane proceeds through macromolecule uptake and pinocytosis. After 21 days, the epithelial lining is "closed," and only small molecules are absorbed by selective transport processes (Henning, 1985). The "closure" occurs first in the distal, then in the proximal part of the intestine, with a timing that coincides with the appearance of α-hydroxy fatty acids in glucosylceramide and G_{M3} (D. Bouhours *et al.,* 1990).

The percentages of hydroxy fatty acids in glucosylceramide, which are specific to each of the three organs at birth, converge toward the same high value found in the adult rat (Fig. 4). By contrast with fatty acids, the sphingoid base composition of glucosylceramide becomes increasingly specific to each organ as development proceeds. As a result, the base composition of glucosylceramide is strikingly different in the three organs of the digestive tract of the adult rat. C_{18}-Sphingenine accounts for 65% of the bases in stomach, 25% in intestine, and 5% in colon; C_{18}-sphinganine content stays at or below 5%; 4*D*-hydroxysphinganines account for the remainder.

C_{18}-4*D*-Hydroxysphinganine is the major base of intestinal glycolipids of many mammals, including humans (Okabe *et al.*, 1968; Smith *et al.*, 1975a,b; Breimer *et al.*, 1981b; Hansson *et al.*, 1982; Sato *et al.*, 1982). It appears that C_{18}-4*D*-hydroxysphinganine and α-hydroxy fatty acids constitute intestine-specific ceramides common to many species, whereas the glycosidic part of glycolipids is species-specific. The primary cause of the presence of 4*D*-hydroxysphinganine in gastrointestinal cell glycolipids is not their exposure to 4*D*-hydroxysphinganine-containing nutrients, since fetus glycolipids already contain it.

Crossman and Hirschberg (1977, 1984) have demonstrated that radiolabeled sphinganine, when injected intravenously in rat, is incorporated into intestinal glycolipids as sphinganine, sphingenine, and 4*D*-hydroxysphinganine, demonstrating that 4*D*-hydroxysphinganine is biosynthesized from sphinganine in rat. If hydroxylation of sphinganine is actually the final step of 4*D*-hydroxysphinganine biosynthesis in intestine, it is not known whether hydroxylation occurs on the free base before ceramide synthesis or on the fatty acid-linked base, as for the dehydrogenation of *N*-acylsphinganine into *N*-acylsphingenine (Merrill and Wang, 1986).

The present review establishes that cells regulate very precisely and in a stage-specific manner during differentiation not only the carbohydrate part of glycolipids, but also their fatty acid and base compositions.

Acknowledgments

This work was supported by the French National Institute of Medical Research (INSERM) and by the Swedish Medical Research Council (MFR, grants 7461 and 8830).

References

Alpers, D. (1972). *J. Clin. Invest.* **51,** 2621–2630.

Angström, J., Breimer, M. E., Falk, K.-E., Griph, I., Hansson, G. C., Karlsson, K.-A., and Leffler, H. (1981). *J. Biochem.* (*Tokyo*) **90,** 909–921.

Angström, J., Falk, P., Hansson, G. C., Holgersson, J., Karlsson, H., Karlsson, K.-A., Strömberg, N., and Thurin, J. (1987). *Biochim. Biophys. Acta* **926,** 79–86.

Ariga, T., Kobayashi, K., Kuroda, Y., Yu, R. K., Suzuki, M., Kitagawa, H., Inagaki, F., and Miyatake, T. (1987). *J. Biol. Chem.* **262,** 14146–14153.

Bouhours, D., and Bouhours, J.-F. (1981). *Biochem. Biophys. Res. Commun.* **99,** 1384–1389.

Bouhours, D., and Bouhours, J.-F. (1983). *J. Biol. Chem.* **258,** 299–304.

Bouhours, D., and Bouhours, J.-F. (1984). *Biochim. Biophys. Acta* **794,** 169–171.

Bouhours, D., and Bouhours, J.-F. (1985a). *J. Biol. Chem.* **260,** 2172–2177.

Bouhours, D., and Bouhours, J.-F. (1985b). *Glycoconjugate J.* **2,** 79–86.

Bouhours, D., and Bouhours, J.-F. (1985c). *J. Biochem.* (*Tokyo*) **98,** 1359–1366.

Bouhours, D., and Bouhours, J.-F. (1988). *J. Biol. Chem.* **263,** 15540–15545.

Bouhours, D., Bouhours, J.-F., and Dorier, A. (1982). *Biochim. Biophys. Acta* **713,** 280–284.

Bouhours, D., Larson, G., Bouhours, J.-F., Lundbladt, A., and Hansson, G. C. (1987). *Glycoconjugate J.* **4,** 59–71.

Bouhours, D., Bouhours, J.-F., Larson, G., and Hansson, G. C. (1990). *Arch. Biochem. Biophys.* **282,** 147–151.

Bouhours, D., Angström, J., Jovall, P.-A., Hansson, G. C., and Bouhours, J.-F. (1991). *J. Biol. Chem.* **266,** 18613–18619.

Bouhours, J.-F. (1980). Thèse de Doctorat d'Etat es-Sciences, N°. 8043. Université Claude Bernard, Lyon, France.

Bouhours, J.-F., and Bouhours, D. (1978). *Biochem. Biophys. Res. Commun.* **85,** 1314–1317.

Bouhours, J.-F., and Bouhours, D. (1989). *J. Biol. Chem.* **264,** 16992–16999.

Bouhours, J.-F., and Glickman, R. M. (1976). *Biochim. Biophys. Acta* **441,** 123–133.

Bouhours, J.-F., and Glickman, R. M. (1977). *Biochim. Biophys. Acta* **487,** 51–60.

Bouhours, J.-F., and Guignard, H. (1979). *J. Lipid Res.* **20,** 897–907.

Bouhours, J.-F., Bouhours, D., and Hansson, G. C. (1987). *J. Biol. Chem.* **262,** 16370–16375.

Brasitus, T. A., Yeh, K.-T., Holt, P. R., and Schachter, D. (1984). *Biochim. Biophys. Acta* **778,** 341–348.

Breimer, M. E., Hansson, G. C., Karlsson, K.-A., and Leffler, H. (1980). *FEBS Lett.* **114,** 51–56.

Breimer, M. E., Hansson, G. C., Karlsson, K.-A., and Leffler, H. (1981a). *Exp. Cell Res.* **135,** 1–13.

Breimer, M. E., Hansson, G. C., Karlsson, K.-A., and Leffler, H. (1981b). *J. Biochem.* (*Tokyo*) **90,** 589–609.

Breimer, M. E., Hansson, G. C., Karlsson, K.-A., and Leffler, H. (1982a). *J. Biol. Chem.* **257,** 50–59.

Breimer, M. E., Hansson, G. C., Karlsson, K.-A., and Leffler, H. (1982b). *Biochim. Biophys. Acta* **710,** 415–427.

Breimer, M. E., Hansson, G. C., Karlsson, K.-A., and Leffler, H. (1982c). *J. Biol. Chem.* **257,** 557–568.

Breimer, M. E., Hansson, G. C., Karlsson, K.-A., and Leffler, H. (1982d). *J. Biol. Chem.* **257,** 906–912.

Ciucanu, I., and Kerek, F. (1984). *Carbohydr. Res.* **131,** 209–217.

Crossman, M. W., and Hirschberg, C. B. (1977). *J. Biol. Chem.* **252,** 5815–5819.

Crossman, M. W., and Hirschberg, C. B. (1984). *Biochim. Biophys. Acta* **795,** 411–416.

Dahiya, R., and Brasitus, T. A. (1986). *Lipids* **21,** 112–116.

Eastwood, G. L., and Trier, J. S. (1974). *Anat. Rec.* **179,** 303–310.

Forstner, G. G., and Wherrett, J. R. (1973). *Biochim. Biophys. Acta* **306,** 446–459.

Fukuda, M. N., and Levery, S. B. (1983). *Biochemistry* **22,** 5034–5040.

Glickman, R. M., and Bouhours, J.-F. (1976). *Biochim. Biophys. Acta* **424,** 17–25.

Hammarström, S. (1970). *J. Lipid Res.* **11,** 175–182.

Hammarström, S., Samuelsson, B., and Samuelsson, K. (1970). *J. Lipid Res.* **11,** 150–157.

Hansson, G. C. (1983). *Biochim. Biophys. Acta* **733,** 295–299.

Hansson, G. C. (1988). *Adv. Exp. Med. Biol.* **228,** 465–494.

Hansson, G. C., Heilbronn, E., Karlsson, K.-A., and Samuelsson, B. E. (1979). *J. Lipid Res.* **20,** 509–518.

Hansson, G. C., Karlsson, K.-A., and Thurin, J. (1980). *Biochim. Biophys. Acta* **620,** 270–280.

Hansson, G. C., Karlsson, K.-A., Leffler, H., and Strömberg, N. (1982). *FEBS Lett.* **139,** 291–294.

Hansson, G. C., Karlsson, K.-A., and Thurin, J. (1984). *Biochim. Biophys. Acta* **792,** 281–292.

Hansson, G. C., Bouhours, J.-F., and Angström, J. (1987). *J. Biol. Chem.* **262,** 13135–13141.

Hansson, G. C., Li, Y.-T., and Karlsson, H. (1989). *Biochemistry* **28,** 6672–6678.

Henning, S. (1985). *Annu. Rev. Physiol.* **47,** 231–245.

Holmes, E. H., and Hakomori, S.-I. (1982). *J. Biol. Chem.* **257,** 7698–7703.

Hübner, C., Lindner, S. G., Stern, M., Claussen, M., and Kohlschütter, A. (1988). *Biochim. Biophys. Acta* **939,** 145–150.

Ito, M., and Yamagata, T. (1989). *J. Biol. Chem.* **264,** 9510–9519.

IUPAC-IUB Commission on Biochemical Nomenclature (1977). *Lipids* **12,** 455–468.

Karlsson, K.-A. (1968). *Acta Chem. Scand.* **22,** 3050–3052.

Karlsson, K.-A., and Larson, G. (1979). *J. Biol. Chem.* **254,** 9311–9316.

Karlsson, K.-A., and Larson, G. (1981). *J. Biol. Chem.* **256,** 3512–3524.

Kraml, J., Kolinska, J., Kadlecova, L., Zakostelecka, M., and Lojda, Z. (1984). *FEBS Lett.* **172,** 25–28.

Larson, G. (1986). *Arch. Biochem. Biophys.* **246,** 531–545.

Larson, G., Karlsson, H., Hansson, G. C., and Pimlott, W. (1987). *Carbohydr. Res.* **161,** 281–290.

Li, S.-C., DeGasperi, R., Muldrey, J. E., and Li, Y.-T. (1986). *Biochem. Biophys. Res. Commun.* **141,** 346–352.

Li, Y.-T., Ishikawa, Y., and Li, S.-C. (1987). *Biochem. Biophys. Res. Commun.* **149,** 167–172.

Lipkin, M. (1985). *Annu. Rev. Physiol.* **47,** 175–197.

Mansson, J.-E., Vanier, M.-T., and Svennerholm, L. (1978). *J. Neurochem.* **30,** 273–275.

Mathan, M., Moxey, P. C., and Trier, J. S. (1976). *Am. J. Anat.* **146,** 73–92.

Merill, A. H., and Wang, E. (1986). *J. Biol. Chem.* **261,** 3764–3769.

Murad, S., Strycharz, G. D., and Kishimoto, Y. (1976). *J. Biol. Chem.* **251,** 5237–5241.

Okabe, K., Keenan, R. W., and Schmidt, G. (1968). *Biochem. Biophys. Res. Commun.* **31,** 137–143.

Palestini, P., Masserini, M., Sonnino, S., Giuliani, A., and Tettamanti, G. (1990). *J. Neurochem.* **54,** 230–235.

Samuelsson, B., and Samuelsson, K. (1969). *J. Lipid Res.* **10,** 41–46.

Sato, E., Uezato, T., Fujita, M., and Nishimura, K. (1982). *J. Biochem.* (*Tokyo*) **91,** 2013–2019.

Smith, E. L., McKibbin, J. M., Karlsson, K.-A., Pascher, I., and Samuelsson, B. E. (1975a). *Biochemistry* **14,** 2120–2124.

Smith, E. L., McKibbin, J. M., Karlsson, K.-A., Pascher, I., Samuelsson, B. E., Li, Y.-T., and Li, S.-C. (1975b). *J. Biol. Chem.* **250,** 6059–6075.

Srivastava, O. P., Steele, M. I., and Torres-Pinedo, R. (1987). *Biochim. Biophys. Acta* **914,** 143–151.

Watanabe, H., and Hakomori, S.-I. (1976). *J. Exp. Med.* **144,** 644–653.

Weiser, M. M. (1973). *J. Biol. Chem.* **248,** 2536–2541.

INDEX

G

S

CONTENTS OF PREVIOUS VOLUMES

ISBN 0-12-024926-X